Global Perspectives on Strategic Storytelling in Destination Marketing

Ana Cláudia Campos
CinTurs, Universidade do Algarve, Portugal

Sofia Almeida
Faculty of Tourism and Hospitality, Universidade Europeia, Portugal

IGI Global
PUBLISHER of TIMELY KNOWLEDGE

A volume in the Advances in Hospitality, Tourism, and the Services Industry (AHTSI) Book Series

Published in the United States of America by
IGI Global
Business Science Reference (an imprint of IGI Global)
701 E. Chocolate Avenue
Hershey PA, USA 17033
Tel: 717-533-8845
Fax: 717-533-8661
E-mail: cust@igi-global.com
Web site: http://www.igi-global.com

Library of Congress Cataloging-in-Publication Data

Names: Campos, Ana Claudia, 1966- editor. | Almeida, Sofia, 1977- editor.
Title: Global perspectives on strategic storytelling in destination marketing / Ana Claudia Campos and Sofia Almeida, editors.
Description: Hershey, PA : Business Science Reference, 2022. | Includes bibliographical references and index. | Summary: "This book presents an updated review of research, as well as of the theoretical frameworks and empirical approaches adopted so far to study storytelling applied to the context of tourism, at the organization and destination levels, and from the perspective of experience providers and customers"-- Provided by publisher.
Identifiers: LCCN 2022007234 (print) | LCCN 2022007235 (ebook) | ISBN 9781668434369 (hardcover) | ISBN 9781668434376 (paperback) | ISBN 9781668434383 (ebook)
Subjects: LCSH: Place marketing. | Tourism--Marketing.
Classification: LCC G155.A1 G4879 2022 (print) | LCC G155.A1 (ebook) | DDC 910.68/8--dc23/eng20220422
LC record available at https://lccn.loc.gov/2022007234
LC ebook record available at https://lccn.loc.gov/2022007235

This book is published in the IGI Global book series Advances in Hospitality, Tourism, and the Services Industry (AHTSI) (ISSN: 2475-6547; eISSN: 2475-6555)

British Cataloguing in Publication Data
A Cataloguing in Publication record for this book is available from the British Library.

For electronic access to this publication, please contact: eresources@igi-global.com.

Advances in Hospitality, Tourism, and the Services Industry (AHTSI) Book Series

Maximiliano Korstanje
University of Palermo, Argentina

ISSN:2475-6547
EISSN:2475-6555

Mission

Globally, the hospitality, travel, tourism, and services industries generate a significant percentage of revenue and represent a large portion of the business world. Even in tough economic times, these industries thrive as individuals continue to spend on leisure and recreation activities as well as services.

The Advances in Hospitality, Tourism, and the Services Industry (AHTSI) book series offers diverse publications relating to the management, promotion, and profitability of the leisure, recreation, and services industries. Highlighting current research pertaining to various topics within the realm of hospitality, travel, tourism, and services management, the titles found within the AHTSI book series are pertinent to the research and professional needs of managers, business practitioners, researchers, and upper-level students studying in the field.

Coverage

- Cruise Marketing and Sales
- Health and Wellness Tourism
- Customer Service Issues
- Service Training
- Leisure & Business Travel
- Tourism and the Environment
- Destination Marketing and Management
- Casino Management
- Travel Agency Management
- Sustainable Tourism

IGI Global is currently accepting manuscripts for publication within this series. To submit a proposal for a volume in this series, please contact our Acquisition Editors at Acquisitions@igi-global.com or visit: http://www.igi-global.com/publish/.

The Advances in Hospitality, Tourism, and the Services Industry (AHTSI) Book Series (ISSN 2475-6547) is published by IGI Global, 701 E. Chocolate Avenue, Hershey, PA 17033-1240, USA, www.igi-global.com. This series is composed of titles available for purchase individually; each title is edited to be contextually exclusive from any other title within the series. For pricing and ordering information please visit http://www.igi-global.com/book-series/advances-hospitality-tourism-services-industry/121014. Postmaster: Send all address changes to above address.

Titles in this Series

For a list of additional titles in this series, please visit: http://www.igi-global.com/book-series/advances-hospitality-tourism-services-industry/121014

Handbook of Research on Urban Tourism, Viral Society, and the Impact of the COVID-19 Pandemic
Pedro Andrade (University of Minho, Portugal) and Moisés de Lemos Martins (University of Minho, Portugal)
Business Science Reference • © 2022 • 430pp • H/C (ISBN: 9781668433690) • US $315.00

Employability and Skills Development in the Sports, Events, and Hospitality Industry
Vipin Nadda (University of Sunderland in London, UK) Ian Arnott (Westminster University, UK) Wendy Sealy (University of Chichester, UK) and Emma Delaney (University of Surrey, UK)
Business Science Reference • © 2022 • 260pp • H/C (ISBN: 9781799877813) • US $215.00

Handbook of Research on Sustainable Tourism and Hotel Operations in Global Hypercompetition
Hakan Sezerel (Anadolu University, Turkey) and Bryan Christiansen (Global Research Society, LLC, USA)
Business Science Reference • © 2022 • 430pp • H/C (ISBN: 9781668446454) • US $315.00

Entrepreneurship Education in Tourism and Hospitality Management
Satish Chandra Bagri (Hemwati Nandan Bahuguna Garhwal University, India) R.K. Dhodi (Hemwati Nandan Bahuguna Garhwal University, India) and K.C. Junaid (Hemwati Nandan Bahuguna Garhwal University, India)
Business Science Reference • © 2022 • 330pp • H/C (ISBN: 9781799895107) • US $230.00

Promoting Social and Cultural Equity in the Tourism Sector
Priscila Cembranel (Universidade Sociedade Educacional de Santa Catarina, Brazil) Jakson Renner Rodrigues Soares (Universidad da Coruña, Spain) and André Riani Costa Perinotto (Parnaíba Delta Federal University, Brazil)
Business Science Reference • © 2022 • 302pp • H/C (ISBN: 9781668441947) • US $240.00

Challenges and Opportunities for Transportation Services in the Post-COVID-19 Era
Giuseppe Catenazzo (ICN Business School, France)
Business Science Reference • © 2022 • 268pp • H/C (ISBN: 9781799888406) • US $250.00

Handbook of Research on Cultural Tourism and Sustainability
Claudia Ribeiro de Almeida (University of the Algarve, Portugal) Joao Carlos Martins (University of the Algarve, Portugal) Alexandra R. Gonçalves (University of the Algarve, Portugal) Silvia Quinteiro (University of the Algarve, Portugal) and Maria Laura Gasparini (University of Bologna, Italy)
Business Science Reference • © 2022 • 482pp • H/C (ISBN: 9781799892175) • US $315.00

IGI Global
PUBLISHER of TIMELY KNOWLEDGE

701 East Chocolate Avenue, Hershey, PA 17033, USA
Tel: 717-533-8845 x100 • Fax: 717-533-8661
E-Mail: cust@igi-global.com • www.igi-global.com

Editorial Advisory Board

Table of Contents

Detailed Table of Contents

Stories have long been a central component to the interpretation of people, culture, and place. Recent research has substantiated the importance of storytelling in heritage tourism settings; however, few studies have provided an overview of storytelling in heritage experience design. This has left academics and service providers alike with little to no guidance as to how to effectively utilize storytelling in their experience offering. This chapter serves to synthetize what is known on the topic of storytelling in heritage experience design. It will explore the relevance of storytelling and experience design in the context of new museology, give an overview of pre-existing frameworks on the subject, and offer a critical analysis highlighting informational gaps and areas of expansion. Beyond guided tours, interpretative signs, and guidebooks, one thing is clear: if they are to facilitate experiences that are more memorable, engaging, and interactive, heritage attractions must utilize storytelling as a design tool.

Cultural heritage interpretation requires mediation processes connecting tangibility with intangible meanings. As a performing narrative, storytelling is an effective communication strategy in cultural and heritage tourism. In particular, digital storytelling allows to segment information and adapt it to the different audience profiles in an interactive and immersive way. Tourists become participants in the mediation process by providing their stories, while local communities enrich the curatorial discourse with their knowledge and memories about the place, culture, and heritage. Tourists and residents' co-creation,

This chapter draws on the case study of aboriginal trackways at World Heritage-listed Willandra Lakes Region of New South Wales, Australia to explore how to re-create a unique, charismatic, and vulnerable heritage site by combining 3D technology and storytelling to assist with the virtual presentation of heritage. This chapter delves into the advantages and challenges of innovative 3D scanning methods available to accurately record fossilised Indigenous footprints, providing best practice guidelines on how to best deliver these outcomes in a form of engaging immersive visitor experience. Findings suggest that storytelling techniques improve intensity of virtual experience recreated using 3D scanning techniques through the process of narrative transportation, which elicits emotional arousal and improves emotional engagement with heritage. These findings contribute to the utilisation of 3D technologies and storytelling for AR/VR to create engaging immersive experiences for visitors.

User-generated videos (UGVs) are stories, and stories reflect how we experience the world. This chapter explores the story-like nature of UGVs by applying narrative analysis to deconstruct a tourism destination's Instagram Reels, revealing their influence on user engagement. Posted comments for each of 20 Reels are coded based on their representation of cognitive, affective, and behavioral dimensions of engagement as they relate to the story elements of plot, character, and verisimilitude. Results indicate that the most engaging Reels feature all story elements, unique locations, and personalized stories. These fulsome video stories are most likely to provoke cognitive engagement, whereas UGVs with fewer elements provoke emotional engagement. This study uniquely connects destination engagement and narrative elements in user-generated storytelling videos to guide destination social media marketing effectiveness.

This chapter reports a study on a religious tourism destination, Fátima, Portugal, with a focus on the contribution of social media to destination communication. The authors analysed the Facebook pages of four entities related to Fátima tourism to examine the number and type of posts and analyse to which extent these entities are sharing destination information or stories and encouraging visitors to tell their own stories about their visit to Fátima. The authors conducted a quantitative and qualitative analysis to Facebook posts during May, August, and October 2021 when Fátima receives the highest number of pilgrims.

Foreword

Stories surround us these days and are not just in forms of entertainment where we become immersed in a plot or characters and events but also in other ways like purchase choices, stamped with marks of authenticity and details of provenance designed to inform about authenticity and develop loyalty. Such narratives hook and transport us to make connections, make sense of the world and help solve problems in times of uncertainty as well as create memories with experiences and escapism. The fusing of stories and their telling in tourism is invaluable and is increasingly recognised as important in the making of place identity and in the formation of destination image. The theme of this book on Global perspectives on strategic storytelling in destination marketing cannot be undervalued and is a timely addition to the literature. Destination management organisations will benefit from the contributions when striving to make connections with visitors from around the world, wanting to connect with them on an emotional level and reading how stories make their destination unique offering a deepening of the visitor experience through an attachment to their narrative.

Recently, while overseas with a mixed group of people, I was surprised to discover a shared enthusiasm for a reality television programme. Once this connection was made, people were drawn to each other and a kinship formed that sustained conversations, perfectly illustrating what unites us, the forming of opinions around an ongoing human interest story. Witnessing group formation around this shared enthusiasm prompted thoughts about exclusion because this was an experience peculiar to place of origin and isolated non-engaged outsiders, of which I was one. Here was a personal reminder of the importance of remembering varying perspectives and interpretations from different voices especially when we know that humour for example, does not always translate across cultures. The inclusion of mixed examples of voices in places within this book will aid understanding and allow empathetic enhancements in the promotion and management of destinations. Destination marketers must be strategic, balancing the drive for revenue with sustaining the destination for a world that flourishes for humans and non-humans alike.

Shifts in thinking in tourism research show the limitations of traditional volume and value statistics and a striving for the personal and qualitative allowing a better understanding of the more ordinary, everyday events occurring in their natural settings. The Pandemic too has heightened such awareness and emphasized the importance of health and wellbeing including the importance of nature and the rural environment. The striving of the individual to cope enhanced interest in self-care and development with an increase in the keeping of diaries and telling of personal stories. This hyper-personalization was already reflected in a move away from the mass packaged product to more niche, self-actualization experiences facilitated by low-cost airlines and ease of access.

Our lives are dominated by images too and these endure over time available in wide-ranging media. Evidence from the past proves this when in the paintings of Renoir we see he successfully showed the

French as leisurely, taking picnics by the Seine when in reality, France was experiencing considerable economic strife. Paintings have been replaced by photographs and images can create a powerful and varied narrative as was seen in the award-winning Incredible India! Marketing campaign. The development of the narrative in this way is inclusive, reflecting input from wide-ranging stakeholders including local residents empowered by networks often formed around and because of a story.

Tour guides are invaluable in their role in taking up the stories and for their interpretation and information providing when visiting a place. They make choices about what they choose to tell and if done well, may offer a far more personal and vivid account of a place visited. A recent visit to the new Burrell collection outside Glasgow was fascinating because of the selection made by the guide on that day, at that time but again I was reminded that such an opportunity is not always available to suit. Similarly, former miners working in a welsh mining museum reveal in powerful detail, what life was like working in a mine. But we may be herded into a large group on such occasions, where we cannot keep up, hear or see and so the development of self-guided tours allows own pace and the luxury to linger. These small compact audio guides are an example of how information and communications technology have transformed accessibility to information and have made dramatic changes to interpretation, interaction and individual experiences.

Social media has democratized the whole travel experience even further by the sharing of tales from around the world in both words and pictures, generating powerful influencing tools affecting choices in purchasing behavior. This development has elevated the role of consumers into co-creators of brands and their communications as well as ambassadors of destination choice. So much information can be overwhelming and difficult to decipher and so the contents of this book is needed and relevant.

Julia Fallon
Cardiff School of Management, Cardiff Metropolitan University, UK

Preface

IMMERSING IN THE WORLD OF STORIES: A LOOK INTO RESEARCH IN TOURISM

Stories contain and tell the ingredients of life: adventure, excitement, encounters, great deeds, herculean challenges, and countless other things. In collective imagination, many stories start with 'once upon a time' and end with 'they lived happily ever after', transporting imagination to a distant past or a blissful and stable future. Story content seems to be as old as human beings. But change, as the overarching principle of life, has created new ways for people to tell stories. Many generations ago, stories were transmitted orally (Yoder-Wise & Kowalski, 2003). With the development of written language, human ingenuity has invented a variety of substrates to use as vehicles for communicating and preserving historical or fictional narratives. Papyruses, parchments, wax tablets, or paper books have allowed the perpetuation of stories in collective life. Nowadays, *storyworlds* are being created with the aid of digital technologies, continuing this process of expanding reality, meaning, and audience reach (Alexander, 2017).

Kreuter and colleagues (2007, p. 222) describe a story "as a representation of connected events and characters that has an identifiable structure, is bounded in space and time, and contains implicit or explicit messages about the topic being addressed". A story has been said "authentic when it is perceived to be 'original' or 'the real thing'" (Almeida, 2022, p. 42). Human beings read, tell, listen to, and stage stories at all hours and places. Stories are acknowledged as a means to communicate a message to an audience. Teachers and lecturers use them in learning activities; journalists find and report events as stories to their readers; people immerse in a story when watching a movie or listening to a music; a casual conversation between two friends is comprised of stories; and readers most often than not grab a book to read one more breathtaking story.

Stories are central to human life and an important part of tourism practice, from the perspectives of both providers and consumers (Lundqvist, Liljander, Gummerus & van Riel, 2012; Nozdreva & Churakova, 2021). On the one hand, tourists are eager to listen to stories about the destination, and also to share with friends and relatives their lived meaningful events during a trip. On the other hand, destinations build identity and distinctiveness by sharing stories with visitors. Stories convey meaning about the destination and are also powerful tools for eliciting emotional responses in visitors (Nabi & Green, 2015). And plenty of evidence shows stories develop emotional bonds between people, and between people and organizations and brands (Escalas, 2004).

Stories, their characteristics, uses and effects have been a topic of research for decades in diverse sciences, such as anthropology and ethnography, history, sociology, linguistics, psychology and more recently neuroscience (Moscardo, 2020b). Storytelling research in tourism draws from various disciplines, however there is need to better address the scientific and theoretical foundations of research on stories in tourism (Moscardo, 2020a). Additionally, exploration of applied sciences and methodologies, such as design thinking, may contribute to a better design of tourism and tourism experiences. Accordingly, progress in innovative approaches in applied areas have been argued as critical to development in tourism (Moscardo, 2020b). Practical issues related to the management and marketing of tourism destination, business, and experience will expectedly benefit from developing knowledge in this area (Campos & Almeida, 2022).

The book *Global Perspectives on Strategic Storytelling in Destination Marketing* focuses on the transformation that hopefully will shape the future of tourism and destination marketing. Its overall aim is to provide readers an overview of storytelling theory and practice in the field of tourism, covering theory, case studies, and applications of interest to both academics (researchers, teachers and students) and practitioners. As a publication dedicated to the theme, it comprises latest research and multiple perspectives on storytelling use across the globe which contribute to the body of knowledge in tourism. The editors believe that the different approaches to stories and storytelling included in this book contribute to a competitive and innovative tourism practice, which increasingly integrates the perspectives of communities and public bodies, businesses, and visitors.

The book is organized in three sections and composed of 13 chapters.

Section 1, "Storytelling, Heritage, and Experience Design and Management", includes six chapters addressing issues that relate storytelling with heritage, and also the design and management of tourism experiences. Chapter 1, "Storytelling and Experience Design in Heritage Tourism", discusses how storytelling is applied in the context of heritage experience design taking into account developments in museology theory and practice. The approach is based on a review of prior work on the subject paying particular attention to frameworks recently developed to support the design of experiences, with a particular focus on the heritage experience. The authors critically analyze these frameworks, reaching the conclusion that storytelling can be considered an element of design, an outcome of design, and a guiding principle for design. In Chapter 2, "Storytelling in Cultural Heritage: Tourism and Community Engagement", the author theoretically explores the paradigm change in museum management and how heritage is currently being proposed to visitors as they become more interested in participative approaches to the heritage experience and also how technologies and co-creation are shaping future developments in the area. Chapter 3, "Autoethnography of Cultural Heritage Tourism Documentary Production: A Story of Tourism Storytelling", reports an autoethnographic research conducted during production of a tourism documentary for television which focuses on cultural heritage stories. This autoethnographic approach to destination storytelling highlights ethical dilemmas involved in story crafting with the aim of sharing inspirational stories about people and places able to support the newly created brand of Kentucky Wildlands. Chapter 4, "The Development of Stories of Bajo Mola in Wakatobi National Park", concentrates on the concept of 'interpretive experience' to approach the design and development of tourism experiences in Indonesia based on stories as interpretation tools of the destination and thoroughly describes the stages of the process. The study highlights the active role played by the local community in this process, but most importantly the values it derived from this involvement, like social and cultural cohesion, sense of pride, or higher awareness of cultural identity and assets. Chapter 5, "From Story-Based to Story-Interactive Experiences: Layers of Narrative Application in Theme Parks and Attractions", revisits theme park literature and research, 'one of the most narrative forms of tourism', to uncover narrative strategies, techniques and styles applied in theme parks. As it happens with many attractions, this chapter is a stimulating ride into the physical spaces of theme parks and how they integrate one or more strata of narrative integration, helping readers to become aware of the diversity and richness of storytelling, drawing from examples from around the world. Chapter 6, "The Influence of Storytelling Strategies on Attitude and Purchase Intention", reports a research which interprets digital storytelling as a set of strategies that tourism companies or destinations can implement to develop proximity with customers by effectively communicate a value proposition. The study concludes that these strategies positively affect individual attitude and purchase intention.

Section 2, "Storytelling and Destination Marketing and Management", comprises of three chapters which provide evidence of the importance of storytelling in the management and marketing of destinations. Studies selected for this section show how stories can be used to develop place branding in close articulation with place history, tradition, and people, as a way, on the one hand, to perpetuate cultural identity and, on the other, to connect at a deeper level the visitor with the destination culture. In Chapter 7, "Promoting Tourism Destinations Through Storytelling", authors emphasize the power of stories in amplifying the quality of the destination's attributes. In their words destination storytelling "is a strategic tool in "wrapping" or promoting destinations". This study highlights the importance of destination promotion as a strategy to gain competitive advantage in a globalized world through communication of distinctive cultural and natural resources and values. Stories are culture-specific and have the power to disseminate local knowledge through the voices of local stakeholders, such as community volunteers' guides, and other local providers. Chapter 8, "Destination Storytelling as a Destination Branding Tool", explores story elements important in knowledge and culture transfer from generation to generation as a way to perpetuate the authenticity of the destination. The study concludes that as stories reflect the authenticity and originality of the destination, storytelling is strongly recommended to build the destination brand. Lastly, in Chapter 9, "Storytelling by the Sea: The Legend of Ílhavo as a Reinvented Tourism Destination", authors report how actors have incorporated stories into the process of marketing the destination to change community's and visitors' perception of its character, which eventually transformed it from a sun and sea destination to a cultural destination.

In Section 3, "Storytelling, Information and Communication Technologies, and Social Media", four chapters discuss the relation between storytelling, ICT, and social media. Currently, digital mobile technologies are part of the overall tourism experience, facilitating global communication among consumers, and accordingly personal stories derived from tourism experiences are common contents found in digital platforms and social media. In Chapter 10, "Cinematic storytelling With Virtual Reality for Cultural Heritage Marketing and Accessibility", explores the application of VR as way to implement experiential marketing and the accessibility of cultural heritage sites. The chapter describes a case study on immersive experiences the Archaeological Park of Cuma (Italy) using cinematic storytelling and exploring its role in the creation of 'invisible' heritage, which is particular important concerning sites that are not fully accessible to visitors. Importantly, authors present the design process used to create a cinematic experience (CVR) based on a user-centered approach. In Chapter 11, "Digital Storytelling and 3D Technologies for Visitor Experience and Heritage Preservation", authors discuss the challenges involved in the preservation of vulnerable heritage through the case of the Willandra 20,000 years old Aboriginal fossil footprint tracks in Mungo National Park, Australia. The case is set to clarify forms of heritage contestation that impact on heritage interpretation and management, and also how heritage digitalization may help in overcoming obstacles in preservation, leading to new opportunities to create immersive experiences by combining 3D technologies and storytelling. Chapter 12, "Show Me a Story:

Social Media-Based User-Generated Videos and Destination Brand Engagement", reports a research adopting the method of narrative analysis to study user-generated storytelling videos, which are seen to increase destination engagement. The interesting perspective is that user-generated videos co-construct places in a dynamic process of storytelling, re-telling, interpretation and re-interpretation which creates visitor engagement in a way that traditional advertising is unable to do. The study concludes that much research is needed to explore the specific methods of research for these new communication media. Chapter 13, "Social Media as Storytelling Destination Communication Channel: The Case of Fátima Sanctuary", discusses how social media, in particular Facebook platform, work as a tool to promote destinations online with a focus on a religious tourism destination. The study concludes that online presence alone is not sufficient to effectively promote the destination and social media platforms are not being used to its full potential in connecting destinations with current and prospect visitors.

With this selection of research on tourism and storytelling, editors hope this book may inspire academic and non-academic readers' future work on the topic. Given that human life is a never-ending story, the best one is not written yet.

Ana Cláudia Campos
CinTurs, Universidade do Algarve, Portugal

Sofia Almeida
Faculty of Tourism and Hospitality, Universidade Europeia, Portugal

REFERENCES

Alexander, B. (2017). *The new digital storytelling: Creating narratives with new media*. Praeger.

Almeida, S. (2022). Stories. In *Encyclopedia of tourism management and marketing*. Edward Elgar Publishing. doi:10.4337/9781800377486.stories

Campos, A. C., & Almeida, S. (2022). Unique stories for unique brands: appealing to the hotel guest experience through digital storytelling. In R. Nunkoo (Ed.), *Handbook on the Tourist Experience: Design, Marketing and Management. Edward Elgar Publishing.*

Escalas, J. E. (2004). Narrative processing: Building consumer connections to brands. *Journal of Consumer Psychology*, *14*(1/2), 168–180. doi:10.120715327663jcp1401&2_19

Kreuter, M., Green, M. C., Cappella, J., Slater, M., Wise, M., Storey, D., & Woolley, S. (2007). Narrative communication in cancer prevention and control: A framework to guide research and application. *Annals of Behavioral Medicine*, *33*(3), 221–235. doi:10.1007/BF02879904 PMID:17600449

Lundqvist, A., Liljander, V., Gummerus, J., & van Riel, A. (2012). The impact of storytelling on the consumer brand experience: The case of a firm-originated story. *Journal of Brand Management*, *20*(4), 283–297. doi:10.1057/bm.2012.15

Moscardo, G. (2020a). The story turn in tourism: Forces and futures. *Journal of Tourism Futures*, *7*(2), 168–173. doi:10.1108/JTF-11-2019-0131

Moscardo, G. (2020b). Stories and design in tourism. *Annals of Tourism Research*, *83*, 102950. doi:10.1016/j.annals.2020.102950

Nabi, R. L., & Green, M. C. (2015). The Role of a Narrative's Emotional Flow in Promoting Persuasive Outcomes. *Media Psychology*, *18*(2), 137–162. doi:10.1080/15213269.2014.912585

Nozdreva, R., & Churakova, A. (2021). Digital Tools in Empirical Marketing Strategies of Global Hotel Companies. In N. Konina (Ed.), *Digital Strategies in a Global Market: Navigating the Fourth Industrial Revolution* (pp. 213–227). Palgrave Macmillan. doi:10.1007/978-3-030-58267-8_16

Skift. (2020). Available at: https://skift.com/destination2020/

Yoder-Wise, P. S., & Kowalski, K. (2003). The power of storytelling. *Nursing Outlook*, *51*(1), 37–42. doi:10.1067/mno.2003.2 PMID:12595827

Acknowledgment

The editors hereby acknowledge the contribution of all colleagues and fellow researchers in the successful achievement of this book project. A special thanks goes to all academia researchers that willingly and enthusiastically participated in the review process and helped achieving the expected standards of quality in research and academic reporting. Without their commitment to the academia, this book would not have been possible.

A special thanks goes to Julia Fallon, who kindly wrote the Foreword to this book.

This book is financed by National Funds provided by FCT, Foundation for Science and Technology through projects UIDB/04020/2020 and UIDB/00295/2020+UIDP/00295/2020..

Section 1

Storytelling, Heritage, and Experience Design and Management

Chapter 1
Storytelling and Experience Design in Heritage Tourism

Meghan Claire Beevor
Faculty of Economics, University of Algarve, Portugal

Ana Cláudia Campos
https://orcid.org/0000-0002-5816-5137
CinTurs, Universidade do Algarve, Portugal

Maria Manuela Guerreiro
https://orcid.org/0000-0002-6398-9712
Faculty of Economics, CinTurs, University of Algarve, Portugal

ABSTRACT

Stories have long been a central component to the interpretation of people, culture, and place. Recent research has substantiated the importance of storytelling in heritage tourism settings; however, few studies have provided an overview of storytelling in heritage experience design. This has left academics and service providers alike with little to no guidance as to how to effectively utilize storytelling in their experience offering. This chapter serves to synthetize what is known on the topic of storytelling in heritage experience design. It will explore the relevance of storytelling and experience design in the context of new museology, give an overview of pre-existing frameworks on the subject, and offer a critical analysis highlighting informational gaps and areas of expansion. Beyond guided tours, interpretative signs, and guidebooks, one thing is clear: if they are to facilitate experiences that are more memorable, engaging, and interactive, heritage attractions must utilize storytelling as a design tool.

INTRODUCTION

Storytelling is part of human nature (Gottschall, 2012; Moscardo, 2020b). It seems humans are compelled to tell stories, whether it be during daily activities, at meals, while travelling, or even while daydreaming (Bassano et al., 2019). Storytelling can be viewed as a core component of human existence, as people use stories to facilitate communication (Moscardo, 2020b), understand the world around them (Dawson

DOI: 10.4018/978-1-6684-3436-9.ch001

& Sykes, 2018), as well as shape their identities (Bassano et al., 2019). It can be defined as "a two-way interaction, written or oral, between someone telling a story and one or more listeners" (Sundin et al., 2018: 1). Stories are a specific type of narrative that must provoke emotional responses (Moscardo, 2020b). They usually tell a sequence of related events for entertainment purposes, including a challenge to the main character(s), the reaction and journey of the character(s) in response to the challenge, as well as some sort of resolution (Moscardo, 2020b; Popova, 2015). They also tend to activate imaginative processes by connecting with people's experiences and emotions (Nielsen, 2017).

Storytelling has been extensively studied in the disciplines of anthropology, psychology, and sociology, but research has just begun exploring its implications in tourism (Moin et al., 2020; Moscardo, 2020b). Recent findings are solidifying storytelling as a crucial component to the design of tourist experiences (Moscardo, 2017a, 2017b, 2018, 2020b). Stories have only recently begun to be examined in more detail in tourism research (Moscardo, 2020b), although some studies have substantiated the importance of stories for tourist experiences (Moscardo, 2010, 2017a, 2017b, 2020a, 2020b). Perhaps stories have not received significant attention in tourism research due to a prevalence of tourism researchers who've adopted modernist and positivist approaches within traditional academic institutions (Munar, 2016; Tribe & Liburd, 2016).

For the managers of heritage attractions, storytelling has presented itself as a co-creative tool that effectively engages audiences and leads to memorable tourist experiences (Bec et al., 2019; Mossberg, 2008; Pine & Gilmore, 1998; Robiady et al., 2020; Ross & Saxena, 2019; Sundin et al., 2018). Moreover, new museums consider elements such as meaning-making, interpretation, and interaction to be central to the modern visitor's experience; yet, there is little guidance for museums as to how to connect these elements with their museological practices (Nielsen, 2017). This is not surprising, as little research has been carried out regarding the implications of storytelling in heritage tourism interpretation and experience design (Moscardo, 2008, 2017a, 2017b, 2020). In fact, the question of how stories might be used to guide the design process of tourist experiences has been left largely unexplored (Moscardo, 2017b), whether it be in heritage contexts or others.

This chapter serves to synthetize what is known on the topic of storytelling in heritage experience design, with the primary objective of assisting academics and practitioners in utilizing storytelling as a design tool for memorable, engaging, and interactive heritage tourism experiences. The chapter will first explore the relevance of storytelling and experience design in the context of the experience economy and new museology. It will then give an overview of pre-existing frameworks on the subject, providing insight on these topics in tourism overall and in heritage tourism, specifically. These frameworks are accompanied by a critical analysis highlighting informational gaps and possible areas of expansion. Finally, the chapter will end with solutions and recommendations, as well as directions for future research for storytelling and experience design in heritage tourism contexts.

BACKGROUND

The Impact of the Experience Economy and Co-Creation on Heritage Tourism

The Experience Economy (Pine and Gilmore, 1998) concept had a profound effect on the tourism industry and brought important changes to the way heritage attractions and museums do business. Heritage attractions – that have long been dominated by a 'curatorial' strategy – had to adjust as experientially-motivated tourists became disinterested in traditional heritage attraction or museum settings (Leighton, 2007; Mijnheer & Gamble, 2019). In addition, increased competition between heritage attractions resulted in a challenging marketplace, ultimately leading to attractions adopting the belief that an experiential approach could set them apart from competitors and improve revenues (Frost et al., 2020; Leighton, 2007). Heritage sites have thus begun transforming their consumer offering into one that is more hospitable, accessible, and engaging (Magnani et al., 2018) by means of co-creation (Frost et al., 2020; Mijnheer & Gamble, 2019).

Co-creation can be broadly defined as the process of creation of products or services in which the customer has invested individual resources such as time, effort, or ability (Prahalad & Ramaswamy, 2004; Sugathan & Ranjan, 2019). This process is inherently interactive in nature, as collaborating parties invest their resources to create value (Buonincontri et al., 2017; Frow et al., 2015). Much research has considered the topic of co-creation in the tourism context (Berrada, 2017; Buonincontri et al., 2017; Campos et al., 2016; Mathis et al., 2016; Prebensen & Xie, 2017; Sugathan & Ranjan, 2019). As recent studies in tourism have focused on the importance of experiences and, more precisely, the role of the tourist in the creation of their own experiences (Prebensen et al., 2013), the concept of co-creation in this context can be understood as "the sum of psychological events a tourist goes through when contributing actively through physical and/or mental participation in activities and interacting with other subjects in the experience environment" (Campos et al., 2018: 391).

The concept of co-creation and value creation have also been explored in archaeological and heritage tourism contexts (Frost et al., 2020; Gao et al., 2020; Ghisoiu et al., 2017; Mijnheer & Gamble, 2019; Ross & Saxena, 2019; Van Der Linde & Mans, 2015). Heritage can be defined as "what is or may be inherited" (Drummond, 2001: 6) or as how modern society utilizes the inherited past (Di Pietro et al., 2018). It can also be described as an important asset that helps us understand the past, face the present, and shape the future (Di Pietro et al., 2018). For tourism purposes, heritage can "be conceptualized as a product, a marketable commodity, although in reality it is multi-faceted, embodying notions of scholarship, culture and personal identity" (Leighton, 2007: 118). Heritage places can include a wide variety of attractions that make for lucrative tourism opportunities, ranging from small-scale local properties, to large attractions that are central to a destination's tourism strategy (Mijnheer & Gamble, 2019).

A co-creative approach has allowed some heritage attractions to regain their competitive edge in the market (Leighton, 2007) by improving customer perceived value of their experience (Meng & Cui, 2020; Prebensen et al., 2013; Prebensen et al., 2014; Prebensen & Foss, 2011). Considering the Service Dominant Logic (SDL) paradigm proposed by Vargo and Lusch (2004), the success of this new approach may be due to the recognition and understanding of how value is co-created between consumer and provider via a collaborative experience. To meet the demands of experientially motivated tourists and remain competitive in a challenging tourism market (Berrada, 2017; Frost et al., 2020; García Haro et al., 2014; Leighton, 2007; Prahalad & Ramaswamy, 2004), heritage attractions need prioritize the implementation of co-creative activities, such as storytelling, in their experience offering (Mijnheer & Gamble, 2019; Wyman et al., 2011).

The rise of storytelling within experience design

Experience design is the ancient practice of designing with the aim of providing a certain kind of experience (McLellan, 2000). It is difficult to give a straight forward definition of the term "design", as it has been used rather loosely in the literature (Tussyadiah, 2014). As Heskett (2005: 3) cleverly puts it: "Design is to design a design to produce a design." In other words, design can be understood at once as a general concept, an activity, a plan, and an outcome (Tussyadiah, 2014). There are three different traditions within design: design science (Fuller, 1975), the science of design (Simon, 1969), and design thinking (Brown, 2009). While these traditions have all developed from distinctive disciplines, they all share a focus on the concepts of stories and experiences (Moscardo, 2017b). Although there is some mention of stories being able to provide a holistic perspective on design (Hatchuel, 2001), as well as stories being considered particularly important within the applied research areas of design science (Holloway et al., 2016) and design thinking (Liedtka, 2018), stories have not been fully analyzed in design literature (Moscardo, 2017b).

Research has only just begun exploring the implications of storytelling in tourism (Moin et al., 2020; Moscardo, 2020b). However, an emerging body of knowledge is beginning to demonstrate its relevance in the "experience tourism" era of the twenty-first century (Bassano et al., 2019; Moin et al., 2020; Moscardo, 2020b; Mossberg et al., 2010; Weiler & Black, 2015). Many researchers have highlighted the potential in further exploring the intersection between experience design and storytelling. In her many works, Moscardo (2010, 2017b, 2018, 2020) has notably addressed this topic the most. "Combining design and systems thinking with the conclusions from social science research highlighting the value of stories as holistic structures for organizing, storing and communicating information, suggests that a focus on stories could provide valuable guidance for innovation in tourism design" (Moscardo, 2020: 2). Some research has stressed the importance of stories as a central element in designing experiences, suggesting that stories provide a memorable and compelling format for transferring information and discoveries (Brown et al., 2016).

According to Moscardo (2020), co-creative storytelling opportunities can include: 1) stories about the destination told to tourists face to face by guides (Weiler & Black, 2015); 2) audio-visual presentations, in text on signs and in guidebooks, as well as through digital platforms, such as mobile apps (Katifori et al., 2018; Page, 2012); 3) activities that are structured to encourage tourists to enact their own personal story (Mossberg et al., 2010); and 4) transmedia storyworlds explored across different media and locations (Hansen et al., 2012). In their study on heritage and storytelling in wine tourism, authors Frost et al. (2020) found that storytelling was an effective way of engaging visitors and creating a competitive advantage. Similarly, Robiady et al. (2020) found that storytelling techniques had a significant positive effect on customer engagement. A study by Ross and Saxena (2019) explored the concept of storytelling, but focused on how creative storytelling can generate meaning-making via participative co-creation at heritage sites where tangible remains have been lost. The authors explain that meaningful tourist experiences are first and foremost determined by 1) the mysticism surrounding "lost" heritage, 2) themed activities and 3) creative storytelling (Ross & Saxena, 2019).

State of the art storytelling experiences in heritage settings often include digital technology (Dal Falco & Vassos, 2017; Gröppel-Wegener & Kidd, 2019; Kidd, 2018). In fact, digital storytelling has taken shape as an effective design tool for creating engaging experiences for visitors (Bassano et al., 2019; Katifori et al., 2018; Katifori, Perry, et al., 2020; Katifori, Tsitou, et al., 2020; Kidd, 2018). Online mobile applications in particular have been the focus of multiple studies (Lombardo & Damiano, 2012), offering a new way to create dialogue between the visitor and the museum (Dal Falco & Vassos, 2017). Therefore, with the emergence of new digital components comes the need to analyze and understand how these can be integrated into the design of tourist experiences, and more specifically digital storytelling experiences in heritage settings.

New museology

Traditionally, museums were places of scholarly devotion that were primarily focused on object-based displays and material authenticity in curation (Baker et al., 2016). However, the 'new museum' "attempts to flip the traditional 'top-down' approach to curation, transforming both the presentation and appeal of these institutions as well as the changing ways in which museum content is perceived by patrons" (Baker et al., 2016: 370). From the visitor's perspective, traditional museums were often perceived as rigid and lacking interactivity (Dal Falco & Vassos, 2017). New museums, on the other hand, share power with communities and recognize the co-construction of knowledge between the museum and the visitor (Marstine, 2006).

Although storytelling has been long employed by museums, the approach has evolved from one of transmitting an authoritative narrative to offering a multi-faced and interactive experience (Wyman et al., 2011). Digital technologies have transformed museums into hybrid and complex spaces by blending the virtual lives of characters and stories with physical (Dal Falco & Vassos, 2017; Gröppel-Wegener & Kidd, 2019; Kidd, 2018). The new museum is primarily interested in visitors *participating* in its stories, and more specifically interacting with artefacts in original and participatory ways (Solima, 2012) using digital technologies (Dal Falco & Vassos, 2017). In other words, the goal is to engage the visitors in memorable and emotionally relatable narrative structures. Visitors are increasingly expecting to be emotionally stimulated by art and artefacts, rather than having a traditional experience based on rational involvement (Dal Falco & Vassos, 2017).

As visitors become less interested in traditional offerings, museums have begun investing in new methods of visitor engagement (Barron & Leask, 2017). Storytelling is now considered as one of the most important tools for creating meaning, as it can ensure emotional engagement among visitors and staff (Nielsen, 2017). Meaning-making can be defined as "the process of making sense of experience, of explaining or interpreting the world to ourselves and others" (Hooper-Greenhill, 1994: 12). However, stories in and of themselves do not necessarily create meaning – it is how stories are understood on an unconscious level that will ultimately make them meaningful to the individual. This process is entirely subjective, as the narrative structure stimulates unique feelings and memories for each person (Nielsen, 2017).

As the process of meaning-making primarily occurs as patrons visit heritage attractions (Smith, 2014), there is a need to better understand how heritage professionals can tailor their experience to facilitate engagement and, consequently, meaning-making. According to Bedford (2001), storytelling is the 'real work' of the museum, as it supports individual interpretation and meaning-making. It is also important to clarify that the process of storytelling is not limited to the stories told by the museum; rather, it includes visitors making up their own stories (Nielsen, 2017). In this way, visitors in and within themselves become central to the experience as they interact with the museum environment and participate first-hand in performances and events (Dal Falco & Vassos, 2017).

The following section of this chapter aims to explore existing frameworks that could assist 'new museums' and academics alike in understanding how storytelling and experience design can be combined to deliver better heritage tourist experiences.

STORYTELLING AND EXPERIENCE DESIGN

An Exploration and Critical Analysis of Existing Frameworks

Storytelling is a powerful means of engaging audiences (Frost et al., 2020; Sundin et al., 2018). It has been identified as a type co-creative experience that offers much potential in terms of visitor engagement, meaning making, and memorability (Frost et al., 2020; Mossberg, 2008; Robiady et al., 2020; Ross & Saxena, 2019; Sundin et al., 2018). In fact, stories have become a central feature of tourism practice, particularly in tourist experience opportunity design (Moscardo, 2017b, 2020b).

There are three main design principles for tourism: creating storyworlds, utilizing specific characteristics that make stories enjoyable, and carefully considering the roles of tourists and the destination community in tourism stories (Moscardo, 2020b). The first design principle for stories in tourism is to establish the fundamental elements of a desirable destination storyworld in which multiple stories are connected by key overarching themes. Studies exploring universal story features have posited that survival, overcoming challenges, the creation and maintenance of relationships, altruism, struggles against injustice, and heroism are likely to be enticing themes to destination communities and tourists (Davis & McLeod, 2003; Moscardo, 2010).

Moreover, different features of stories can be linked to different audience reactions (Moscardo, 2020b). Four concepts have been consistently linked to audience attention, engagement, and positive responses to stories, namely: perceived realism, narrative transportation, emotional engagement, and character identification or admiration (Green & Dill, 2013; Nabi & Green, 2015). Perceived realism refers to the belief that a story is authentic. It is based on the perception that the characters and their reactions are comparable to real people, and that the sequence of causal events makes sense (Cho et al., 2012). Narrative transportation refers to the idea that the audience becomes so involved and immersed in the story that they become part of the storyworld itself. Emotional engagement requires that stories have strong emotional elements (Nabi & Green, 2015). Finally, characters should be described in sufficient detail to be accepted and understood by the audience (Green & Dill, 2013). Story applications in tourism should, therefore, follow these design principles: authentic credible characters described in detail (and who are similar to and/or likely to be admired by the targeted tourists), consistent and plausible plots, and strong expressions of emotion (Moscardo, 2020b).

It is also important to think carefully about the roles of tourists and the destination community in the story (Moscardo, 2020b). It is crucial to engage tourists in stories that give them greater control (i.e., give them choices and challenges in stories and variability in how they experience them). Studies have found that, overall, fostering tourist agency in a story is likely to result in more positive experiences, greater learning, and increased likelihood of sharing the story with others (Moscardo, 2017a, 2017b). For the destination community, offering opportunities to share their personal stories, as well as engage in story co-creation with tourists, can be an important way of encouraging the community's participation in tourism (Moscardo, 2020b).

Studies have, therefore, solidified the relevance of storytelling within the experience design of memorable and enjoyable tourist experiences. However, what frameworks exist to guide academics and practitioners in understanding and implementing storytelling as a design tool? In this section, we will explore this question on two levels: experience design frameworks for storytelling in tourism (in general) and experience design frameworks for storytelling in heritage tourism (specifically). A critical analysis will also be provided, addressing any gaps or problems that these frameworks bring to light.

Experience Design Frameworks for Storytelling in Tourism

Tussyadiah (2014) provides a theoretical framework for experience design in tourism. Drawing from a multi-disciplinary review of the literature, human-centeredness, an iterative designing process, and a holistic experience concept are proposed as pillars of tourism experience design. Human-centeredness design is based on the idea that the needs, wants, expectations, and limitations of the end users should be carefully considered at every stage of the design process. An iterative designing process refers to the acknowledgement of service providers that designing is a cyclical process that aims to continuously improve the quality and functionality of the design via adaptive learning. Lastly, a holistic experience concept means designing in a way that considers the complexity and variability of human interaction within an experience environment where meanings and values of the tourist emerge (Tussyadiah, 2014).

At the intersection of these three pillars are the approaches and methodologies for tourism experience design (Tussyadiah, 2014). At the intersection of human-centeredness and the holistic experience concept is naturalistic inquiry, which aims to understand tourist behavior and experiences (including meanings and values) by considering socio-cultural contexts. This requires the gathering of information about tourists in natural experience settings or, more precisely, capturing the experience narratives of tourists to provide a holistic view of these experiences. At the intersection of human centeredness and the iterative designing process is participatory design, or the active engagement of users at all stages of the design process. The goal of this approach is to generate ideas and concepts relevant to tourist experiences (e.g., storyboarding). Finally, at the convergence of iterative designing process and the holistic human concept is integrative research, which refers to the integration of explorative, generative, and evaluative research into the design cycle to test, provide feedback, and monitor the design concepts and systems being created (Tussyadiah, 2014).

Tussyadiah (2014) proceeds to present the tourism experience as an operational and meta-concept. The meta-concept of tourism experience highlights strategic positioning, value propositions, and opportunities for meaning as elements that must come together to form core and peripheral experiences. These are supported by the storytelling of these experiences or, in other words, by the meanings and values associated with these experiences. The operational concept, on the other hand, focuses on the facilitation of these experiences through meaningful application of design principles and the storytelling aspects of these experiences. It should also consider the factors of interactivity and triggers (i.e., how all interactions between the tourist and physical and social elements, as well as the media, may shape their experience).

Although Tussyadiah's (2014) framework provides a much needed theoretical and multi-disciplinary perspective on experience design in tourism, it does not have a central focus on storytelling. Rather, storytelling is interpreted as an outcome (i.e., the product of a successful participatory design) or as a supportive element (i.e., something that supports experiences in their strategic positioning, value propositions, and opportunities for meaning). Therefore, this framework alone cannot fully answer the primary questions of this chapter, as it does not explore the role of storytelling as a design tool or, in other words, as a guiding principle of effective experience design overall. However, it does assist in the understanding of storytelling in its different forms. Like the word "design", the concept of storytelling can often adopt different definitions and meanings depending on the context and field of study in which it is used.

Moscardo (2020), on the other hand, proposes a conceptual framework for stories and storytelling in tourism experience design. Based on this author's previous research (Moscardo, 2017b), five dimensions of stories are highlighted: 1) the level of the story within a story hierarchy; 2) when the story is told or received; 3) the functions the story can fulfill for different stakeholders; 4) whose story is told; and 5) the role of the tourist in the story. These five dimensions are combined into an overall framework and two levels of tourist experience opportunity design are proposed: design for experiences within specific activities, place visits, and attractions, and design for experience of destinations as a whole. Moscardo (2017b) also proposes three categories of stories: 1) those that exist before the tourist experience; 2) those that emerge or unfold during the experience; and 3) those that are told after the experience. Pre-experience stories are those that tourists already have with them when they arrive, while emerging stories are those that are presented as part of the experience or that are created through the experience. The stories told after the experience, for their part, are the stories that tourists share with others and that tourism organizations may share afterwards with tourists (Moscardo, 2017b).

Moscardo's (2020) conceptual story model for design in tourism highlights these three types of stories, namely: the pre-experience, emerging experience, and post-experience story. Three intersecting elements are highlighted for each type. At the center of pre-experience stories are the stories that are desirable to tourists, while the actual tourist experience is central to emerging experience stories. Lastly, at the core of post-experience stories is the meaning and value of the experience for the tourist, the destination, and the tourism provider (Moscardo, 2020b). The three main components for pre-experience stories are the tourist's life story, the destination storyworld, and the tourism provider storyworlds (Moscardo, 2020b). The tourist's life story includes all the stories that the tourist associates with the destination (including their own previous experiences, stories from other tourists and from travel writers, and stories from fiction and popular culture). The destination's storyworld, on the other hand, includes all the stories about that destination (including the stories of residents, of its history, culture, and environment). The storyworlds of tourism providers, for their part, include the stories of their staff and their history with the destination (Moscardo, 2020b). The three central elements for emerging experience stories are tourist agency, experience opportunity management, and features of the story that are incorporated into these opportunities (Moscardo, 2020b). Tourist agency refers to the tourist's role in, and engagement with, the experience opportunities provided, while experience opportunity management are the features, activities, and information provided to the tourist. Story features include the decisions made surrounding the story that was told (Moscardo, 2020b). Finally, the three essential dimensions of post-experience stories are the same as those in pre-experience stories, except now they are all now changed due to tourists being in the destination (Moscardo, 2020b).

Moscardo's (2020) framework clearly acknowledges that effective experience design is centered around stories and that tourist experiences are intrinsically linked to stories. However, it does lack some substance from an experience design perspective, as it addresses the topic on a purely theoretical level rather than a practical one. In their earlier work, Moscardo (2017b) does provide guidance to designers by suggesting twelve design principles within their story framework: 1) Determine the key story dimensions, 2) Consider story content carefully, 3) Offer surprise, build suspense and encourage curiosity, 4) Plan story pacing around a main event, 5) Establish authentic characters, 6) Support physical orientation and access, 7) Provide good cognitive orientation, 8) Provide choice and control for tourists, 9) Include appropriate challenges for tourists, 10) Encourage learning, 11) Build in connections, 12) Be consistent.

However, there is something to be said about the difference between *designing stories* and *designing experiences using storytelling as a design tool*. For example, what steps can designers practically take to favour the creation of post-experience stories? What opportunities need to be made available to tourists for these stories to come to be? The tourism provider could include steps within their experience design that address questions such as these. In this scenario, a potential practical design step could be to plan a point of contact with the tourist after the experience to encourage them to share their story on social media (e.g., by reaching out via email, Instagram, etc.). These are impactful design considerations that can have a significant influence on the number, quality, and impact of the stories being told.

Overall, both Tussyadiah's (2014) and Moscardo's (2017b, 2020) frameworks are useful in providing a foundation on which we can begin to understand the purpose of storytelling and stories in tourist experience design. This analysis provides two key takeaways considering the primary question of this section *(i.e., What frameworks exist to guide academics and practitioners in understanding and implementing storytelling as a design tool?)*. First, no framework can address the concept of storytelling in all sense of the word within the context of experience design. The author has no choice but to decide what storytelling means within their theoretical framework to provide insight on the topic of interest. This will ultimately lead to the framework having some blind spots when it comes to understanding a conceptual phenomenon. Second, theoretical frameworks may sometimes lack applicability in their goal to link concepts on a more abstract level. Providing concrete examples for realizing the framework in a real service context will assist designers in their task and should be more common in tourism literature. In addition, neither framework specifically addresses experience design for storytelling in heritage tourism contexts specifically, leaving a gap for the special considerations that might exist within these circumstances.

Experience Design Frameworks for Storytelling in Heritage Tourism

Tussyadiah (2014) and Moscardo (2017b, 2020) supply us with some answers regarding the question of storytelling within tourist experience design. However, if we are to understand this topic within the specific context of heritage tourism, further exploration is needed. This section serves to present the existing frameworks that address experience design for storytelling in heritage tourism, as well as provide a continuation of the critical analysis.

Chu and Mazalek (2019) present a Tangible and Embodied Narrative Framework (TENF) to design digitally-enhanced experiences that engage museum visitors with heritage artifacts. Inspired by the works of Ryan (2015) and Murray (2017), the framework developed by Chu and Mazalek (2019) is based on three spectra: 1) diegetic vs. non-diegetic, 2) internal vs. external, 3) ontological vs. exploratory. These spectra are meant to guide experience designers in integrating the design elements of physical engagement, narrative role, and narrative consequence in various ways. The diegetic vs. non-diegetic dimension refers to the visitor's position as they interact with the experience environment. Said differently, it determines whether a visitor can physically situate themselves within a story. In a diegetic experience, the interaction of the visitor motivates them to immerse themselves in the context of the story. On the other hand, non-diegetic experience is one where the visitor is limited to interacting with symbolic representations of narrative elements to interpret the story. The internal vs. external spectrum is focused on the visitor's narrative perspective in the story. An internal experience describes an experience where the visitor is part of the virtual story world, whether it be from a first-person or third person perspective. Alternatively, an external experience is one where the visitor situates themselves outside the virtual story world and interacts with the story like a god would. Lastly, the ontological vs. exploratory captures how visitors interact with the story's plot. An experience is ontological when a visitor's actions alter the events of a story (Ryan, 2015), while an exploratory experience is when the visitor can choose to navigate through a story in various ways by adopting different perspectives, but cannot alter the ending of the story (Chu & Mazalek, 2019).

Storytelling projects at museums or heritage sites can use the TENF to guide the design of exhibits that promote experiences where visitors can contextualize the meaning of artifacts and make personalized connections with them (Chu & Mazalek, 2019). In this way, the TENF aids practitioners with the integration of cultural heritage, digital media, and storytelling in a cohesive design plan that supports visitor engagement with cultural materials. However, this framework needs to be further explored and developed in collaboration with museum professionals to identify its value, limitations, and applicability (Chu & Mazalek, 2019; Petrelli et al., 2016). In addition, this framework is very specific in that it is meant to aid in the design of digitally-enhanced exhibits. For these reasons, it may be difficult to implement the TENF without the use of digital media and interactive technologies, which limit its applicability to experience design for storytelling in heritage tourism overall.

Similarly to tangible interactions, blended spaces seamlessly couple the digital and the physical (O'Keefe et al., 2014). Fauconnier and Turner (2002) are the firsts to introduce blending theory, arguing that cognition involves blending mental domains together to create a new domain that has its own structure. Today, the Blended Spaces Framework can be used to help experience designers make decisions that favor the careful blending of digital and physical spaces (Benyon, 2012; Benyon et al., 2013; Jetter et al., 2012). In their work, O'Keefe et al. (2014) explore Blended Theory as a framework for designing heritage storytelling experiences. Blended spaces can be conceptualized in terms of four main characteristics: ontology, topology, agency, and volatility. Ontology of spaces refers to the POIs (Points of Interest) at a particular site. The topology characteristic is focused on how POIs relate to one another, while agency is concerned about the opportunities for action in that space for the visitor. Finally, the volatility characteristic describes how digital content and physical interaction change over time (Benyon, 2012).

Using the Blended Spaces Framework, O'Keefe et al. (2014) designed a heritage story as part of a broader project of developing of an app that aims to enhance the visitor experience to a heritage destination in New York State. In the context of blended spaces, a heritage story can be understood as follows: 1) On an ontological level, it is a collection of pre-packaged POIs that is designed to be attractive to visitors, 2) Considering the topological characteristic, a heritage story is a reason for two or more POIs to be related, 3) From the agency perspective, it is a digital agent that motivates users to navigate in the space, and 4) Factoring in volatility, a story can lead to unexpected changes. Overall, the developed app and heritage storytelling experience were positively received by the participants of the study. Considering the first two characteristics, almost all participants agreed that the heritage story was fun and engaging (ontology) and facilitate navigation around the site (topology). Considering the last two characteristics, almost all participants agreed that the characters added mystery to the visit, which made the visit more interesting overall (agency), and that they would be interested in visiting again should new stories become available (volatility). Therefore, the study concluded that the Blended Spaces Framework provided an efficient way of developing and designing heritage storytelling experiences in mixed reality.

Dal Falco and Vassos (2017) suggest that storytelling experiences in museums should adopt an interaction design, interactive storytelling, and artificial intelligence to promote innovative opportunities in gaining knowledge. An interaction design refers to a design that integrates interactive technologies that offer new ways of experiencing the world around us (e.g., Augmented Reality, Virtual Reality, etc.). Interactive storytelling is storytelling that adapts to the visitor based on their interactions (i.e., personalized stories according to the decisions of the user). Artificial intelligence, in this context, is often used to generate a conversational component to the experience that enables user to ask and receive responses to questions (e.g., 'I want to know more about the owner of this building') or chat for fun using Machine Learning technology. The authors go on to propose four ingredients for design methods that can deliver storytelling experiences. First, it is necessary to gather information on the historical figures, as well as the historical sites and buildings from museum experts. Second, based on this information, authored stories must be created and specifically tailored for interactive experiences. This will require the involvement of experienced authors who know how to generate engaging narratives. Third, these narratives must be implemented in the form of interactive experiences, such as games, within a physical, digital, or hybrid space. This step necessitates the involvement of game designers and interaction design specialists, as well as application developers and engineers for scenario implementation. Finally, the fourth ingredient concerns the museum's branding identity and communication strategy, which should captivate the target audience and bring the museum experience to their attention (Dal Falco & Vassos, 2017).

O'Keefe et al.'s (2014) Blended Theory Framework, Chu and Mazalek's (2019) TENF, and the design principles suggested by Dal Falco and Vassos' (2017) are similar in that they are centered on digital storytelling in the design of heritage tourism experiences. In fact, most of the recent research concerning the topics of experience design and storytelling in heritage tourism contexts tend to focus solely on digital storytelling. Although digital storytelling is a very relevant topic in heritage tourism, there is an academic gap for newer conceptual frameworks for storytelling in the design of heritage tourism experiences that are more holistic in their approach. For example, the creation of an app or interactive digital exhibits may be out of reach for some heritage attractions or museums, rendering these frameworks ineffective for these institutions. Digital technologies can certainly enhance tourist experiences; however, it is important to remember that they are not necessarily a prerequisite to impactful storytelling experiences in heritage tourism contexts.

Analysis: How to compare, summarize, and draw conclusions from these frameworks?

The aforementioned frameworks vary in scope and focus, making their comparison and summation with regards to experience design for storytelling a challenge. While some frameworks are relevant to tourism experiences overall and others to heritage experiences specifically, all can be categorized in terms of how storytelling is conceptualized and positioned within experience design (Table 1).

Table 1. Categorization of frameworks based on the role of storytelling within experience design

Storytelling as a design element	Storytelling as an outcome of a design	Storytelling as a guiding principle for a design
Tussyadiah, 2014 Dal Falco & Vassos, 2017	O'Keefe et al., 2014	Moscardo 2017, 2020b Chu & Mazalek, 2019

The first category consists of frameworks that view storytelling as a component within an overall design. Tussyadiah's (2014) meta concept positions storytelling as a supportive element to the core and peripheral tourist experience. In other words, storytelling is simply viewed as the meanings and values that tourists associate with the overall experience. Similarly, Dal Falco & Vassos, (2017) view storytelling as a necessary 'ingredient' to effective museum experience design, alongside an interaction design and artificial intelligence.

The second category includes those frameworks that consider storytelling as an outcome of a design. Said differently, if an experience design was a recipe, storytelling would be the final dish. In the case of O'Keefe et al. (2014), the goal was to utilize the Blended Spaces Framework to design a heritage storytelling experience. By considering the characteristics of ontology, topology, agency, and volatility, the authors demonstrated how blended spaces can lead to a framework for storytelling experiences in historical sites.

Lastly, the third category groups those frameworks where storytelling is used as a guiding principle for a design. To continue the analogy of cooking a recipe, storytelling in this case is neither an ingredient nor the final product, but the overall inspiration that guides the chef's every decision. Moscardo (2017, 2020b) of designing experiences that take into consideration pre-, emerging-, and post-experience stories and the many factors that come to play within each of these story categories. Chu & Mazalek (2019) TENF aims to aid in the design of digitally-enhanced experiences that engage museum visitors with artifacts. The parameters of this framework are each focused on a specific dimension of storytelling within the visitor experience, allowing designers to make decisions based on the kind experience they intend to create.

SOLUTIONS AND RECOMMENDATIONS

The presented frameworks and their analysis can bring value to heritage experience design by offering a foundation on which experiences can be built on. As of yet, little research has been dedicated to summarizing relevant information within the fields of experience design and heritage experience design. The frameworks identified in this chapter all have their strengths and limitations; however, together they provide a solid base on which the elaboration of new frameworks can take place. Special attention should be paid to the definition of storytelling within the theoretical design, as well as to the applicability of the framework to real service contexts. Researchers should focus their attention on developing a holistic experience design framework for storytelling in heritage contexts specifically, as this feat has not yet been achieved.

Museum professionals and heritage experience designers can implement any of the presented frameworks, or a combination thereof, depending on their goal and how they conceptualize storytelling within the design of their heritage experience. Considering storytelling as a component of experience design, designers can ask themselves how they are shaping the meanings and values that will ultimately support the visitor experience (e.g., brand communication and marketing). If a 'storytelling experience' is the intended outcome of a design, designers can ask themselves how they can offer opportunities for visitor storytelling within the experience environment. Finally, if using storytelling as a guiding principle of the design, designers must consider the stories that take place at every stage of the visitor experience and design to effectively manage and optimize these. Regardless of how designers may conceptualize storytelling, what is most important is that storytelling be regarded as a crucial factor to consider within the practice of heritage experience design.

Future frameworks should be co-designed with museum professionals for better integration of interactive technologies in heritage tourism settings. As museums and heritage attractions increasingly seek to design exhibits that favor the visitor's active participation and stimulate their imagination, experience designers should adjust frameworks to align with these goals. It is also important to evaluate the effectiveness and utility of these frameworks in real service contexts to identify new issues and challenges surrounding the use of technology in museums and heritage attractions (Chu & Mazalek, 2019). There is also some concern surrounding the fact that experience design frameworks for storytelling in heritage tourism contexts tend to focus solely on digital storytelling. Although digital storytelling is currently a very relevant topic for experience designers, frameworks that are broader in their perspective are important when considering factors of accessibility, applicability, and inclusion.

FUTURE RESEARCH DIRECTIONS

Although it is known that many tourists listen to, create, and tell stories about their experiences, very few studies have thoroughly investigated storytelling and even fewer have examined how stories can be used to design better tourist experiences (Moscardo, 2018). Future research opportunities could include a systematic analysis and evaluation of the use of stories in various aspects of tourism. This could guide actual practice and understanding the role of stories for tourists and tourism. Future research papers should also be wary in utilizing the word 'story' without providing a clear definition of the term and an extensive literature review on stories from a multi-disciplinary perspective (Moscardo, 2020a). In addition, the development of case studies and applications will be crucial to the field of experience design moving forward (O'Leary & Fesenmaier, 2017).

Little research has been done to introduce digital storytelling in the context of heritage tourism, despite early evidence that its implementation can lead to improved experiences that are more engaging to visitors. As heritage tourists increasingly seek more interaction in their experiences, new frameworks should explain how to design experiences that use digital storytelling to increase visitor interaction. This research could provide guidance to experts as they attempt to design meaningful digital storytelling experiences in heritage tourism settings (Katifori et al., 2018). However, these frameworks should be inclusive in their approach, accommodating different levels of technological capability to maintain relevance to most museums and heritage sites, regardless of their current state of digital adoption.

Future research directions should include finding ways of using tourism to tell stories of sustainability that can have a greater influence on individuals beyond their experiences as tourists (Moscardo, 2020a). This means designing experiences that enable tourists to enact or tell stories that encourage sustainable living. From an ethical perspective, it is equally important for future studies to consider questions surrounding authenticity and authority (i.e., *Whose stories are told to tourists? Who has the authority to tell these stories or alter these stories?*). Experience designers should also take into account the influence of social media in tourism contexts, and that it is becoming increasingly necessary to design experiences that meet tourists' social media requirements (Moscardo, 2017b).

CONCLUSION

The elaboration of more holistic and multi-disciplinary frameworks that assist practitioners in harnessing storytelling as a design tool, coupled with case studies that demonstrate their applicability in real service contexts, will pave the way for museums and heritage attractions to succeed in the competitive tourism market of the twenty-first century. Tourism design will continue to evolve as we expand our understanding of the elements that impact experience, such as storytelling (O'Leary & Fesenmaier, 2017). It is likely that stories will become prevalent in experience opportunity design in the future. Consequently, experience designers are expected to prioritize increasing tourist engagement and interaction, as well as focus on tourists creating their own stories (Moscardo, 2020a). Beyond guided tours, interpretative signs, and guidebooks, one thing is clear: If they are to facilitate experiences that are more memorable, engaging, and interactive, heritage attractions must utilize storytelling as a design tool.

ACKNOWLEDGMENT

This research was supported by National Funds provided by the FCT - Foundation for Science and Technology (Portugal) through project UIDB/04020/2020.

REFERENCES

Baker, S., Istvandity, L., & Nowak, R. (2016). Curating popular music heritage: Storytelling and narrative engagement in popular music museums and exhibitions. *Museum Management and Curatorship, 31*(4), 369–385. doi:10.1080/09647775.2016.1165141

Barron, P., & Leask, A. (2017). Visitor engagement at museums: Generation Y and 'Lates' events at the National Museum of Scotland. *Museum Management and Curatorship, 32*(5), 473–490. doi:10.1080/09647775.2017.1367259

Bassano, C., Barile, S., Piciocchi, P., Spohrer, J. C., Iandolo, F., & Fisk, R. (2019). Storytelling about places: Tourism marketing in the digital age. *Cities (London, England), 87*, 10–20. doi:10.1016/j.cities.2018.12.025

Bec, A., Moyle, B., Timms, K., Schaffer, V., Skavronskaya, L., & Little, C. (2019). Management of immersive heritage tourism experiencs: A conceptual model. *Tourism Management, 72*, 117–120. doi:10.1016/j.tourman.2018.10.033

Bedford, L. (2001). Storytelling: The Real Work of Museums. *Curator (New York, N.Y.), 44*(1), 27–34. doi:10.1111/j.2151-6952.2001.tb00027.x

Benyon, D. (2012). Presence in blended spaces. *Interacting with Computers, 24*(4), 219–226. doi:10.1016/j.intcom.2012.04.005

Benyon, D., Mival, O., & Ayan, S. (2013). Designing Blended Spaces. *The 26th BCS Conference on Human Computer Interaction (HCI)*, 398–403. 10.14236/ewic/HCI2012.1

Berrada, M. (2017). Co-Creation of the Tourist Experience via Internet: Towards Exploring a New Practice. *Journal of International Business Research and Marketing, 2*(5), 18–23. doi:10.18775/jibrm.1849-8558.2015.25.3003

Brown, J. S., Collins, A., & Duguid, P. (2016). Situated Cognition and the Culture of Learning. *Educational Researcher, 18*(1), 32–42. doi:10.3102/0013189X018001032

Brown, T. (2009). *Change by Design*. Harper Collins. https://www.amazon.com/Change-Design-Transforms-Organizations-Innovation/dp/0061766089

Buonincontri, P., Morvillo, A., Okumus, F., & van Niekerk, M. (2017). Managing the experience co-creation process in tourism destinations: Empirical findings from Naples. *Tourism Management, 62*, 264–277. doi:10.1016/j.tourman.2017.04.014

Campos, A. C., Mendes, J., do Valle, P. O., & Scott, N. (2018). Co-creation of tourist experiences: A literature review. *Current Issues in Tourism, 21*(4), 369–400. doi:10.1080/13683500.2015.1081158

Campos, A. C., Mendes, J., Oom do Valle, P., & Scott, N. (2016). Co-Creation Experiences: Attention and Memorability. *Journal of Travel & Tourism Marketing,* 1–28. doi:10.1080/10548408.2015.1118424

Cho, H., Shen, L., & Wilson, K. (2012). Perceived Realism: Dimensions and Roles in Narrative Persuasion. *Communication Research, 41*(6), 828–851. doi:10.1177/0093650212450585

Chu, J. H., & Mazalek, A. (2019). Embodied Engagement with Narrative: A Design Framework for Presenting Cultural Heritage Artifacts. *Multimodal Technologies and Interaction, 3*(1), 1. Advance online publication. doi:10.3390/mti3010001

Dal Falco, F., & Vassos, S. (2017). Museum experience design: A modern storytelling methodology. *The Design Journal, 20*(sup1), S3975–S3983. doi:10.1080/14606925.2017.1352900

Davis, H., & McLeod, S. L. (2003). Why humans value sensational news: An evolutionary perspective. *Evolution and Human Behavior, 24*(3), 208–216. doi:10.1016/S1090-5138(03)00012-6

Dawson, P., & Sykes, C. (2018). Concepts of Time and Temporality in the Storytelling and Sensemaking Literatures: A Review and Critique. *International Journal of Management Reviews, 21*(1), 97–114. doi:10.1111/ijmr.12178

Di Pietro, L., Guglielmetti Mugion, R., & Renzi, M. F. (2018). Heritage and identity: technology, values and visitor experiences. In Journal of Heritage Tourism (Vol. 13, Issue 2, pp. 97–103). Taylor and Francis Ltd. doi:10.1080/1743873X.2017.1384478

Drummond, S. (2001). Introduction to quality. In *Quality Issues in Heritage Visitor Attractions* (pp. 5–15). Butterworth-Heinemann. doi:10.1016/B978-0-7506-4675-8.50007-1

Fauconnier, G., & Turner, M. (2002). *The way we think: conceptual blending and the mind's hidden complexities*. Basic Books.

Frost, W., Frost, J., Strickland, P., & Smith Maguire, J. (2020). Seeking a competitive advantage in wine tourism: Heritage and storytelling at the cellar-door. *International Journal of Hospitality Management, 87*(102460), 1–9. doi:10.1016/j.ijhm.2020.102460

Frow, P., Nenonen, S., Payne, A., & Storbacka, K. (2015). Managing Co-creation Design: A Strategic Approach to Innovation. *British Journal of Management, 26*(3), 463–483. doi:10.1111/1467-8551.12087

Fuller, R. B. (1975). *Synergetics: Explorations in the Geometry of Thinking*. Macmillan. https://philpapers.org/rec/FULSEI

Gao, J., Lin, S., & Zhang, C. (2020). Authenticity, involvement, and nostalgia: Understanding visitor satisfaction with an adaptive reuse heritage site in urban China. *Journal of Destination Marketing & Management*, *15*(100404), 1–10. doi:10.1016/j.jdmm.2019.100404

García Haro, M. Á., Martínez-Ruiz, M. P., & Martínez-Canas, R. (2014). The Effects of the Value Co-Creation Process on the Consumer and the Company. *Expert Journal of Marketing*, *2*(2), 68–81.

Ghisoiu, M., Bolan, P., Gilmore, A., & Carruthers, C. (2017). "Conservation" and co-creation through film tourism at heritage sites: An initial focus on Northern Ireland. *Revista Turismo & Desenvolvimento*, *1*(27/28), 2125–2135.

Gottschall, J. (2012). *The Storytelling Animal: How Stories Make Us Human*. Houghton Mifflin Harcourt.

Green, M. C., & Dill, K. E. (2013). Engaging with Stories and Characters: Learning, Persuasion, and Transportation into Narrative Worlds. In K. Dill (Ed.), *Oxford Handbook of Media Psychology* (pp. 449–461). Oxford University Press. doi:10.1093/oxfordhb/9780195398809.013.0025

Gröppel-Wegener, A., & Kidd, J. (2019). Introduction: The seductive power of immersion. In *Critical Encounters with Immersive Storytelling*. https://www.taylorfrancis.com/books/mono/10.4324/9780429055409/critical-encounters-immersive-storytelling-alke-gröppel-wegener-jenny-kidd

Hansen, F. A., Kortbek, K. J., & GrønbÆk, K. (2012). Mobile Urban Drama: Interactive storytelling in real world environments. In *New Review of Hypermedia and Multimedia* (Vol. 18, Issues 1–2, pp. 63–89). Taylor & Francis Group. doi:10.1080/13614568.2012.617842

Hatchuel, A. (2001). Towards design theory and expandable rationality: The unfinished program of Herbert Simon. *The Journal of Management and Governance*, *5*(3–4), 260–273. doi:10.1023/A:1014044305704

Heskett, J. (2005). *Design: A very short introduction*. Oxford University Press. doi:10.1093/actrade/9780192854469.001.0001

Holloway, S. S., van Eijnatten, F. M., Romme, A. G. L., & Demerouti, E. (2016). Developing actionable knowledge on value crafting: A design science approach. *Journal of Business Research*, *69*(5), 1639–1643. doi:10.1016/j.jbusres.2015.10.031

Hooper-Greenhill, E. (1994). *The Educational role of the museum*. Routledge.

Jetter, H.-C., Geyer, F., Schwarz, T., & Reiterer, H. (2012). Blended Interaction: Toward a Framework for the Design of Interactive Spaces. *AVI 2012 : International Working Conference on Advanced Visual Interfaces*. https://kops.uni-konstanz.de/handle/123456789/43527

Katifori, A., Karvounis, M., Kourtis, V., Perry, S., Roussou, M., & Ioanidis, Y. (2018). Applying interactive storytelling in cultural heritage: opportunities, challenges and lessons learned. In R. Rouse, H. Koenitz, & M. Haahr (Eds.), *Interactive Storytelling. ICIDS 2018* (Vol. 11318, pp. 603–612). Springer. doi:10.1007/978-3-030-04028-4_70

Katifori, A., Perry, S., Vayanou, M., Antoniou, A., Ioannidis, I.-P., McKinney, S., Chrysanthi, A., & Ioannidis, Y. (2020). "Let Them Talk!": Exploring Guided Group Interaction in Digital Storytelling Experiences. *Journal on Computing and Cultural Heritage*, *13*(3), 21. doi:10.1145/3382773

Katifori, A., Tsitou, F., Pichou, M., Kourtis, V., Papoulias, E., Ioannidis, Y., & Roussou, M. (2020). Exploring the Potential of Visually-Rich Animated Digital Storytelling for Cultural Heritage. In *Springer Series on Cultural Computing* (pp. 325–345). Springer., doi:10.1007/978-3-030-37191-3_17

Kidd, J. (2018). "Immersive" Heritage Encounters. *The Museum Review, 3*(1). http://orca.cf.ac.uk/id/eprint/110788

Leighton, D. (2007). "Step back in time and live the legend": Experiential marketing and the heritage sector. *International Journal of Nonprofit and Voluntary Sector Marketing, 12*(2), 117–125. doi:10.1002/nvsm.288

Liedtka, J. (2018). Why design thinking works. *Harvard Business Review, 96*(5), 72–79. https://designatdarden.org/app/uploads/2017/07/DSWP13-01rev2-1.docx

Lombardo, V., & Damiano, R. (2012). Storytelling on mobile devices for cultural heritage. *New Review of Hypermedia and Multimedia, 18*(1–2), 11–35. doi:10.1080/13614568.2012.617846

Magnani, M., Guttorm, A., & Magnani, N. (2018). Three-dimensional, community-based heritage management of indigenous museum collections: Archaeological ethnography, revitalization and repatriation at the Sámi Museum Siida. *Journal of Cultural Heritage, 31*, 162–169. doi:10.1016/j.culher.2017.12.001

Marstine, J. (2006). Introduction. In J. Martsine (Ed.), *New Museum Theory and Practice: An Introduction* (pp. 1–36). Blackwell. doi:10.1002/9780470776230.ch

Mathis, E. F., Kim, H. L., Uysal, M., Sirgy, J. M., & Prebensen, N. K. (2016). The effect of co-creation experience on outcome variable. *Annals of Tourism Research, 57*, 62–75. doi:10.1016/j.annals.2015.11.023

McLellan, H. (2000). Experience design. *Cyberpsychology & Behavior, 3*(1), 59–69. doi:10.1089/109493100316238

Meng, B., & Cui, M. (2020). The role of co-creation experience in forming tourists' revisit intention to home-based accommodation: Extending the theory of planned behavior. *Tourism Management Perspectives, 33*(100581), 1–13. doi:10.1016/j.tmp.2019.100581

Mijnheer, C. L., & Gamble, J. R. (2019). Value co-creation at heritage visitor attractions: A case study of Gladstone's Land. *Tourism Management Perspectives, 32*(100567), 1–12. doi:10.1016/j.tmp.2019.100567

Moin, S. M. A., Hosany, S., & O'Brien, J. (2020). Storytelling in destination brands' promotional videos. *Tourism Management Perspectives, 34*, 100639. doi:10.1016/j.tmp.2020.100639

Moscardo, G. (2008). Understanding tourist experience through mindfulness theory. In M. Kozak & A. DeCrop (Eds.), *Handbook of Tourist Behavior* (pp. 99–115). Routledge. doi:10.4324/9780203881804.pt4

Moscardo, G. (2010). The Shaping of Tourist Experience: The Importance of Stories and Themes. In M. Morgan, P. Lugosi, & J. Ritchie (Eds.), *The Tourism and Leisure Experience* (pp. 43–58). Channel View. doi:10.21832/9781845411503-006

Moscardo, G. (2017a). Exploring mindfulness and stories in tourist experiences. *International Journal of Culture, Tourism and Hospitality Research, 11*(2), 111–124. doi:10.1108/IJCTHR-11-2016-0108

Moscardo, G. (2017b). Stories as a Tourist Experience Design Tool. In D. R. Fesenmaier & Z. Xiang (Eds.), *Design Science in Tourism: Foundations of Destination Management* (pp. 97–124). Springer International Publishing. doi:10.1007/978-3-319-42773-7_7

Moscardo, G. (2018). Tourist Experience Design: A Storytelling Framework. In Quality Services and Experiences in Hospitality and Tourism (pp. 93–107). Emerald. doi:10.1108/S2042-144320180000009007

Moscardo, G. (2020a). The story turn in tourism: Forces and futures. *Journal of Tourism Futures*, *7*(2), 168–173. doi:10.1108/JTF-11-2019-0131

Moscardo, G. (2020b). Stories and design in tourism. *Annals of Tourism Research*, *83*, 102950. doi:10.1016/j.annals.2020.102950

Mossberg, L. (2008). Extraordinary experiences through storytelling. *Scandinavian Journal of Hospitality and Tourism*, *8*(3), 195–210. doi:10.1080/15022250802532443

Mossberg, L., Therkelsen, A., Huijbens, E. H., Björk, P., & Olsson, A. K. (2010). *Storytelling and Destination Development*. Nordic Council of Ministers. www.handels.gu.se

Munar, A. M. (2016). The house of tourism studies and the systemic paradigm. In *Tourism Social Science Series* (Vol. 22, pp. 131–153). Emerald Group Publishing Ltd. doi:10.1108/S1571-504320150000022014

Murray, J. H. (2017). *Hamlet on the Holodeck: The Future of Narrative in Cyberspacr*. MIT Press. https://mitpress.mit.edu/books/hamlet-holodeck-updated-edition

Nabi, R. L., & Green, M. C. (2015). The Role of a Narrative's Emotional Flow in Promoting Persuasive Outcomes. *Media Psychology*, *18*(2), 137–162. doi:10.1080/15213269.2014.912585

Nielsen, J. K. (2017). Museum communication and storytelling: Articulating understandings within the museum structure. *Museum Management and Curatorship*, *32*(5), 440–455. doi:10.1080/09647775.2017.1284019

O'Keefe, B., Benyon, D., Chandwani, G., Menon, M., & Duke, I. I. R. (2014). A Blended Space for Heritage Storytelling. *Proceedings of the 28th International BCS Human Computer Interaction Conference: Sand, Sea and Sky - Holiday HCI, HCI 2014*, 120–129. 10.14236/ewic/HCI2014.13

O'Leary, J. T., & Fesenmaier, D. (2017). Concluding Remarks: Tourism Design and the Future of Tourism. In *Design Science in Tourism: Foundations of Destination Management* (pp. 265–272). Springer. doi:10.1007/978-3-319-42773-7_16

Page, R. E. (2012). Stories and social media: Identities and interaction. In *Stories and Social Media: Identities and Interaction*. Taylor and Francis. doi:10.4324/9780203148617

Petrelli, D., Dulake, N., Marshall, M., Kockelkorn, H., & Pisetti, A. (2016, April 7). Do it together: the effect of curators, designers, and technologists sharing the making of new interactive visitors' experiences. *Museums and the Web 2016*. https://mw2016.museumsandtheweb.com/paper/do-it-together-the-effect-of-curators-designers-and-technologists-sharing-the-making-of-new-interactive-visitors-experiences/

Pine, J., & Gilmore, J. H. (1998). Welcome to the experience economy. *Harvard Business Review*, *76*(4), 97–105. PMID:10181589

Popova, Y. B. (2015). *Stories, meaning, and experience: Narrativity and enaction* (1st ed.). Routledge., doi:10.4324/9781315880488

Prahalad, C., & Ramaswamy, V. (2004). Co-creation experiences: The next practice in value creation. *Journal of Interactive Marketing, 18*(3), 5–14. doi:10.1002/dir.20015

Prebensen, N. K., & Foss, L. (2011). Coping and co-creating in tourist experiences. *International Journal of Tourism Research, 13*(1), 54–67. doi:10.1002/jtr.799

Prebensen, N. K., Vittersø, J., & Dahl, T. I. (2013). Value Co-creation significance of tourist resources. *Annals of Tourism Research, 42*, 240–261. doi:10.1016/j.annals.2013.01.012

Prebensen, N. K., Woo, E., & Uysal, M. S. (2014). Experience value: Antecedents and consequences. *Current Issues in Tourism, 17*(10), 910–928. doi:10.1080/13683500.2013.770451

Prebensen, N. K., & Xie, J. (2017). Efficacy of co-creation and mastering on perceived value and satisfaction in tourists' consumption. *Tourism Management, 60*, 166–176. doi:10.1016/j.tourman.2016.12.001

Robiady, N. D., Windasari, N. A., & Nita, A. (2020). Customer engagement in online social crowdfunding: The influence of storytelling technique on donation performance. *International Journal of Research in Marketing*. Advance online publication. doi:10.1016/j.ijresmar.2020.03.001

Ross, D., & Saxena, G. (2019). Participative co-creation of archaeological heritage: Case insights on creative tourism in Alentejo, Portugal. *Annals of Tourism Research, 79*(102790), 1–14. doi:10.1016/j.annals.2019.102790

Ryan, M.-L. (2015). *Narrative as Virtual Reality II: Revisiting Immersion and Interactivity*. JHU Press. https://www.academia.edu/16494139/Narrative_as_Virtual_Reality_II_Revisiting_Immersion_and_Interactivity

Simon, H. A. (1969). *The sciences of the artificial*. MIT Press.

Smith, L. (2014). Visitor Emotion, Affect and Registers of Engagement at Museums and Heritage Sites. *Conservation Science in Cultural Heritage, 14*(2), 125–131. doi:10.6092/issn.1973-9494/5447

Solima. (2012). *The listening museum. New communications strategies for public museums*. Rubbettino.

Sugathan, P., & Ranjan, K. R. (2019). Co-creating the tourism experience. *Journal of Business Research, 100*, 207–217. doi:10.1016/j.jbusres.2019.03.032

Sundin, A., Andersson, K., & Watt, R. (2018). Rethinking communication: Integrating storytelling for increased stakeholder engagement in environmental evidence synthesis. In Environmental Evidence (Vol. 7, Issue 6, pp. 1–6). BioMed Central Ltd. doi:10.118613750-018-0116-4

Tribe, J., & Liburd, J. (2016). The tourism knowledge system. *Annals of Tourism Research, 57*, 44–61. doi:10.1016/j.annals.2015.11.011

Tussyadiah, I. P. (2014). Toward a Theoretical Foundation for Experience Design in Tourism. *Foundations of Tourism Research: A Special Series, 53*(5), 543–564. doi:10.1177/0047287513513172

Van Der Linde, S., & Mans, J. (2015). Visualising values in the caribbean: A creative approach to value assessment. *Conservation and Management of Archaeological Sites*, *17*(3), 257–269. doi:10.1080/13505033.2015.1129800

Vargo, S. L., & Lusch, R. F. (2004). Evolving to a New Dominant Logic for Marketing. *Journal of Marketing*, *68*(1), 1–17. doi:10.1509/jmkg.68.1.1.24036

Weiler, B., & Black, R. (2015). The changing face of the tour guide: One-way communicator to choreographer to co-creator of the tourist experience. *Tourism Recreation Research*, *40*(3), 364–378. doi:10.1080/02508281.2015.1083742

Wyman, B., Smith, S., Meyers, D., & Godfrey, M. (2011). Digital Storytelling in Museums: Observations and Best Practices. *Curator (New York, N.Y.)*, *54*(4), 461–468. doi:10.1111/j.2151-6952.2011.00110.x

ADDITIONAL READING

Chu, J. H., & Mazalek, A. (2019). Embodied Engagement with Narrative: A Design Framework for Presenting Cultural Heritage Artifacts. *Multimodal Technologies and Interaction*, *3*(1), 1. Advance online publication. doi:10.3390/mti3010001

Dal Falco, F., & Vassos, S. (2017). Museum experience design: A modern storytelling methodology. *The Design Journal, 20*(sup1), S3975–S3983. doi:10.1080/14606925.2017.1352900

Moscardo, G. (2017). Stories as a Tourist Experience Design Tool. In D. R. Fesenmaier & Z. Xiang (Eds.), *Design Science in Tourism: Foundations of Destination Management* (pp. 97–124). Springer International Publishing., doi:10.1007/978-3-319-42773-7_7

Moscardo, G. (2018). Tourist Experience Design: A Storytelling Framework. In Quality Services and Experiences in Hospitality and Tourism (pp. 93–107). Emerald. doi:10.1108/S2042-144320180000009007

Moscardo, G. (2020). Stories and design in tourism. *Annals of Tourism Research*, *83*, 102950. doi:10.1016/j.annals.2020.102950

O'Keefe, B., Benyon, D., Chandwani, G., Menon, M., & Duke, I. I. R. (2014). A Blended Space for Heritage Storytelling. *Proceedings of the 28th International BCS Human Computer Interaction Conference: Sand, Sea and Sky - Holiday HCI, HCI 2014*, 120–129. 10.14236/ewic/HCI2014.13

Tussyadiah, I. P. (2014). Toward a Theoretical Foundation for Experience Design in Tourism. *Foundations of Tourism Research: A Special Series, 53*(5), 543–564. doi:10.1177/0047287513513172

KEY TERMS AND DEFINITIONS

Co-Creation: The joint creation of value during experiences involving two or more parties. It is an inherently interactive process that necessitates the active contribution of all parties to an outcome. In tourism studies, it is often used to describe a consumer-provider relationship that acknowledges the active role of the consumer in service contexts.

Digital Storytelling: A specific type of storytelling that is enhanced by the use of digital media and interactive technology. It is also commonly referred to as "immersive storytelling" or "interactive storytelling".

Experience Design: Can be understood as a concept, an activity, a plan, or an outcome. It is the act of intentionally and mindfully planning, organizing, and implementing all components of an experience within a service context with the goal of providing enjoyable and memorable tourist experiences.

Heritage Tourism: A specific type of tourism relating to tourist activities occurring within the heritage sector, such as museums, heritage attractions, or heritage sites.

New Museology: A management approach that has emerged within the museum sector aiming to provide visitors with more participative, interactive, and engaging museum experiences. This approach runs counter to the traditional approach, which favors a one-directional transmission of information to visitors and the continuation of an authorized heritage discourse.

Storytelling: The practice and art of telling a story to an audience (called listener, reader, or viewer) deeply rooted in human cultures and taking many different forms (everyday life stories, myths, fables or other).

Tourist Experience: A phenomenon occurring within the minds of individuals before, during, and after they have participated in activities as a tourist.

Chapter 2
Storytelling in Cultural Heritage:
Tourism and Community Engagement

Maria Isabel Roque
https://orcid.org/0000-0002-2258-8904
Universidade Católica Portuguesa, Portugal

ABSTRACT

Cultural heritage interpretation requires mediation processes connecting tangibility with intangible meanings. As a performing narrative, storytelling is an effective communication strategy in cultural and heritage tourism. In particular, digital storytelling allows to segment information and adapt it to the different audience profiles in an interactive and immersive way. Tourists become participants in the mediation process by providing their stories, while local communities enrich the curatorial discourse with their knowledge and memories about the place, culture, and heritage. Tourists and residents' co-creation, contributing to strengthening perception of local authenticity and sense of belonging, enhances the experience and engagement with the place. The literature review aims to analyse the impact of storytelling and co-creation on cultural heritage mediation and the tourist experience and community engagement.

INTRODUCTION

Studies on cultural heritage emphasise the relevance of storytelling and digital technologies in the understanding of tangible and intangible components of heritage in their relation to community involvement and local perception. This phenomenon is recent, however, practices involving processing and display of digital heritage have increased, reducing fear over destruction of cultural authenticity.

Due to the fluidity and uncertainty of current societies, characterised by Zygmunt Bauman (2000) as the "liquid modernity" of our time, cultural heritage and values of the past are connected to the perception of security and timelessness (Prentice, 2005). Cultural heritage provides communities with a sense of belonging and continuity, which is strengthened through preserved meanings. By stabilising the representation of memories, patrimonialization increases place and artefacts depth, which attracts people's attention and subsequently motivates them to travel.

DOI: 10.4018/978-1-6684-3436-9.ch002

Storytelling is present in everyday life by telling and listening to stories of lived or imaginary experiences (Langellier, & Peterson 2004). Storytelling mediates narratives associated with cultural heritage and creates public engagement with it. Recent technological developments lead to digital storytelling, defined as "the practice of using computer-based tools to write and tell stories" (Stanley, 2018, p. 1), maintaining the story analogic features and combining them with hypertext, still and moving images, video, sound, animation, and other forms of media, such as virtual reality (VR) and augmented reality (AR). In the last decades, digital storytelling has been used to disseminate knowledge, stimulate creative thinking and learning, and engage audiences. As digital storytelling became a universal way of communicating, monuments, heritage sites, and museums use this tool to inspire visitors to connect with places, collections, and artefacts based on their own experiences and history (Robin, & McNeil, 2019, p. 3).

Therefore, an increasingly popular aim of heritage organisations and managers is to explore the potential of digital storytelling to connect both local communities and visitors with the tangible and intangible facets of the heritage experience by using interactive and immersive approaches to interpret the cultural identity of a place. The assumption is that knowledge and understanding of place by the local communities and visitors enhances a sense of belonging, a non-hegemonic acceptance of other cultures, and a personal involvement in interpretation, communication, and protection processes. This chapter explores storytelling as a tool to experience and communicate heritage in an interactive and immersive way, taking into account the personal interests, tastes, needs, and skills of heritage users. Following an overview of the literature, the role of mediation is discussed as the process to that facilitates understanding of cultural tangible and intangible heritage, and storytelling the mediation strategy that leads to a co-creative heritage experience.

BACKGROUND

There is extensive research on heritage (Howard, 2003; Logan, Nic, & Kockel, 2015; Messenger, & Smith, 2014) and the different narratives about the past (Hanna et al., 2017). The promotion of cultural values inherent to heritage has transformed heritage sites into touristic attractions (Alvarez et al., 2016; Prentice, 1993). However, adapting heritage sites for international visitors provoked a progressive distancing of local communities (Court, & Gamini, 2015; Waterton, & Smith, 2010).

Following the growth of an industry focused on heritage, both UNESCO (1976) and the World Tourism Organization (UNWTO, 1995), as well as other specialist organisations (Wells, & Stiefel, 2019) highlighted the need of preserving and managing heritage sites, included in the broader frame of sustainable tourism (Robinson, & Picard, 2006) and community's engagement (Court, & Gamini, 2015). Case studies have been published in recent years, demonstrating how residents' engagement creates advantages to the preservation of heritage and the empowerment of communities (Chinyele, & Lwoga, 2018; Ferreira, 2018; Li et al., 2020; Stephens, & Tiwari, 2015).

The use of digital mediation tools aims to connect visitors with places through the understanding of intangible components and their implicit meanings (Díaz-Andreu, 2017). Acknowledging the importance of technology in the processes of knowledge acquisition, visualisation and promotion of tangible and intangible heritage (Guazzaroni, & Pillai, 2020), studies have been conducted on digital humanities, giving rise to a new field of research called "digital heritage" (Kremers, 2020). Digital humanities are seen as the digital convergence that resets "the domain of art, culture and the humanities in digital contexts (Giannini & Bowen, 2019, p. 10). Digital humanities and heritage "share concepts such as the idea of spatialisation, rich information about an object as research base and a strong link to the creation and perception of visualization and imagery" (Muenster et al, 2019, p. 818). However, digital technology-based developments, such as 3D visualisation and printing, virtual and augmented reality, or holography, have been accepted essentially as sporadic endeavours, requiring crowdsourcing in the digital humanities for heritage visualisation and analysis (Yuxiang, & Jingwen, 2021).

Digital storytelling is used to understand, promote and protect tangible and intangible heritage (Udeaja et al., 2021; Vrettakis et al., 2019). Cultural mediation is currently being re-conceptualised as a storytelling activity, engaging "a multitude of voices" (Sylaiou, & Dafiotis, 2020), allowing heritage curators and communities to participate as narrators (Djabarouti, 2020; Palombini, 2017; Wilson, & Desha, 2016). Although the implementation of a verification process is required to prevent narrative freedom from misrepresenting or distorting historical truth (Palombini, 2017), storytelling-based activities contribute to a community-centred approach to cultural heritage.

STORYTELLING AS MEDIATION TO CULTURAL HERITAGE

Cultural Heritage Mediation

Cultural heritage comprehends *tangible immovable* (buildings and cultural landscapes) and *movable productions* (artistic, ethnographic, scientific or technical, archaeological, archival, bibliographical), *intangible creations* (oral expressions and traditions, cults, social practices, handicrafts), as well as *natural resources* (with particular physical, biological, and geological features). All these elements contribute to identity building of a place, a region, a nation, or a civilisation. Cultural heritage involves a process of selection based on what, according to time, place and culture, is recognised as valuable or representative of the identity of an individual or a social group. It is considered a form of social memory (Hanna *et al.*, 2017), based on a sense of ownership and belonging, which constructs identity. Intercultural dialogue contributes to strengthening cultural identity, which is constantly redefined and negotiated, as it promotes recognition of *Otherness*, both with its similarities and differences. The concept of 'cultural heritage' is based on this shared bond to a community, representing the history of a cultural group and collective knowledge. The sociocultural system, authorities and communities, when dealing with collective memories, implicitly select which heritage is to be preserved and how it is communicated to the next generations. Commemorative, historical, antiquity, aesthetic or the correspondence to functional use are values assigned to cultural heritage (Riegl, 1903) and, by the process of patrimonialisation, places and sites gain prominence in cultural representation.

Authenticity has been considered a Western European concept (Lowenthal, 1995, pp. 125-127) that recognizes cultural values as relative to the cultures where they emerge. The *Nara Document on Authenticity* states that "all judgements about values attributed to cultural properties as well as the credibility of related information sources may differ from culture to culture, and even within the same culture" (ICOMOS, 1994, article 11). Conversely, if the valuation of the past is based on the discontinuity between a monument and contemporaneousness, the concept of 'living heritage site' suggests that communities and heritage are continuous and indivisible entities, although considering different bond strength levels (Poulious, 2014). Also, it is assumed that heritage must be taken as evidence of the past as living in the present. Thus its significance is not based on historical facts but on contemporary meanings and values (Wells, & Stiefel, 2019). These highly affect how heritage is appropriated in various contexts and perceived by audiences.

Heritage values have stimulated the desire to travel and experience the places and artefacts representing other civilisations and cultures. Heritage tourism, described as the movement toward attractions such as monuments and historical sites to obtain knowledge in response to cultural needs (Richards, 1996), relies on the tourist perception of heritage attributes (Poria, Butler, & Airey, 2003). Moreover, "heritage tourism is tied to myth making and stories, creative content that can be shared, stored, combined and manipulated in different ways to present and remember place associations embedded in the local community" (Go, Alvarez, & Yüksel, 2016, p. 186). On the other hand, heritage resources are supported by stimulating narratives drawn on local culture that attract tourists, emphasising the connection with local communities and driving self-projection towards the past and place values.

In his seminal book *The past is a foreign country*, Lowenthal observed that if the difference associated with the past stimulates heritage preservation, preserving enhances the differences: "preservation has deepened our knowledge of the past but dampened creative use of it" (1985, p. 17). When revisiting the book, Lowenthal maintained that "our precursors identified with a simplified unitary antiquity whose vestiges became models for their own creations. Our own more numerous and exotic pasts, prized as vestiges, are divested of the iconographic meanings they once embodied" (Lowenthal, 1985, p. 17; Lowenthal, 2015, p. 595). However, while in the book's first edition he claimed that the past was venerated as a fount of communal identity" (Lowenthal, 1985, p. 17), in the revisited edition, he instead says that "no longer revered or feared, the past is swallowed up by the ever-expanding present; we enlarge our sense of the contemporary at the expense of dissevering it from the past" (Lowenthal, 2015, p. 595). He also criticises a new "*admiration* of ignorance" (Lowenthal, 2015, p. 590), referring to contemporary societies marked by a vague understanding of both geography and history, allowing the past to be "swallowed up by the ever-expanding present" (Lowenthal, 2015, p. 595). Also, it has been noted that excessive, uninformed or unaware visitors cause severe damage to heritage sites and artefacts and deteriorate their authenticity. "It is a paradox that participants in [tourism] are drawn to attractive environments, whether natural or built, but that their mere presence is likely to result in the modification of those environments." (Wall, 1989, p. 10). In this process, heritage attractions lose their value and become less attractive to tourists who provoked the site modification (Timothy, 1994). All these changes induce a new perspective on heritage and impose the need for mediation as a contributing factor to heritage values preservation running against cultural commodification.

Heritage has evolved from traditional practices, centred on passive visitation to monuments, historical sites and museums, to others more dynamic and participatory. Simultaneously, cultural and heritage tourism are no longer exclusive to the social elites or the intellectuals and experts, involving all potential visitors. Heritage tourism evolved from an early vision limited to remnants of the past (Poria et al., 2003) to include a broader range of cultural, historical and natural assets, intangible elements of culture, places, attractions and events (Di Pietro et al., 2018). A wide range of cultural heritage consumers are visitors and ubiquitous tourists, but they are also locals and insiders. Currently recognised as an industry that manages historical sites, monuments and museums, "heritage has many markets, only one of which is the tourist" (Howard, 2003, p. 25). Even though cultural consumers usually have high-quality demands in their leisure time (Piscitelli, 2019, p. 88), heritage audiences, regarded as plurals and heterogeneous, also have different expertise, skills, and purposes, demanding a complex and segmented mediation process.

The relationship between tourism, heritage and culture requires a rigorous approach to management so that sustainable objectives can be achieved, involving the integration and promotion of cultural assets, activities and traditions. Cultural mediation must include the representation and mostly the interpretation of the place, presenting the features of its original functionality and meanings. Tilden defined 'interpretation' as "the work of revealing [...] something of the beauty and wonder, the inspiration and spiritual meaning that lie behind what the visitor can with his senses perceive" (1977, 3-4), performing "the function of the custodians of our treasures" (1977, 4). Cultural mediation to tourists demands interpretation, including heritage polysemic meanings. Heritage meanings are preserved through the community's collective memory, understood as a group that shares values, beliefs, and interests, even considering their contradictions and disruptions (Waterton, & Smith, 2010). Despite fluidity, the community contains information and capacities that outlast political structures and complement experts' knowledge about the place. Thus, community narratives are currently implemented in the cultural heritage mediation.

The introduction of mediation tools to interpret and understand the place has been assumed as a strategy of sustainability (Alvarez et al., 2016), as the local community and visitors are more encouraged to preserve and respect a place in so far as they understand its meanings, which increases the feeling of belonging. However, heritage curators and professionals advertise that introducing data in analogical supports can generate an excessive visual impact, disturbing the site's observation and experience. Hence creating non-invasive cultural mediation strategies is suggested to contextualise architectural, archaeological or artistic items in their functional and symbolic components.

Technological data communication methods have led to profound changes in how postmodern societies organise, access, and retrieve information in the different knowledge domains, heritage included. "The digital revolution has not only impinged on how heritage management is organised at an institutional level, but also on how heritage data is organised to facilitate access to systematised information" (Díaz-Andreu, 2017, p. 404). Digital technology provides tools and strategies to process, visualise, and disseminate a large amount of data, adapting them to the skills and expectations of the different users through the possibility of information segmentation and access personalisation. Moreover, it allows cross-referencing over tangible and intangible heritage, developing an integrated knowledge of these areas and allowing local communities and visitors to connect, relate and unify concepts and enhance a holistic experience of the place.

Storytelling as Cultural Mediation in Creative Tourism

Storytelling crosses all times and spaces of humanity: "people have always told each other stories about the places they are from, the places they live in, and the places they have visited. A sense of place is essential to human life" (Bassano et al., 2019, p. 11). With the advent of the digital age, storytelling benefited from technological advances and became an innovative communication modality about heritage, assuming the main components to building a narrative: the theme and correlated issues, the setting to define the time and geographic location within it; the characters and the relations between them; the plot that comprises the diverse events that construct the story; the narrative arch which may be synthesised as the sequence of those events, describing the path that unfolds throughout the story.

Digital storytelling enhances the affective experience of interpretation by evoking personal experiences, expressions and emotions (Cameron, 2012). The first practices go back to the early 1990s, when digital technologies provided new information, presentation, and communication tools. A group of media artists and designers at the Center for Digital Storytelling (CDS), called StoryCenter since 2015, began to explore how to create personal stories using the new digital media technology, allowing Joe Lambert, CDS co-founder and executive director, to define seven elements of digital storytelling: (i) point of view (of the author), (ii) a dramatic question, (iii) emotional content, (iv) the gift of your voice (narration), (v) the power of the soundtrack (to support the storyline), (vi) economy (of length), and (vii) pacing of the story (2003). These elements were later updated to the seven steps of digital storytelling as follows: (i) owning your insight, (ii) owning your emotions, (iii) finding the moment, (iv) seeing your story, (v) hearing your story, (vi) assembling your story, and (vii) sharing your story (Lambert, 2010).

While connecting narratives to heritage and enhancing experiences and motivations to visit cultural sites, virtual storytelling is also a powerful instrument for cultural inclusion and accessibility, approaching the places "too expensive to visit, too inhospitable, too far away, too fragile, too dangerous, or [that] no longer exist" (Kasemsarn, & Nickpour, 2015). Online applications and media also allow ubiquitous and permanent access to heritage sites and museum collections, complemented with history, stories and interpretative information as it commonly happens in physical places. Despite the potential benefits, heritage organisations do not address the issue of motivating potential visitors to access them, nor do they intensively use interactive technologies, e.g. digital storytelling, to present items, collections and data in a more appealing way. In the cultural heritage context, storytelling is defined as "an experience created around a coherent narrative that leverages the interpretation of the available cultural heritage artefacts to develop the essential elements of storytelling: setting, characters, plot, conflict, theme, and a satisfying narrative arc" (Vrettakis et al., 2019, p. 1).

Knowledge transmission and cultural communication by curators and other heritage professionals used to be passive, neutral, or objective, but it has become more active, appealing to emotional memories without jeopardising data accuracy. "The role of the curator is not only to develop a creative story of place, but to highlight particular elements of culture and creativity that are of importance for local and global actors" (Richards, 2020, p. 9). Heritage curacy becomes a creative process to deal with a wider range of visitors and to increase interest in local cultures based on accurate perceptions and uses. "The countless possibilities ICT offer to cultural organisations have generated *lines of flight* into new methodological approaches favoring investment in affective, multimodal and immersive experiences, personalised interactivity and even co-authorship of the narratives" (Sylaiou, & Dafiotis, 2020, p. 370). Digital information and communication technologies may enrich, transform, and enhance the heritage experience and accordingly the features of cultural tourism.

Storytelling is applied to enrich visitors' and local communities' experiences at heritage sites and museums (Guazzaroni, & Pillai, 2020). By designing immersive experiences, storytelling contributes to the transition from cultural tourism toward creative tourism, characterised by users' active participation and involvement and by the convergence of tourism and the creative economy. As described in *A research agenda for creative tourism* (Duxbury & Richards, 2019), creative tourism activities incorporate four dimensions: active participation, learning, creative self-expression, and community engagement. Although requiring literacy skills to research data and technological skills to use computer-based applications, creative tourism stimulates new consumption patterns and more skilled activities. "As a new approach to tourism development and marketing, creative tourism expands the very concept of tourism as a whole, driving a shift from conventional models of heritage-based cultural tourism to new models of creative tourism centred on contemporary creativity, innovation and intangible content" (OECD, 2014, p. 7). The expansion of creative tourism generates "knowledge based creative activities that link producers, consumers and places by utilising technology, talent or skill to generate meaningful intangible cultural products, creative content and experiences" (OECD, 2014, p. 14). Even considering criticism on the risks of asymmetrical benefits and intruding in the host communities' private domains (Richards, 2018), creative tourism, which can be developed with relatively limited resources, has a relevant potential to create meaning and sensory experiences through storytelling.

In reaction to a growing mass cultural tourism, creative tourism is concerned with placemaking (Richards, 2020), described as the intentional and collaborative process of activating new or existing public spaces (Ball, 2014; Richards, 2021). The process uses arts, culture, and heritage to create emotional connections, establish a sense of place, and enhance shared values. Thus, it plays a crucial role in assembling resources, meanings and creativity, and, conversely, creative tourism may be used to preserve the values that local communities might otherwise be in danger of losing (Richards, 2021). Even if it is mainly concerned with promoting better urban design, placemaking aims to create innovative patterns of heritage places' uses by boosting the cultural identities, which sustainably support their ongoing development.

The *Sustainable Cultural Tourism Open Method of Coordination* (SCT OMC) working group for the European Union defined sustainable cultural tourism as "the integrated management of cultural heritage and tourism activities in conjunction with the local community, creating social, environmental and economic benefits for all stakeholders in order to achieve tangible and intangible cultural heritage conservation and sustainable tourism development" (European Commission, 2019, p. 8). Sustainability deals with the cultural and creative tourism benefits to local communities by attracting and retaining visitors, alongside the disadvantages of the potential degradation of cultural sites due to overuse and commodification. On the other hand, as sustainable cultural tourism is also concerned with the involvement of both local communities and tourists and with internal and external stakeholders, heritage storytelling can contribute to long-term cultural, environmental, and economic outcomes. So, the SCT OMC working group advises sustainable cultural tourism relating to both the tangible and intangible cultural heritage to emphasise concepts such as "slow" tourism, "authenticity", "well-being", "contact with locals", and "storytelling", which "can be a powerful way to convey the significance of cultural heritage" (European Commission, 2019, p. 19). Also, "through narrative, storytelling is one way of making cultural heritage more accessible with the aim of creating qualified and differentiated experiences for the visitors to a place" (European Commission, 2019, p. 19). At the same time, tangible historic places begin to have meaning and be correlated with their intangible components through interpretative narratives.

While digital technologies potential is accepted to connect heritage with its visitors and locals, "storytelling has the potential to revolutionise the way we engage with cultural heritage and has been widely recognised as an important direction for attracting and satisfying the audience of museums and other cultural heritage sites" (Vrettakis et al., 2019, p. 1). Digital storytelling engages a wide range of audiences that would not otherwise participate, resulting in strong, sensitive and holistic experiences in cultural and heritage tourism. As it enhances visitors' experience, it may also attract virtual users and motivate them to visit real cultural sites. To achieve this, storytelling must involve a coherent narrative that interprets the cultural heritage sites and artefacts' morphologies, uses, and meanings in increasingly general and flexible models.

Storytelling to a Co-Creative Heritage Experience

Sense of place includes "the experiential and expressive ways places are known, imagined, yearned for, held, remembered, voiced, lived, contested and struggled over; and the multiple ways places are metonymically and metaphorically tied to identities" (Feld, & Basso, 1996, p. 11). The connection to heritage and local-based culture and traditions is made through doing, making, sensing, experiencing, and creating processes as defined in creative tourism (Bakas et al., 2020), reported to support the trend towards co-creation, in which visitors and residents become co-producers and co-consumers of experiences (Richards, 2020).

It has been observed that heritage tourism used to be connected with elitist consumption, involving mainly cultured tourists (Timothy & Boyd, 2003), suggesting that a lack of cultural knowledge could inhibit heritage experiences (Davis & Prentice, 1995). To prevent this, cultural and heritage tourism aims to involve a broader range of resources and experiences (Di Pietro et al., 2018). Heritage tourism experiences includes the interactive participation in knowledge acquisition and interpretation of aspects related to historical sites, monuments, museum collections and artefacts, material history, past values, functionalities, relationships through the use of original objects, and their correlated meanings and beliefs, as well as learning about descriptions and interpretations produced in academic circles (Tilden, 1977). Heritage interpretation, conceived as an informal learning activity, was in the past focused on experts' explanations, but current practices allow the inclusion of other stories and expressions about the past and more creative experiences.

Creative tourism aims to develop more active involvement between tourists and local communities' culture, traditions, and creativity. "Through digital tools, storytelling strategies allow heritage curators to listen, incorporate and encourage the voices of their audiences to enter into a fruitful dialogue with them (Sylaiou, & Dafiotis, 2020). Heritage sites and museums transform from passive places and repositories of collections into interactive, dialogical, and participatory spaces. Users may achieve segmented data based on their personal interests and select the preferred channel, and the information level they want to achieve (Biedermann, 2020), but they also are contributors, verbalising their memories, practices, beliefs, and emotions. Participative heritage interpretation is a form of co-creation by both tourists and local communities (Ross, 2020). When tourists become co-producers, providing data related to cultural heritage and participating in the patrimonialisation process, perceived values arise, improving the experience quality. "Experience value becomes an integrated process between host and guests in a certain atmosphere where their respective meanings of value are shared and recognised" (Prebensen et al., 2018, p. 3). However, co-creation surpasses the sensorial contact with heritage time and place, as it may happen before, during and after staying at the destination (Campos et al., 2018; Prebensen et al., 2018), crossing all travelling procedures, from preparation to memorialisation (Hung, 2016). Before the visit, tourists are exposed to stories disseminated through diverse media channels, which influence the destination image, forging a preconception about it, and after the visit they return to the home environment with constructed new stories at the destination. "During interaction processes, tourists' involvement and co-creation behaviours influence their interpretation and reconstruction of the destination brand stories, and their creation of destination brand meaning, which enhances their experienced value" (Mathisen, 2018, p. 141). Value creation in cultural and heritage tourism includes tourists' participation and involvement (Prebensen et al., 2013).

Stories constructed by tourists are signs of their active participation and interaction in the physical, sensorial, cognitive, emotional and social experience at the destination. Besides, those stories express the perception, interpretation, and evaluation of the cultural heritage found there, creating consecutive layers of meanings to the progressive construction of its identity and sense. Tourists give an additional sense to heritage, providing evidence of how it is used and perceived, expressed through their co-created stories. By constructing a narrative based on their values, knowledge, memories, beliefs and motivations, tourists play an active role in interpreting historical sites and artefacts (Ross, 2020), crossing their stories with the curators and cultural mediator's discourses. In the myriad of individual stories and interpretations produced by tourists, there are variations and divergences from the academic narrative elaborated by researchers and transmitted by heritage sites and museum channels of communication. As a result, curators and cultural mediators face the challenge of managing the subjectivities of common knowledge with rigorous and objective scientific data whilst simultaneously providing a personalised and meaningful tourism experience (Hung, 2016).

Tourists' participation in co-creative activities and co-preservation of local heritage and traditions are viewed as participatory opportunities to perceive and understand and interpret their meanings, learn about cultures, and engage with communities while developing creative skills and self-expression activities (Bakas et al., 2020). Creative tourism is also related to the link between residents, places and cultural heritage. Communities well integrated into their homeplace and concerned with local cultural heritage may lively introduce it to visitors, enhancing the connection with the place. Therefore, "the most meaningful memories originate in the destination environment, when tourists find themselves in a new environment, undertaking activities and interacting with people, all of which increase the strength of emotions and feelings and acquisition of new knowledge" (Campos et al., 2018, p. 26). Thus, value creation enhances by using storytelling during tourists' and residents' interactions (Mathisen, 2018).

In general, local communities are the successors and heirs of the heritage founder cultural group and maintain the memory of its original function, use and meanings. Listening to their stories and integrating them into the heritage site narrative strengthens the community's sense of belonging and enhances tourists' authenticity perception (Roque & Forte, 2017a, 2017b). While tourism in heritage sites generates a risk of inauthenticity, requiring a balanced curatorship to provide an accurate cultural mediation, the integration of community storytelling is a decisive factor in values preservation. Heritage values are also promoted "if the core society is the main instrument of socialisation, assimilating 'outsiders' into the values of the core while continually reasserting and reinforcing it to insiders and outsiders alike" (Ashworth et al., 2007, p. 91). So, residents begin to be called to participate in the design, production and delivery of creative tourism experiences. Communities' role surpasses the experience provision sphere as they participate in a co-creative tourism development process. Host community involvement may contribute to solving emerging challenges in cultural and heritage tourism by linking creativity and experimentation, and combining immersion in local culture with active learning and creative self-expression (Emmendoerfer, 2020). Also, community involvement in the relationship between heritage and tourism, by integrating their stories in the authorised heritage narratives, strengthens the sense of belonging, local identity, social cohesion and development (Di Pietro et al., 2018); additionally, creativity helps engagement in the local regeneration process (Trunfio, & Della Lucia, 2018).

Storytelling practices use digital technology for data dissemination in a reactive, interactive and performative way, as they offer unprecedented opportunities to update the processes of use, preservation, management and enhancement of cultural tourism and to heritage interpretation and mediation. Under the effect of the Covid-19 pandemic, which highlighted the vulnerability of tangible and intangible cultural heritage (Clini, & Quattrini, 2021), changing the paradigm of mass tourism and overcrowded spaces towards a sustainable and proximity tourism focused on co-creative activities (Romagosa, 2020), engaging tourists and residents, digital storytelling gains prominence as a communication tool for cultural heritage.

FUTURE RESEARCH DIRECTIONS

Research gaps and directions for future empirical research include the analysis of the COVID-19 pandemic impact on heritage tourism and digital acceleration. Therefore, it is proposed to pursue multidisciplinary research on available methods in the field of digital storytelling to improve tangible and intangible heritage awareness and preservation.

CONCLUSION

Prior research emphasises the potential of storytelling to interpret tangible and intangible heritage and promote local community's cultural identity. Telling or listening to a story may confer authenticity to places or contribute to its perception, as storytelling conveys knowledge and memories associated with the heritage experience. Through storytelling strategies, currently with the support of digital technologies, both tourists and residents approach heritage in an interactive and immersive way to better understand its multiple meanings. Storytelling strategies enhance the sense of belonging together with the acceptance of a different culture.

Research in digital humanities and heritage studies directs the development of strategies for cultural mediation towards tourists and local communities in a segmented way, adapting them to the diverse profiles of audiences, personal interests, tastes, needs, and skills. Mediation projects through information and communication technologies also allow different modalities of visualisation and experience, from the more intellectual and rational to the more emotional and sensitive. Heritage mediation has evolved towards more participative behaviour and co-creative activities before, during and after the visit. The engagement of visitors and other heritage users is a critical factor in placemaking and sustainability. Also, as communities act as co-creators, their engagement through storytelling may effectively contribute to a greater societal value of cultural heritage.

REFERENCES

Alvarez, M. D., Go, F. M., & Yüksel, A. (Eds.). (2016). *Heritage tourism destinations: Preservation, communication and development*. CABI. doi:10.1079/9781780646770.0000

Ashworth, G. J., Graham, B. J., & Tunbridge, J. E. (2007). *Pluralising pasts: Heritage, identity and place in multicultural societies*. Pluto Press.

Bakas, F. E., Duxbury, N., Silva, S., & Castro, T. G. (2020). Connecting to place through creative tourism. In S. M. Cabeça, A. R.Gonçalves, J. F.Marques, M. Tavares (Eds.), Creative tourism dynamics: connecting travellers, communities, cultures and places (pp. 119-139). Grácio Editor.

Ball, R. (2019, September-October). Economic development: It's about placemaking. *The European Business Review*. https://www.europeanbusinessreview.com/economic-development-its-about-placemaking/

Bassano, C., Barile, S., Piciocchi, P., Spohrer, J. C., Iandolo, F., & Fiske, R. (2019). Storytelling about places: Tourism marketing in the digital age. *Cities*, *87*, 10–20. doi:10.1016/j.cities.2018.12.025

Bauman, Z. (2000). *Liquid modernity*. Polity Press.

Biedermann, B. (2020). The potentials of the digital cultural heritage: A best practice example of presentation and use: The digital representation of the Hans Gross Kriminalmuseum. In H. Kremers (Ed.), Digital cultural heritage (pp. 195-207). Springer.

Cameron, E. (2012). New geographies of story and storytelling. *Progress in Human Geography*, *36*(5), 573–592. doi:10.1177/0309132511435000

Campos, A. C., Mendes, J., Valle, P. O., & Scott, N. (2018). Co-creation of tourist experiences: A literature review. *Current Issues in Tourism*, *21*(4), 369–400. doi:10.1080/13683500.2015.1081158

Chinyele, B. J., & Lwoga, N. B. (2018). Participation in decision making regarding the conservation of heritage resources and conservation attitudes in Kilwa Kisiwani, Tanzania. *Journal of Cultural Heritage Management and Sustainable Development*, *9*(5), 184–5198. doi:10.1108/JCHMSD-05-2017-0019

Clini, P., & Quattrini, R. (2021). Editorial: Digital cultural heritage, arts reproduction and museums systems: Languages and techniques in a Covid and post-Covid scenario for new forms of heritage against the silence of a fragile culture. *SCIRES-IT: Scientific Research and Information Technology*, *11*(1), 1–10. doi:10.2423/i22394303v11n1p1

Court, S., & Gamini, W. (2015). *People-centred approaches to the conservation of cultural heritage: Living heritage*. ICCROM. https://www.iccrom.org/sites/default/files/PCA_Annexe-2.pdf

Davis, A. J., & Prentice, R. C. (1995). Conceptualising the latent visitor to heritage attractions. *Tourism Management*, *16*(7), 491–500. doi:10.1016/0261-5177(95)00066-W

Di Pietro, L., Mugion, R. G., & Renzi, M. F. (2018). Heritage and identity: Technology, values and visitor experiences. *Journal of Heritage Tourism*, *13*(2), 97–103. doi:10.1080/1743873X.2017.1384478

Díaz-Andreu, M. (2017). Introduction to the themed section' digital heritage and the public.'. *International Journal of Heritage Studies*, *23*(5), 404–407. doi:10.1080/13527258.2017.1286780

Djabarouti, J. (2021). Stories of feelings and things: Intangible heritage from within the built heritage paradigm in the UK. *International Journal of Heritage Studies*, *27*(4), 391–406. doi:10.1080/13527258.2020.1798271

Duxbury, N., & Richards, G. (Eds.). (2019). *A research agenda for creative tourism*. Edward Elgar Publishing. doi:10.4337/9781788110723

Emmendoerfer, M. L., Olavo, A. V. A., Silva, A. C., Jr., & Mediotte, E. J. (2020). Innovation lab in the touristic development context: Perspectives for creative tourism. In S. M. Cabeça, A. R.Gonçalves, J. F. Marques & M. Tavares (Eds.), Creative tourism dynamics: Connecting travellers, communities, cultures and places (pp. 87-101). Grácio Editor.

European Commission, Directorate-General for Education, Youth, Sport and Culture. (2019). *Sustainable cultural tourism*. Publications Office. https://data.europa.eu/doi/10.2766/400886

Feld, S., & Basso, K. H. (1996). *Senses of place*. School of American Research Press.

Ferreira, T. C. (2018). Bridging planned conservation and community empowerment: Portuguese case studies. *Journal of Cultural Heritage Management and Sustainable Development*, *8*(2), 179–193. doi:10.1108/JCHMSD-05-2017-0029

Georgescu, M., & Popescul, D. (2015). The impact of new Information and Communication Technologies on the creation and dissemination of knowledge inheritance. *Procedia: Social and Behavioral Sciences*, *188*, 122–129. doi:10.1016/j.sbspro.2015.03.347

Giannini, T., & Bowen, J. (2019a). Digital culture. In T. Giannini & J. Bowen (Eds.), *Museums and digital culture: New perspectives and research* (pp. 3–26). Springer. doi:10.1007/978-3-319-97457-6_1

Go, F. M., Alvarez, M. D., & Yüksel, A. (2016). Conclusions and implications for heritage tourism destinations. In M. D. Alvarez, F. M. Go, & A. Yüksel (Eds.), *Heritage tourism destinations: Preservation, communication and development* (pp. 186–193). CABI.

Guazzaroni, G., & Pillai, A. S. (2020). *Virtual and augmented reality in education, art, and museums*. IGI Global. doi:10.4018/978-1-7998-1796-3

Hanna, S. P., Potter, A. E., Modlin, E. A., Carter, P., & Butler, D. L. (2017). *Social memory and heritage tourism methodologies*. Routledge.

He, Y., Ma, Y. H., & Zhang, X. R. (2017). "Digital heritage": Theory and innovative practice. *The International Archives of the Photogrammetry, Remote Sensing and Spatial Information Sciences*, *42*(2), 335–342. doi:10.5194/isprs-archives-XLII-2-W5-335-2017

Howard, P. (2003). Heritage: Management, interpretation, identity. *Continuum*.

Hung, W.-L., Lee, Y.-J., & Huang, P.-H. (2016). Creative experiences, memorability and revisit intention in creative tourism. *Current Issues in Tourism*, *19*(8), 763–770. doi:10.1080/13683500.2013.877422

ICOMOS. (1994). *The Nara document on authenticity*. https://www.icomos.org/charters/nara-e.pdf

Kasemsarn, K., & Nickpour, F. (2015). The challenge of digital storytelling for cultural tourism. *Journal of the Faculty of Architecture King Mongkut's Institute of Technology Ladkrabang*, *20*(1), 154–164.

Kremers, H. (2020). *Digital cultural heritage*. Springer. doi:10.1007/978-3-030-15200-0

Lambert, J. (2003). *Digital storytelling cookbook and traveling companion*. Digital Diner Press.

Lambert, J. (2010). *The digital storytelling cookbook*. Digital Diner Press.

Langellier, K., & Peterson, E. (2004). *Storytelling in daily life: Performing narrative*. Temple University Press.

Li, J., Krishnamurthy, S., Roders, A., & Wesemael, P. (2020). Community participation in cultural heritage management: A systematic literature review comparing Chinese and international practices. *Cities (London, England)*, *96*, 102476. doi:10.1016/j.cities.2019.102476

Logan, W. S., Nic, C. M., & Kockel, U. (Eds.). (2015). *A companion to heritage studies*. Wiley-Blackwell. doi:10.1002/9781118486634

Lowenthal, D. (1985). *The past is a foreign country*. Cambridge University Press.

Lowenthal, D. (2015). *The past is a foreign country: Revisited.* Cambridge University Press. doi:10.1017/CBO9781139024884

Mathisen, L. (2018). Storytelling in a co-creation perspective. In N. K. Prebensen, J. S. Chen, & M. S. Uysal (Eds.), *Creating experience value in tourism* (pp. 137–147). CABI. doi:10.1079/9781786395030.0137

Messenger, P. M., & Smith, G. S. (Eds.). (2014). *Cultural heritage management: A global perspective.* University Press of Florida.

Muenster, S., Apollonio, F. I., Bell, P., Kuroczynski, P., Di Lenardo, I., Rinaudo, F., & Tamborrino, R. (2019). Digital cultural heritage meets digital humanities. In D. Gonzalez-Aguilera, F. Remondino, I. Toschi & ICOMOS/ISPRS Committee for Documentation of Cultural Heritage (Eds.), *27th CIPA International Symposium "Documenting the past for a better future" 1–5 September 2019, Ávila, Spain* (pp. 812-820). Copernicus Publications.

OECD. (2014). *Tourism and creative economy.* OECD. doi:10.1787/9789264207875-

Palombini, A. (2017). Storytelling and telling history. Towards a grammar of narratives for cultural heritage dissemination in the digital era. *Journal of Cultural Heritage, 24*, 134–139. doi:10.1016/j.culher.2016.10.017

Piscitelli, M. (2019). Promotion of heritage tourism through branding and graphic communication. In S. Srivastava (Ed.), *Conservation and promotion of heritage tourism* (pp. 87–106). IGI Global. doi:10.4018/978-1-5225-6283-2.ch004

Poria, Y., Butler, R., & Airey, D. (2003). The core of heritage tourism. *Annals of Tourism Research, 30*(1), 238–254. doi:10.1016/S0160-7383(02)00064-6

Poulios, I. (2014). *Past in the present: A living heritage approach.* Ubiquity Press.

Prebensen, N. K., Chen, J. S., & Uysal, M. S. (2018). Co-creation of tourist experience: Scope, definition and structure. In N. K. Prebensen, J. S. Chen, & M. S. Uysal (Eds.), *Creating experience value in tourism* (pp. 1–10). CABI. doi:10.1079/9781786395030.0001

Prebensen, N. K., Vittersø, J., & Dahl, T. I. (2013). Value co-creation significance of tourist resources. *Annals of Tourism Research, 42*, 240–261. doi:10.1016/j.annals.2013.01.012

Prentice, R. (1993). *Tourism and heritage attractions.* Routledge.

Prentice, R. (2005). Heritage: The key sector in the 'new' tourism. In G. Corsane (Ed.), *Heritage, museums and galleries: An introductory reader* (pp. 267–278). Routledge. doi:10.4324/9780203326350_chapter_19

Richards, G. (1996). *Cultural tourism in Europe.* Cab International.

Richards, G. (2020). Designing creative places: The role of creative tourism. *Annals of Tourism Research, 85*, 102922. Advance online publication. doi:10.1016/j.annals.2020.102922

Richards, G. (2021). Making places through creative tourism? In N. Duxbury (Ed.), *Cultural sustainability, tourism and development: (Re)articulations in tourism contexts* (pp. 36–48). Routledge. doi:10.4324/9780367201777-4

Richards, G., & Cooper, C. (2018). The creative economy, entertainment and performance. In C. Cooper, S. Volo, W. C. Gartner, & N. Scott (Eds.), *SAGE handbook of tourism management: Applications of theories and concepts to tourism* (Vol. 2, pp. 315–327). SAGE Reference. doi:10.4135/9781526461490.n22

Robin, B. R., & McNeil, S. G. (2019). Digital storytelling. In *The international encyclopedia of media literacy,* 1-8. doi:10.1002/9781118978238.ieml0056

Robinson, M., & Picard, D. (2006). *Tourism, culture and sustainable development.* UNESCO.

Romagosa, F. (2020). The COVID-19 crisis: Opportunities for sustainable and proximity tourism. *Tourism Geographies: An International Journal of Tourism Space, Place, and Environment, 22*(3), 690–694. doi:10.1080/14616688.2020.1763447

Roque, M. I., & Forte, M. J. (2017a). Digital strategies to a local cultural tourism development: Project e-Carnide. In V. Katsoni, A. Upadhya, & A. Stratigea (Eds.), *Tourism, culture and heritage in a smart economy* (pp. 365–383). Springer International Publishing. doi:10.1007/978-3-319-47732-9_24

Roque, M. I., & Forte, M. J. (2017b). The festivities of the Light of Carnide: A persistent authenticity. In S. Lira, R. Amoêda & C. Pinheiro (Eds.), *Sharing cultures 2017: Proceedings of the 5th international conference on intangible heritage* (pp. 375-383). Green Lines Institute for Sustainable Development.

Ross, D. (2020). Towards meaningful co-creation: A study of creative heritage tourism in Alentejo, Portugal. *Current Issues in Tourism, 23*(22), 2811–2824. doi:10.1080/13683500.2020.1782355

Snyder, H. (2019). Literature review as a research methodology: An overview and guidelines. *Journal of Business Research, 104*, 333–339. doi:10.1016/j.jbusres.2019.07.039

Stanley, N. (2018). Digital storytelling. *The TESOL encyclopedia of English language teaching*, 1–7. doi:10.1002/9781118784235.eelt040

Stephens, J., & Tiwari, R. (2015). Symbolic estates: Community identity and empowerment through heritage. *International Journal of Heritage Studies, 21*(1), 99–114. doi:10.1080/13527258.2014.914964

Sylaiou, S., & Dafiotis, P. (2020). Storytelling in virtual museums: Engaging a multitude of voices. In F. Liarokapis, A. Voulodimos, N. Doulamis, & A. Doulamis (Eds.), *Visual computing for cultural heritage* (pp. 369–388). Springer. doi:10.1007/978-3-030-37191-3_19

Tilden, F. (1977). *Interpreting our heritage.* University of North Carolina Press.

Timothy, D. J. (1994). Environmental impacts of heritage tourism: Physical and sociocultural perspectives. *Manusia dan Lingkungan, 11*(4), 37-49.

Timothy, D. J., & Boyd, S. W. (2003). *Heritage tourism.* Prentice Hall.

Torraco, R. J. (2005). Writing integrative literature reviews: Guidelines and examples. *Human Resource Development Review, 4*(3), 356–367. doi:10.1177/1534484305278283

Trunfio, M., & Della Lucia, M. (2018). Engaging destination stakeholders in the digital era: The best practice Of Italian regional DMOs. *Journal of Hospitality & Tourism Research (Washington, D.C.), 43*(3), 349–373. doi:10.1177/1096348018807293

Udeaja, C., Mansuri, L. E., Makore, B. C. N., Awuah, K. G. B., Patel, D. A., Trillo, C., & Jha, K. N. (2021). Digital storytelling: The integration of intangible and tangible heritage in the city of Surat, India. In M. Rauterberg (Ed.), *Culture and computing; Interactive cultural heritage and arts* (pp. 152–168). Springer. doi:10.1007/978-3-030-77411-0_11

UNESCO. (1976). *Charter of cultural tourism*. http://www.lacult.unesco.org/lacult_en/docc/ChartCulturTour1976.doc

UNWTO. (1995). *Global code of ethics for tourism*. https://cf.cdn.unwto.org/sites/all/files/docpdf/gcetbrochureglobalcodeen.pdf

Vrettakis, E., Kourtis, V., Katifori, A., Karvounis, M., Lougiakis, C., & Ioannidis, Y. (2019). Narralive: Creating and experiencing mobile digital storytelling in cultural heritage. *Digital Applications in Archaeology and Cultural Heritage*, *15*, e0120513. Advance online publication. doi:10.1016/j.daach.2019.e00114

Wall, G. (1989). An international perspective on historic sites, recreation, and tourism. *Recreation Research Review*, *14*(4), 10–14.

Waterton, E., & Smith, S. (2010). The recognition and misrecognition of community heritage. *International Journal of Heritage Studies*, *16*(1-2), 4–15. doi:10.1080/13527250903441671

Wells, J. C., & Stiefel, B. L. (Eds.). (2019). *Human-centered built environment heritage preservation: Theory and evidence-based practice*. Routledge.

Wilson, K., & Desha, C. (2016). Engaging in design activism and communicating cultural significance through contemporary heritage storytelling. *Journal of Cultural Heritage Management and Sustainable Development*, *6*(3), 271–286. doi:10.1108/JCHMSD-10-2015-0039

Yuxiang, Z., & Jingwen, L. (2021). Review of cultural heritage crowdsourcing in the domain of digital humanities. *Data Analysis and Knowledge Discovery*, *5*(1), 36–55. doi:10.11925/infotech.2096-3467.2020.0906

ADDITIONAL READING

Cabeça, S. M., & Gonçalves, A. R., Marques, J. F., & Tavares, M. (Eds.). (2020). Creative tourism dynamics: Connecting travellers, communities, cultures and places. Grácio Editor.

Duxbury, N. (2021). *Cultural sustainability, tourism and development: (Re)articulations in tourism contexts*. Routledge. doi:10.4324/9780367201777

Lambert, J., Hessler, H. B., & ProQuest (Firm). (2018). *Digital storytelling: Capturing lives, creating community*. Routledge.

Prebensen, N. K., Chen, J. S., & Uysal, Muzaffer, S. (Eds.). (2018). *Creating experience value in tourism*. CABI.

Prebensen, N. K., Chen, J. S., & Uysal, M. (2017). *Co-creation in tourist experiences*. Routledge. doi:10.4324/9781315645919

Srivastava, S. (Ed.). (2019). *Conservation and promotion of heritage tourism.* IGI Global. doi:10.4018/978-1-5225-6283-2

Timothy, D. J. (2021). *Cultural heritage and tourism: An introduction.* Channel View Publications.

KEY TERMS AND DEFINITIONS

Co-Creation: Collaborative development of new values, contents, products, and services together with experts and outsiders (consumers).

Creative Tourism: Extension of cultural tourism that engages local communities to provide interactive and immersive experiences for tourists.

Cultural Heritage: The legacy of tangible and intangible assets of a cultural group, expressing their value systems, beliefs, traditions and lifestyles through times and spaces.

Cultural Mediation: The process of interpreting the meaning of cultural works and products.

Digital Storytelling: Digital media tools to tell stories combining text, sound and still and moving imagery with interactive features.

Heritage Tourism: Branch of tourism centred on the destination's cultural heritage and on the experience of historical sites, artefacts, and performances that represent its culture and past.

Placemaking: Collaborative and multi-faceted design and management process of public spaces, reflecting local community shared values and local place identity.

Chapter 3
Autoethnography of Cultural Heritage Tourism Documentary Production:
A Story of Tourism Storytelling

Jason R. Swanson
University of Kentucky, USA

Taylor Johannigman
University of Kentucky, USA

ABSTRACT

Using an auto-ethnographic approach, this chapter describes aspects of how a story about cultural and heritage tourism is produced as a documentary for public television. A documentary storytelling production about a proposed national heritage area in the United States provides a case study for the chapter. Included is a discussion of ethical dilemmas that arose during documentary production and how the dilemmas were addressed. Ethical dilemmas included dealing with concerns about protecting people and property, telling stories about subjects that could be viewed as controversial, covering narratives that have been neglected through history, and avoiding advancing stereotypes about people in the subject region. Solutions to dilemmas were found in how stories were crafted for the documentary production.

INTRODUCTION

Stories have stories to tell. This chapter tells the story of how tourism stories have been produced as a video documentary for public television in the United States, using a particular tourism documentary as a case study. I, as the documentary producer, drafted the chapter while also producing the tourism documentary. This simultaneity allows for a real-time reporting of challenges and successes associated with how ethical dilemmas were addressed as the production team worked through them.

DOI: 10.4018/978-1-6684-3436-9.ch003

A myriad of challenges exist when creating a video story designed to advance understanding of under-told stories in a way that entertains and encourages people to visit places featured in the documentary. Challenges described in this chapter that arose during production include protecting people and property, telling stories that could be seen as controversial, covering historically neglected topics, and advancing stereotypes in the stories presented in the documentary. Each of these ethical dilemmas are detailed following a brief description of the purpose and setting of the documentary.

This chapter is autoethnographic, as I recount some of the challenges in telling cultural heritage stories in a way to entice interest in tourism visitation to a region. The chapter also includes an account of how I became involved with documentary television production as a professor of tourism. A key to autoethnography is the author's use of self-reflection while exploring personal experiences and connecting with larger social phenomena and understanding (Ellis, 2004). Through autoethnography, the chapter presents a connection between tourism and art, in this case documentary television production.

Autoethnography has been used in research that analyzes the impacts on travel and tourism derived from other artforms such as literature, poetry, and television dramas (Best, 2017; Brown, 2016; & Noy, 2007). Noy (2007), through innovative analysis of poetry, used autoethnography to capture tourist experiences in more realistic ways that may be more vibrant and more uncomfortable than traditional research methods. Underscoring the importance of cultural preservation, Brown (2016) created an autoethnographic account showing the linkages between travelers' desire to admire a place of literary importance while in the place where books were written. Best (2017) presented a personal account of a visit to a cultural heritage museum spurred by a lifetime of fascination with American cars. He realized the museum experience, having been sanitized like tourism marketing tends to be, lacked authenticity that was presented better by the television script writers and novelists that originally spurred his interest in the subject. Similar to the authors of these works, I am also an autoethnographic traveler using art to frame this research.

Documentary Setting and Purpose

The subject of the tourism documentary used as a case for this chapter is a 41-county region in southern and eastern Kentucky, along with the region's communities, cultures, and characters. Kentucky is the northernmost southern state and the easternmost mid-western state in the United States. Kentucky was the unexplored backyard at America's inception and became the front door of westward expansion as the nation grew beyond the original 13 states. Kentucky remains an American cross-roads, of sorts, where some people pass through briefly, while others stay for generations. Southern and eastern Kentucky reflect this diversity of culture as the region is home to some of the poorest counties in the country while also home to owners of million dollar race horses and yacht-like houseboats.

The United States government has invested heavily in creating economic opportunity in this region. One of the funded projects was a region-wide tourism branding initiative that set out to position the region as the *Kentucky Wildlands* and collectively promote the region's natural beauty, forests, mountains, cliffs, lakes, and outdoor recreation. Separate from – but complementary to – the branding initiative, legislation was introduced in late 2019 to designate the region as a National Heritage Area (NHA) through the National Park Service. An NHA provides a region with technical and funding assistance to promote the natural, cultural, and historic resources that combine to form a nationally important lived-in landscape. In announcing the legislation United States Senator from Kentucky Mitch McConnell stated, "The rich history and scenic beauty of Kentucky's Appalachian region deserve to be preserved for visitors, families and communities to enjoy. Promoting the unique culture of the Commonwealth's wilderness as our first National Heritage Area can encourage new tourism and economic activity in Southern and Eastern Kentucky" (Rogers, 2019).

The impetus for my tourism documentary project was the struggle for funding faced by smaller, rural destination management organizations (DMOs) in the early stages of the COVID-19 pandemic as travel slowed. Nearly all DMOs in Kentucky are funded by hotel occupancy or local restaurant taxes. Both revenue sources ceased when travel stopped in Spring 2020. To further exacerbate funding challenges, since these quasi-governmental organizations are not private-sector small businesses nor non-profits, they were not eligible for COVID-related economic relief from the federal government for qualifying organizations.

Despite funding setbacks, opportunities arose out of pandemic challenges for smaller, rural communities. Travelers were changing their plans in ways that would benefit communities in places like the Kentucky Wildlands. Approximately 26% of travelers changed their travel plans to destinations to which they could drive, and 10% replaced international destinations with domestic destinations in their travel plans. While planning trips to large cities dominated travel plans at the end of 2019, 40% of travelers reported changing the type of destination to which they planned travel in 2020. Nearly 56% reported they hoped to avoid crowded places and more than half (50.5%) planned to avoid destinations hit hardest by COVID-19 (Destination Analysts, 2020). Small towns offering nature-based outdoor experiences benefitted from new demand. The emerging shift from international to domestic travel was part of the push toward localism, as travelers looked for authenticity about a place and its people while also helping local regions as they recover (Spinks, 2020). While rural Kentucky fits the profile of the sought-after destination as demand changed, Kentucky tourism organizations were still limited in capturing this new demand because of tourism marketing funding constraints.

As many people did during the Spring of 2020, when the world shut down, I reflected on my contributions to society. In talks with state tourism leaders about how I could help, I realized I could use my research endowment and my experience in producing programs for public television to create a documentary that might help the region tell tourism stories to support its newly created brand. After learning about the National Heritage Area initiative for the Kentucky Wildlands, I decided to reframe a grant-funded project that was designed for a pre-Covid world. Further, my own reinvented normal – including not being able to travel internationally during the pandemic – forced me to look at, and more deeply into, the physical and cultural landscape that surrounds me.

I designed a documentary for public television that tells inspirational stories about people and places, along with community-based tourism activities in destinations within the Kentucky Wildlands. Storytelling benefits communities as it plays a strategic role in tourism marketing and economic development (Bassano, Barile, Piciocchi, et al, 2019). The hour-long production is centered around themes discovered during research that characterize the culture and heritage of the region. Themes include music, industry, women influencers, pioneers, migration, and Native American heritage, among others cultural heritage topics.

The broadcast outlet for the production is the Public Broadcasting Service (PBS) network affiliate in the state of Kentucky. PBS is similar to the Canadian Broadcasting Corporation, the British Broadcasting Company, and the Special Broadcasting Service in Australia. In Kentucky, the state-wide PBS affiliate is Kentucky Educational Television (KET). KET was founded in 1962 and currently reaches over five million potential viewers in Kentucky and parts of seven surrounding states. Programs I have produced for KET match the organization's mission to make communities stronger through education, inspiration, and information.

Television Producer as Researcher and Teacher

Autoethnographic research incorporates the researcher's (i.e., the storyteller's) sense of self (Marechal, 2010). I am not a television producer, but I identify as one. My university supervisors see me as a tourism scholar and teacher who has an outside interest in making television shows. The project that is the subject of this manuscript and my previous television production projects stemmed from my research and teaching background.

I am a qualitative researcher by training. While all good research should tell a story, qualitative methods are better suited for storytelling than quantitative, particularly in tourism, recreation, and leisure fields (Henderson, 2006). Techniques such as in-depth personal interviews, content analysis of documents, and video analysis are activities of qualitative scholarly research as well as research that goes in to producing documentary television. Prior to academia, as a tourism development consultant, I learned to analyze a destination from a variety of perspectives far beyond how the typical tourist might view a place during vacation. Through these analytical skills I am able to look broadly at a place or experience to extract a deeper meaning. Both my scholarly and consulting research training enable me to tell tourism stories in ways that are not common in traditional travel media.

Teaching has also been a contributing factor to my involvement in television production. I have taught several education away courses – courses with domestic or international travel components. After leading my second group of students in an education abroad course that traveled to Panama, I began to think about the transformative nature that travel can have on people. I ultimately created an education abroad course devoted to the study of travel's transformative effects on travelers. The first course was titled Transformative Travels: Exploring Ikaria, during which the students and I traveled to a remote Greek island for a short-term immersion in a place where 80-year-old residents outnumber tourists (at least at the time of our visit). Research about this travel experience showed the transformative nature of travel centers around the traveler's willingness to allow themselves to be affected, the challenge involved in the travel experience, and spending time to get to know the destination (Cavender, Swanson, & Wright, 2020).

After teaching that class, I wondered how the impact on travelers might be different or similar if they were to travel closer to home – instead of boarding an airplane, hitting the road on a motorcoach. I designed a course that would take students on an exploration of Kentucky, the state in which they studied. While planning this course, which was titled Transformative Travels: Exploring Kentucky, the motorcoach operator suggested it could make an interesting TV show for our local public television affiliate. This passing comment lit a spark. I spent a year developing the idea, then pitched the story of ten University of Kentucky students venturing out to small Kentucky communities to local television producers and television network executives. We produced six 30-minute episodes that aired in late 2018 and early 2019. The TV show carried the same name as the course: *Transformative Travels: Exploring Kentucky*. We then produced a second season of three 30-minute episodes that featured three University of Kentucky students who were part of a term-long education abroad program in London during spring of 2019. This show also had the same title as the course I taught during the London semester: *Transformative Travels: Exploring London*.

The current production, which is the third season of my Transformative Travels series, features communities, cultures, and characters as I wander through the Kentucky Wildlands. The show is titled, *Transformative Travels: Wandering the Wildlands*. One of the major differences in this season compared to the first two seasons is the show does not feature students. Students were not available because production started during the COVID-19 pandemic when the university taught all courses online after sending students home. Without students in the story, the success of the production depends more on my research, writing, and presenting skills than previous seasons.

During production for the Wildlands documentary, I taught a class on the topic of community-based tourism storytelling. The purpose of the class was to teach students to research, write, and produce multimedia travel stories about people and places. Selected student assignment submissions became part of the documentary script and video segments. Students benefitted from a real-world application and the production benefitted from the content provided by the students. One of the students in the storytelling course is a co-author on this chapter.

STORY CRAFTING AROUND ETHICAL DILEMMAS

With limited documentary production experience and no formal training in the field, I have had to learn how to address ethical dilemmas as they present themselves. Confronting ethical dilemmas is a standard challenge of documentary production, as the final product is a convergence of subjects creating an impression on viewers that depends on the context presented, truthfully or untruthfully, by the producer (Nash, 2011). Analyzing ethical dilemmas is part of the autoethnographic research process to uncover how observations and experiences contrast with greater social and cultural issues (Ellis, 2004). Four ethical issues I faced during production of the Kentucky Wildlands tourism documentary are protecting people and property featured in the program, stories that could be viewed as controversial by tourism promoters and pro-tourism viewers, topics that are typically neglected or passed over quickly, and unintentionally advancing stereotypes about people who live in the region. Each are described below along with a discussion of how challenges were address in the production.

Protecting People and Property

Documentary programs in general bring an audience's attention to real people and their real issues. Often, these issues come from dark places. However, tourism documentaries should be lighter, telling happier stories that lead potential visitors to want to can spend their leisure time learning about people and experiencing places. Still, protocols to protect subjects and property should be followed when telling tourism stories.

The Kentucky Wildlands tourism documentary includes a story about the array of cultures that exist in a region with a population that is often viewed as homogenous. One population I found surprisingly abundant was Amish people. Amish culture includes a religious tradition that favors simple living, pacificism, reluctance to use modern technology, and a desire for self-sufficiency. Amish people are immediately recognizable by their plain dress and favored mode of transportation – a horse-drawn buggy. The Amish have a strong aversion to being photographed (or videoed) because such recordings would violate the second of the Ten Commandments: "Thou shalt not make unto thee any graven image…" (Exodus 20:4, KJV). Photographs and videos are seen as graven images by Amish people.

Armed with this knowledge, I cultivated a relationship with a non-Amish community member who had strong ties to the community in one of the counties. I wanted to get as close as possible to Amish life. My ultimate goal was to earn their trust so that I could shoot a scene of me driving an Amish horse and buggy – an ethical dilemma in itself as it could be a form of cultural appropriation common in many tourism activities (Gertner, 2019). After several weeks, my local contact secured a tour of an Amish buggy maker's workshop. The Amish buggy maker willingly obliged, as long as we did not film his face or his house.

Another dilemma arose when we noticed a field, in which a group of Amish school children and their teacher were playing baseball during a school recess. Although the outdoor recreation activity probably happens everyday, it was a rare scene to outsiders as we were a 30-minute drive from the end of the backroad where most people would turn around. Being a rare scene, it was a shot we wanted. But again, we wanted to avoid creating a "graven image" that could distress and disrespect our hosts. Still, it is hard for a seasoned videographer to not casually lean a monopod-mounted camera on a fencepost and let it roll while he tends to other matters. And when the editor and producer discover this 'graven image' that already exists, the question arises about if and how to include the video in the production. All the while knowing, those in the video or their family members would likely never see the show, since their avoidance of modern technology includes not watching public television.

The solution for this dilemma was in the editing. We placed short b-roll clips of the unidentifiable children playing baseball at points in the program that were not next to other segments that included the Amish storyline. Since the Amish story included clips from multiple counties, we were able to use a mashed-up approach for the video sequence so that the viewer should not be able to identify the private location where we filmed. However, being a documentary about how viewers can visit and experience the place (i.e., a tourism documentary), we included the public places where non-Amish regularly mingle with the Amish – produce auction houses, food markets, furniture stores, and other Amish-run businesses.

Amish storylines were not the only ones to present ethical dilemmas about protecting people and property. For decades, farmers in the region have turned up historical artifacts on their land while tending to farm fields. A variety of Civil War and Native American relics presented in county museums were found in this way. Other items get passed down from one generation to the next. We came across such a collector while filming a story about Native American heritage.

He wanted to show us some of the interesting artifacts his grandfather had found over the years. He took us to an old school building the community hopes to repurpose as a community museum. While lack of funds have limited development of the larger museum, one classroom space served as a showcase of the community's history with the typical exhibits – news stories about a famous son, foreign war memorials, and black and white photos of the town square in its heyday.

But it was a room down the hall we came to see. The classroom turned storage room contained roughly one hundred large plastic bins, stacked three high in four 25-foot-long rows. Each bin contained multiple Native American artifacts – likely more than 1,000 pieces, some allegedly dating back 14,000 years and worth thousands of dollars. The entire collection belonged to one man who indicated he pulled the pieces out of his grandparents' house over the years.

Questions arose immediately. Are they authentic? Or was his grandpa just an avid stone carver and good at making arrowheads smoking pipes, statuettes, bowls, and necklace charms? If they were worth so much, why would they not sell one or two to collectors and raise funds to properly preserve and showcase them? Why were none of these items highlighted in the community's small visitor center museum on the town square? How do I ask these questions to someone who takes pride in having and is willing for us to film the items?

To add to the story and to answer some of these questions, we invited them to bring selected artifacts to the university campus where they could be evaluated by a heritage science lab to determine details about their origin and other information that would be of interest to the collector and viewers. The lab was happy to have their services featured in the television program. We were able to showcase the artifacts in the field – in their storage place and in locations where some of them could be found – and also in a scientific lab to give new dimension to the story.

However, the scientific analysis of artifacts did not address potential issues associated with cultural appropriation since the collector and his family were not Native American. This would be addressed, in part, by interviewing a Native American leader in the community about how the items should be preserved and presented in ways that honored the Native American past so that visitors could learn from the vast collection of Native American artifacts.

Potentially Controversial Subjects

Tourism can be detrimental to the natural and cultural landscape in communities in which tourism is developed (Choi & Murray, 2010; Streimikiene, Svagzdiene, et. al., 2021). Present-day travelers may likely be unaware of the previous natural state of the destination to which they travel and in which they recreate. This can be particularly true in outdoor areas that appear to be natural. Such is the case in one of the Kentucky Wildlands' most popular tourism destination – Lake Cumberland.

Lake Cumberland showcases the stunning nature for which Kentucky is known. The lake encompasses 102 square miles (265 square km) within 1,255 miles (2,020 km) of shoreline, covering portions of six counties in southern Kentucky. Each year, more than 5,000,000 people visit Lake Cumberland, the 9th largest reservoir in the United States. The lake is a nature lover's and visitor's paradise with serene hiking trails and boating opportunities on the glistening water. In addition to its immensity and unending natural beauty, the unnatural nature of Lake Cumberland's creation is also captivating.

Steamboats used the Cumberland River regularly for transportation south into Tennessee. Although this was an excellent resource, regular natural flooding made the river dangerous. In the 1920s, a plan was developed to build a massive concrete-heavy and earth-filled dam that would become the 25th largest dam in the United States. The Wolf Creek Dam now controls flooding and creates electricity for homes and businesses in the region.

Constructing a dam to block a major section of the Cumberland River was anything but easy. World War II caused great delays and eventually brought a complete halt to construction. Yet, the complications did not end when construction resumed. The dam was built to control flooding, not to eliminate it. Through controlled flooding, a dam is how a river becomes a lake. As water rose it quickly and permanently overcame obstacles in its path.

After completing the dam in 1950, entire towns had to adapt before the flooding water from the creation of the lake destroyed everything they once knew. Burnside, Kentucky was a small town just off of the Cumberland River in Pulaski County that completely relocated to higher ground in order to avoid destruction by the creation of the lake. The U.S. Army Corp of Engineers expected the water within the lake to reach full capacity in two years, giving people time to relocate; however, there were unprecedented amounts of rainfall and flooding that caused it to fill in under a year. Structures that were not destroyed prior to the filling of the lake are now situated at the bottom of the lake. Reports indicate 1,649 structures were affected as a result of the Wolf Creek Dam and creation of Lake Cumberland including seven schools, three churches, and 17 cemeteries.

An analysis of local newspaper reporting from the time uncovered no mentions of local sentiment regarding the creation of the lake and the loss of homes and properties. The only negative remark recorded at the time came from a speaker at a church men's club who lamented the soon arrival of tourists by saying, "Oftentimes people on vacation will do things they might not do at home where they are more vulnerable to public opinion" (Mardis, 2007). Indeed, more could be covered on the topic of the creation of this lake and the many similar lakes and the negative impacts caused by their creation. However, this tourism documentary is not the place. But I would be happy to explore this further with another documentarian with a focus other than tourism.

In my view, we cannot tell a story about Lake Cumberland without acknowledging the social impacts resulting from its creation. But doing so could alienate the tourism professionals who supported the project and also some of the viewers who are more concerned about enjoying a houseboat on the lake then reminiscing about the houses on the lake bottom. We addressed this issue by featuring the stunning beauty of the lake and many of the boats on the lake. We shot an interview with a boat manufacturing executive on the deck of a 105-foot luxury houseboat. I asked the boat builder a question about how the lake was formed. After his answer, we cut to a story about the lake's creation and included a brief interview with an older couple who lived along the river when it was flooded. We included their real impression of the lake at the time of the creation and how their impression evolved over time. After the brief history lesson provided by the couple, we returned our focus to the boats that are built in factories surrounding the lake.

Memories for most people subside faster than floodwaters. For a lake that permanently flooded more than 100 square miles of livelihoods only 70 years ago, few in the region will object to what Lake Cumberland provides in terms of year-round boating, fishing, and other outdoor recreation activities in what is now perceived as a natural setting. The economic impact extends beyond direct visitor spending as the community surrounding the lake is now known as the Houseboat Capital of the World, not only because of the number of houseboats on the lake, but also because of several prominent houseboat manufacturers nearby that deliver boats to other prominent houseboating lakes around the country.

Historically Neglected Topics

Interestingly, but not surprisingly, the Native American heritage of the region has been largely overlooked by existing storytelling. The central portion of the Kentucky Wildlands region is best known for being the gateway to the pioneering west as thousands of early American pioneers traveled through the Cumberland Gap, a natural break in the mountains that made passage easier. The negative impact on the Native American population is well-documented, which may be the reason many Native American stories have been historically neglected.

Presenting the dissonant heritage of native or aboriginal people at interpretive sites can heal past hurts and empower communities, as prescribed by Lemelin, Whyte, et. al. (2013). Showcasing Native American heritage also allows destinations to diversify tourism products. Upon realizing the significant Native American heritage and associated physical elements that still exist in the Wildlands region, I made a point to feature these under-told stories while touching on traditional pioneer stories more lightly than other cultural heritage accounts that I reviewed.

Native American tourism attractions are not featured in the region's tourism marketing. Categories under *Heritage & Culture* on the regional DMO's website included Culture, Music, Art, Coal, Civil War History, Legends, and Pioneering Education. Some of these sub-categories may contain links to Native American attractions. This lack of prominence for Native American heritage at the regional level may exist because the individual communities within the region do not typically focus tourism development and marketing efforts on Native American heritage.

A handful of native American attractions, such as burial mounds and museum exhibits, do exist in the region. During a visit to a well-established community museum, where we recorded an interview about a certain music pioneer from the county, the tour guide we interviewed insisted we take a tour of the entire museum. She was particularly interested in showing us an exhibit on dresses worn at Kentucky Derby parties. Enroute to another exhibit, we breezed through the Native American room, where she paused only long enough to say that she did not know about it, but the school children she leads through the room during tours find it interesting. This museum is in a city that has burial mounds preserved in a public park. This lack of awareness by tourism employees about the Native American attractions in their own communities made me realize the need to feature the Native American heritage in the region was even greater than I originally thought.

The lack of focus on Native American heritage in tourism is comparable to how the same heritage has been treated in programming on public television the United States. The lack of Native American storytelling is reflected in the major programming PBS has shown over time. The series, *Native America*, which debuted in 2018, was the first prime-time nationally-aired show chronicling Native history since 2009 on PBS. Network executives sought to feature programs reflective of the nation's diversity, while realizing opportunities to tell under-told stories on television (Goldsmith, 2018).

This dearth of focus on Native American heritage in the region may change as there is a movement – with Native American involvement – to re-create a pathway that Cherokee and Shawnee tribes used as trading routes for millennia prior to European arrivals. The pathway is known as *Athiamiowee*, or Warrior's Path. The plan is to develop the pathway throughout the region and surrounding regions as a mixed-used recreational trail system that would cater to hikers, bikers, horseback riders, kayakers, and motorists. This project also has support from the National Park Service.

To address issues of under-representation of Native American heritage in the region and on public television, the Warrior's Path and other Native American heritage and attractions in the region are a prominent featured storyline early in the production. References to Native American heritage appear throughout the document, where appropriate, particularly when talking about the cultural diversity of the region. We hope the production can shed some light on the contributions Native Americans have had, and are having, on the region. While this production is expected to have limited impact, I believe each contribution to highlighting under-represented groups and communities can contribute to a larger, cumulative effect. Or, perhaps will inspire others to learn more about these groups beyond what we can cover in the production. Above all, this is a tourism documentary with the aim of inspiring people to explore more for themselves.

Avoiding Advancing Stereotypes

The people in the larger Appalachian region, in which the Kentucky Wildlands is situated, have been subject to negative stereotypes for more than a century. The most common ethnophaulic term to describe Appalachian people is *hillbilly*. A hillbilly was first described in a 1900 *New York Journal* article as an unrestrained, white man who talks, drinks, dresses, and shoots his gun as he pleases (Jones, 2019). The term has evolved to be applied, at times, to all working-class white people from the southern United States. The stereotype has been perpetuated for decades by a variety of television programs, Hollywood movies, and other media (Guy, 2019).

When I initially approached the director of the Kentucky Wildlands DMO with my idea to do a television documentary about the region, her first concern was ensuring the production would portray the region's people in a positive light. This was an easy assurance to make, knowing my intentions, but the region's stereotype surfaced several times during pre-production research. I decided on three approaches to mitigate advancing this stereotype in the production.

The first approach was to highlight the diversity of people who currently live in the region or who had historically lived in the region. The area has been home to communities of Swiss, German, and Hungarian immigrants, among people from other European nations. The region was home to a large freed slave population following the American Civil War because of Kentucky's role as a border state – situated in between the northern and the southern states. The Amish and Native Americans, previously described in this chapter, are other groups that add to the diversity of the region. With several cultures to select from, we decided to introduce those listed in this paragraph while telling the story of a community that has preserved parts of an Austrian settlement, including a church that is still used by descendants of the original Austrian settlers. Interviews were also conducted with individuals who could speak to the African-American heritage in the region in a deliberate attempt to be more inclusive with the documentary.

The second approach to mitigate stereotype advancement was to contribute energy to busting the myth by going to the roots of what many people believe is the origin of the hillbilly saga – the early 20th century dispute between the Hatfield family of West Virginia and the McCoy family of Kentucky. The Hatfield and McCoy feud lasted at least three generations as family members continued to steal, sabotage, and occasionally murder. The disagreement caught the attention of national media. When reporters would visit the area, they would report on the general lawlessness of the mountain region, portraying all of the people with the hillbilly stereotype. I learned about a community museum in the heart of where the feud took place that has created exhibits showing direct evidence that the people of the region, and particularly the McCoy family, were not as unrestrained as the media reports and subsequent interpretation in the popular culture has led us to believe. In the production, instead of recounting the well-known events of the feud, we focus on this historical evidence refuting the hillbilly stereotype.

The third approach to combat the stereotype of the hillbilly, which is an unrestrained white man, was to focus on women who contributed to the cultural development of the region. Women's contribution to the region included the packhorse librarians who would carry library books on horseback for several days over treacherous paths from town libraries to remote homes without access to books. All the adventurous librarians were women. Another woman important to the region was Jean Thomas, a court reporter known as the Traipsin' Woman, who realized the broad presence of music in mountain communities as she followed a circuit court judge through the region, while also documenting mountain life. She started the American Folk Song Festival and helped to bring bluegrass music to the mainstream. We also focused on young women entrepreneurs operating tourism and hospitality businesses. These women who grew up in the area left home but eventually returned and are having important impacts on their communities. In many cases, it was the women who built and are continuing to build the region.

FUTURE RESEARCH DIRECTIONS

Further research on ethical dilemmas in tourism documentary production is warranted, particularly as documentary production is becoming easier. Video cameras with increased recording quality are becoming less expensive and easier to operate. The same is true for editing and graphics design software. Further, the influx of distribution channels – online streaming services and video sharing social media platforms – where production standards may be lower than required by public television networks could lead producers to be more careless when faced with ethical dilemmas. A valuable research study would be an analysis of existing tourism documentaries distributed across a variety of platforms to identify how ethical dilemmas were addressed. Expanding this autoethnographic analysis into a larger empirical study, using focus groups, in-depth personal interviews, or surveys could inform future documentary production and provide a framework for dealing with ethical challenges during pre-production.

Continuing the analysis related to the distribution of documentaries could also provide valuable insights on the resulting impacts that tourism documentaries may have on the subject destination. Positive impacts could include increased visibility and subsequent visitation demand induced by the production. An example analysis could investigate if comparable numbers of viewers on social media platforms and on broadcast television networks induce demand for tourism visitation to a subject destination in different ways. Research analyzing the impacts of tourism documentaries related to variables in addition to distribution channel – such as intended documentary audience, budget, or length – would also be valuable. Knowing this, tourism documentary producers could allocate resources to develop the most effective distribution channels to best assist tourism marketers promote their destinations using tourism documentary storytelling.

CONCLUSION

Telling under-told cultural heritage stories about places can help viewers (or readers) see the place in new ways that might encourage visitation. But as these ethical dilemmas illustrate, the storyteller must proceed cautiously, while still taking risks, to present the place and people in ways that boost the viewers' appreciation and entices them to learn more by visiting. Solutions presented to the ethical dilemmas described could be viewed as simply rationalizing my desired outcomes in order to advance my documentary goals. To some extent this is true, as it would be for anyone with a professional agenda while also trying to make fair and balanced choices between multiple alternatives. However, it is my hope that discussing the solutions to these challenges could inform others faced with decisions related to how tourism development and visitor experiences should be presented to not only documentary viewers from afar, but also to visitors in the destination.

The easiest way to mitigate challenging dilemmas that arise in documentary production is to avoid the stories that create challenging dilemmas. However, in doing so, story producers run the risk of telling stories that are overtold, sanitized, or just less interesting. In my work on the documentary project, I realized that very few stories remain untold. The public television network to which the documentary will be submitted has been telling stories of Kentucky culture for nearly six decades. At this point, most of the cultural heritage stories have been told. However, these stories can still be presented in innovative ways and from different perspectives. That is a challenge for any documentary producer but may be a challenge that is easier to overcome for someone who is new to the activity and still learning how to produce tourism documents.

Above all, I attribute my success in getting my documentaries on public television air to incorporating my whole self into the documentary. Despite my urge to identify otherwise, I am a tourism researcher and teacher who happens to make tourism documentaries. Bringing my tourism researcher, teacher, and consulting skills and abilities to a television production automatically makes my tourism documentary projects unique. I am also a wanderer. I prefer a saunter to a brisk walk. While wandering through a place, I am able to use my professional training to understand tourism destinations different than the typical tourist or documentary producer, I believe. I am the only professor of tourism in the state of Kentucky, and probably within a larger region, who makes tourism documentaries for television. As such, I hope to produce a travel show that offers viewers some 'why' about tourism, instead of just seeing the 'what' and 'how' that typically comes from traditional travel media. What is true for me is true for other storytellers – incorporating oneself into any story should differentiate the final product and help the storyteller tell a better story.

Final notes: While this book chapter was written during post-production and completed prior to the final version of the documentary that aired, some of the footage described in this manuscript may not appear in the final production. Editing for content and quality dictate what stories are included in the final documentary production. This research received no specific grant from any funding agency in the public, commercial, or not-for-profit sectors.

REFERENCES

Bassano, C., Barile, S., Piciocchi, P., Spohrer, J., Iandolo, F., & Fisk, R. (2019). Storytelling about places: Tourism marketing in the digital age. *Cities (London, England)*, *87*, 10–20. doi:10.1016/j.cities.2018.12.025

Best, G. (2017). Media memories on the move: Exploring an autoethnographic heritage of automobility and travel. *Journal of Heritage Tourism*, *12*(1), 52–66. doi:10.1080/1743873X.2016.1201088

Brown, L. (2016). Treading in the footsteps of literary heroes: An autoethnography. *European Journal of Hospitality. Tourism and Recreation*, *7*(2), 135–145.

Cavender, R., Swanson, J., & Wright, K. (2020). Transformative travel: Transformative learning through education abroad in a niche tourism destination. *Journal of Hospitality, Leisure, Sport and Tourism Education*, *27*, 100245. Advance online publication. doi:10.1016/j.jhlste.2020.100245

Choi, H., & Murray, I. (2010). Resident attitudes toward sustainable community tourism. *Journal of Sustainable Tourism*, *18*(4), 575–594. doi:10.1080/09669580903524852

Destination Analysts. (2020, May 11). *Nature-Based Destinations and the Future of Travel. Destination Analysts*. https://www.destinationanalysts.com/blog-nature-based-destinations-and-the-future-of-travel/

Ellis, C. (2004). *The ethnographic I: A methodological novel about autoethnography*. AltaMira Press.

Gertner, R. (2019). The impact of cultural appropriation on destination image, tourism, and hospitality. *Thunderbird International Business Review*, *61*(6), 873–877. doi:10.1002/tie.22068

Goldsmith, J. (2018, July 27). *Undertold history of Native America lands on PBS with a big splash*. Current-News for people in public media. https://current.org/2018/07/undertold-history-of-native-america-lands-on-pbs-with-a-big-splash/

Guy, R. (2019). Will the real hillbilly please stand up? Urban Appalachian migration and culture seen through the lens of *Hillbilly Elegy*. In A. Harkins & M. McCarroll (Eds.), *Appalachian Reckoning: A Region Responds to Hillbilly Elegy* (pp. 86–104). West Virginia University Press.

Henderson, K. (2006). *Dimensions of Choice: Qualitative Approaches to Parks, Recreation, Tourism, Sport, and Leisure Research* (2nd ed.). Venture Publishing, Inc.

Jones, J. (2019). Notes on a Mountain Man. In A. Harkins & M. McCarroll (Eds.), *Appalachian Reckoning: A Region Responds to Hillbilly Elegy* (pp. 270–277). West Virginia University Press.

Lemelin, R. H., Whyte, K., Johansen, K., Higgins Desbiolles, F., Wilson, C., & Hemming, S. (2013). Conflicts, battlefields, indigenous peoples and tourism: Addressing dissonant heritage in warfare tourism in Australia and North America in the twenty-first century. *International Journal of Culture, Tourism and Hospitality Research*, *7*(3), 257–271. doi:10.1108/IJCTHR-05-2012-0038

Mardis, B. (2007, May 17). Looking back at the birth of Lake Cumberland. *Commonwealth Journal*. https://www.somerset-kentucky.com/news/looking-back-at-the-birth-of-lake-cumberland/article_35f3a55c-93b1-5030-8283-421648164e48.html

Marechal, G. (2010). Autoethnography. In A. Mills, G. Durepos, & E. Wiebe (Eds.), *Encyclopedia of case study research* (Vol. 2, pp. 43–45). Sage Publications.

Nash, K. (2011). Documentary-for-the-Other: Relationships, Ethics and (Observational) Documentary. *Journal of Mass Media Ethics*, *26*(3), 224–239. doi:10.1080/08900523.2011.581971

Noy, C. (2007). The Poetics of Tourist Experience: An Autoethnography of a Family Trip to Eilat. *Journal of Tourism and Cultural Change*, *5*(3), 141–157. doi:10.2167/jtcc085.0

Rogers, H. (2019, December 17). *Rogers, McConnell Secure Kentucky Wildlands National Heritage Area Study Act in Final Federal Funding Package for 2020* [Press Release]. United States House of Representatives, Office of Congressman Hal Rogers. https://halrogers.house.gov/2019/12/rogers-mcconnell-secure-kentucky-wildlands-national-heritage-area-study-act-in-final-federal-funding-package-for-2020

Spinks, R. (2020, May 21). *What is the future of vacations?* Skift. https://skift.com/2020/05/21/what-is-the-future-of-vacations/

Streimikiene, D., Svagzdiene, B., Jasinskas, E., & Simanavicius, A. (2021). Sustainable tourism development and competitiveness: The systematic literature review. *Sustainable Development*, *29*(1), 259–271. doi:10.1002d.2133

ADDITIONAL READING

Alamouti, D. (2020). The digital ethical space: Towards a transnational documentary ethics, a filmmaker's point of view. *Transnational Screens*, *11*(2), 103–120. doi:10.1080/25785273.2020.1734305

An, G., & Witt, C. (2022). Ethics of care in documentary filmmaking since 1968. *French Screen Studies*, *22*(1), 1–5. doi:10.1080/26438941.2021.2003551

Bassano, C., Barile, S., Piciocchi, P., Spohrer, J., Iandolo, F., & Fisk, R. (2019). Storytelling about places: Tourism marketing in the digital age. *Cities (London, England)*, *87*, 10–20. doi:10.1016/j.cities.2018.12.025

Domínguez-Quintero, A. M., González-Rodríguez, M., & Paddison, B. (2020). The mediating role of experience quality on authenticity and satisfaction in the context of cultural-heritage tourism. *Current Issues in Tourism*, *23*(2), 248–260. doi:10.1080/13683500.2018.1502261

Grant, B., & Sloniowski, J. (Eds.). (1998). *Documenting the Documentary: Close Readings of Documentary Film and Video*. Wayne State University Press.

Lewis, P. J. (2011). Storytelling as Research/Research as Storytelling. *Qualitative Inquiry*, *17*(6), 505–510. doi:10.1177/1077800411409883

Liestol, G., Ritter, C. S., & Ibrus, I. (2019). Audiovisual Industries and Tourism: Forms of Convergence. In I. Ibrus (Ed.), *Emergence of Cross-innovation Systems* (pp. 165–172). Emerald Publishing Limited. doi:10.1108/978-1-78769-977-920191018

Vázquez-Herrero, J., & Gifreu-Castells, A. (2019). Interactive and Transmedia Documentary: Production, Interface, Content and Representation. In M. Túñez-López, V. Martínez-Fernández, X. López-García, X. Rúas-Araújo, & F. Campos-Freire (Eds.), *Communication: Innovation & Quality. Studies in Systems, Decision and Control* (pp. 113–127). Springer.

KEY TERMS AND DEFINITIONS

Autoethnography: A type of qualitative research that includes a reflection of the author's personal experience with socio-cultural issues, in the context of the research topic.

Documentary Producer: The person who oversees production of a video program that presents a factual record. For small-scale documentaries, the producer may perform many functions including scheduling interviews, researching locations, scheduling and assembling the crew, running interviews, videography, and script writing.

Ethical Dilemmas: A situation requiring a choice between two or more competing alternatives that may violate a moral code.

Kentucky Wildlands: The tourism brand for a 41-county region in southern and eastern Kentucky. Work is underway to designate the region as a National Heritage Area.

National Heritage Area: A lived-in region designated by the United States Congress in which historic preservation and promotion is encouraged.

Public Television: Broadcast television programming funded by government or philanthropic sources that focuses on educational content.

Story Crafting: To produce an informative and entertaining narrative with skill, creativity, and resourcefulness.

Tourism Documentary: A video program that presents a factual record about a community with the intent to encourage visitation to the subject community.

Under-Told Stories: Factual accounts that have been previously communicated infrequently or presented with details lacking.

Chapter 4
The Development of Stories of Bajo Mola in Wakatobi National Park

Wiwien Tribuwani Wiyonoputri
Indonesian Society for Interpretation, Indonesia

Buntu Marannu Eppang
Politeknik Pariwisata Makassar, Indonesia

Noel Scott
University of the Sunshine Coast, Australia

ABSTRACT

Tourism is part of the experience economy, and more and more governments are interested in improving their tourism experiences. This improvement can add value for visitors as well as increasing income for local people. There are several techniques that are used in developing experiential tourism products, and this chapter focuses on development through the use of storytelling, particularly applying interpretation approach, principles, and techniques to create a story-based experiences well-known as interpretive experience. This chapter provides a case study of the development of interpretive experiences by the Bajo sea-farer, the largest sea nomad descendant in the world, who live in Mola, Wakatobi National Park, Indonesia, a UNESCO Man and Biosphere site and listed in their World Heritage Tentative List. Through series of workshops, community investigation, simulation, and tryouts, they developed four interpretive experiences strongly related to Bajo culture, namely Dolphin Watching Tour, Cultural Walking Tour, Village Canoeing Tour, and Star-Telling.

DOI: 10.4018/978-1-6684-3436-9.ch004

INTRODUCTION

Tourist destination managers are tasked with increasing visitor expenditure. To do this they may increase the average length of stay, expenditure per visitor or the number of visitors. One important strategy used to achieve these aims is to improve tourism product quality. A tourism product combines elements of tangibility and intangibility. The United Nation World Tourism Orgnasization (UNWTO) includes natural, cultural and man-made resources as well as attractions, facilities, services and activities around a specific center of interest that made up a tourism product and it represents the core of the destination marketing mix and creates an overall visitor experience including emotional aspects for the potential customers (UNWTO, 2019, para. 1). As a result, tourism managers often try to improve the quality of their tourism products through experiences. Medlik and Middleton (1973) consider that a tourism product involves the creation of experiences for tourists.

Scholars such as Prebensen et al. (2017) and Moscardo (2020) have discussed how to create experiences. Creation of experiences requires tourists to be actively or passively involved (Prebensen et al., 2017). This involvement can be facilitated by telling stories (Moscardo, 2020). A story about a destination or a place told to a visitor is mentally engaging and more memorable than just objective facts or statistics. When listening to an interesting story, a visitor creates a picture of the events of the story in their mind. This mental image allows a story to be better understood, the meaning interpreted, and its relevance identified. This process can lead to stronger emotional arousal and a memorable and meaningful experience. Hence, stories are an important tool used to improve experience quality.

There are a number of ways to improve visitor experiences. Physical activities (Scott et al., 2017) attract visitors' attention as can meaningful and interactive activities. Stories and storytelling also attract a visitor's attention as well as mental engagement (Li, 2014; Prebensen et al., 2017). Stories are utilized in experiences in a destination and have a focus for academic research. Research into stories has examined their symbolic and cultural components, stories as communication and learning tools, their role in dramatic performances and leading to emotion, and into their structures (McCabe & Foster, 2006). "[S] tories are central to the way tourists organize and talk about their travel experiences and .... may be a critical element of tourism (Moscardo, 2020: 4). The creation of meaning in the story is important to deliver value and emotion (Li, 2014). Li (2014) considers that stories can help visitors understand life experiences; master surrounding events and life situations; assist people in perceiving, recognizing, and developing their self-image; and encourage people to take action. Moscardo (2020) and Kim & Youn (2017) suggest that the main role of stories is provoking emotion. Moscardo (2020) also argues that successful stories have a clear beginning, setting, and ending. Binkhorst & Dekker (2009: 313) also emphasize that tourists search for "psychological needs such as inspiration, authenticity, belonging to a meaningful community, value and meaning".

The general term used to describe communication of information about a place is "interpretation". Interpretation of places sometimes only involves a history lesson or a series of facts. More recently, good practice interpretation links the communication to creation of an experience. For example, Wiyonoputri, Anindita, and Masita (2021) believe that interpretation is more effective if it includes a story-based experience. This is called an interpretive experience.

Such interpretive experiences are useful in improving the tourist experiences in Indonesian communities. There is an increasing number of community-based tourism initiatives across the country, in part because of the number and variety of ethnic groups in the country. One such initiative involved the Bajo ethnic group living in Mola, Wakatobi National Park. Bajo Mola is located in west Wangi-Wangi

Island and has a population of around 7,000 residents. The Bajo Mola people live in wooden stilt houses built over the sea. They are descendants of sea nomads, with cultural knowledge and skills about living on and from the sea inherited from previous generations. Their culture is perceived as exotic for those who live on land and this allows creation of unique tourist experiences.

This chapter discusses the process of designing and developing interpretive tourist experiences as part of a Bajo Mola community-based tourism initiative. A number of tours were developed, based on the resources both physical and intangible. The intangible resources were identified from the community's way of life and their cultural practices. The process aimed to benefit the community by providing an additional source of income for the community as well as providing social and cultural benefits.

INTERPRETIVE EXPERIENCES

Interpretation has long been used in various tourist attractions in many destinations to create meaningful visitation and, therefore, quality experience. The term 'interpretation' was first defined and used by Tilden in 1957 when he worked for the US-National Park Service. Tilden (1977: 8) defines interpretation as "an educational activity which aims to reveal meanings and relationships through the use of original objects, by the first experience, and by illustrative media, rather than simply to communicate factual information." The US National Association for Interpretation (NAI) is continuously improving its working term of interpretation and they recently revisited their definition in 2021 as "a purposeful approach to communication that facilitates meaningful, relevant, and inclusive experiences that deepen understanding, broaden perspectives, and inspire engagement with the world around us" (NAI, 2021, para. 1). Their new definition explicitly states the link of interpretation with experience. In the Indonesian context, drawing on several definitions of interpretation and field works in several areas of the country, Wiyonoputri, Anindita and Masita (2021) considered that interpretation creates a story-centric/-based visitor experience.

Tilden (1977) identified several principles for the delivery of interpretation: relevance, revelation, art of story, provocation, whole, and appropriateness to the audience age. First, the interpretation must be related to the visitor's personality or experience (Relevance). Second, information is not interpretation. Interpretation is disclosure based on information (Revelation). Third, interpretation is an art that combines various arts, whether the material presented is scientific, historical, or architectural (Art of story). Next, the main purpose of interpretation is not instruction but provocation (Provocation). Furthermore, interpretation should aim to present the whole rather than a part (Whole) in delivering both its story content and experience. Finally, the interpretation must consider the capacity of the audience so that the interpretation addressed to children must be different from the delivery to adults (Appropriateness).

A number of models have been used in development of interpretation. One of the models offered in the interpretation development is the TORE model presented by Ham (2013). Interpretation places great importance on communication, therefore Ham (2013) sees that communication will be successful if it is thematic, organized, relevant, and enjoyable. The thematic feature of interpretation means that interpretation has the main idea that will be communicated to visitors. The theme is the overarching conclusion about a topic that the interpreters would like the audience to think about (Ham, 2013). An organized interpretation is an interpretation in which a story is easy to follow and its success will be seen when visitors understand what is conveyed through a well-structured flow. It will look at what information need to be included and what is not. Relevancy in interpretation occurs when the interpretation is meaningful and personal. The meaning of interpretation will emerge when the ideas conveyed are connected with the minds of visitors based on their previous experiences and knowledge. Interpretation is personal if the interpretation can connect visitors with their concerns about the theme being discussed. Interpretation will be enjoyable when interpretation provides mental pleasure and the opportunity to learn about what is presented which will ultimately give satisfaction to visitors. Mei et al. (2020) argue that entertainment value is needed to gain visitors' interest and keep their attention in conveying stories. Therefore, interpretation employs various techniques to make the process of communicating ideas engaging. This is the reason interpretation is more than simply creating programs and media (Gross & Zimmerman, 2002). Interpretation enhances the experience (Ham, 2013).

Ludwig (2021) believes that turning experiences at a site into a whole interpretive experience significantly increases the chance of memories staying in visitors' minds. Whole interpretive experiences include the outer (sensory) experiences and inner (mental) experiences. Therefore, he suggested five aspects of a whole interpretive experience as:

- **Physical**: experiences using external sensory elements - sight, hearing, touch, taste, smell - and internal ones - position, balance, temperature, and pain.
- **Intellectual**: a process that links the sensory experiences with resources and pieces of information about the resources with ideas and values.
- **Emotional**: an experience that influences the quality of its memorization and can bring people into a constructive motion.
- **Spiritual**: experience to search for personal meaning, experiencing a larger truth behind the facts.
- **Social**: an opportunity to exchange about issues that are critical for themselves and for their shared future.

Given the importance of interpretation, developing interpretive experiences in tourist destinations with great significance, such as national parks, are imperative. Thus, this chapter provides an example of creating interpretive experiences in Wakatobi National Park.

Wakatobi National Park

The Tukang Besi Islands were allocated as a national park in 1996 and, finally, declared as Wakatobi National Park by the Minister of Forestry in 2002 covering an area of 13,900 km^2 (Figure 1). The name Wakatobi is an acronym taken from the names of the four largest islands in the area, namely **Wa**ngi-Wangi, **Ka**ledupa, **To**mia, and **Bi**nongko. The Government of Indonesia designated Wakatobi as a regency in 2003. Wakatobi Regency is responsible of land management within the national park which covers an area of 559.5 km^2 and is inhabited, today, by more than one hundred thousand inhabitants.

Figure 1. Wakatobi National Park Area
Source: University of Wellington (n.d.)

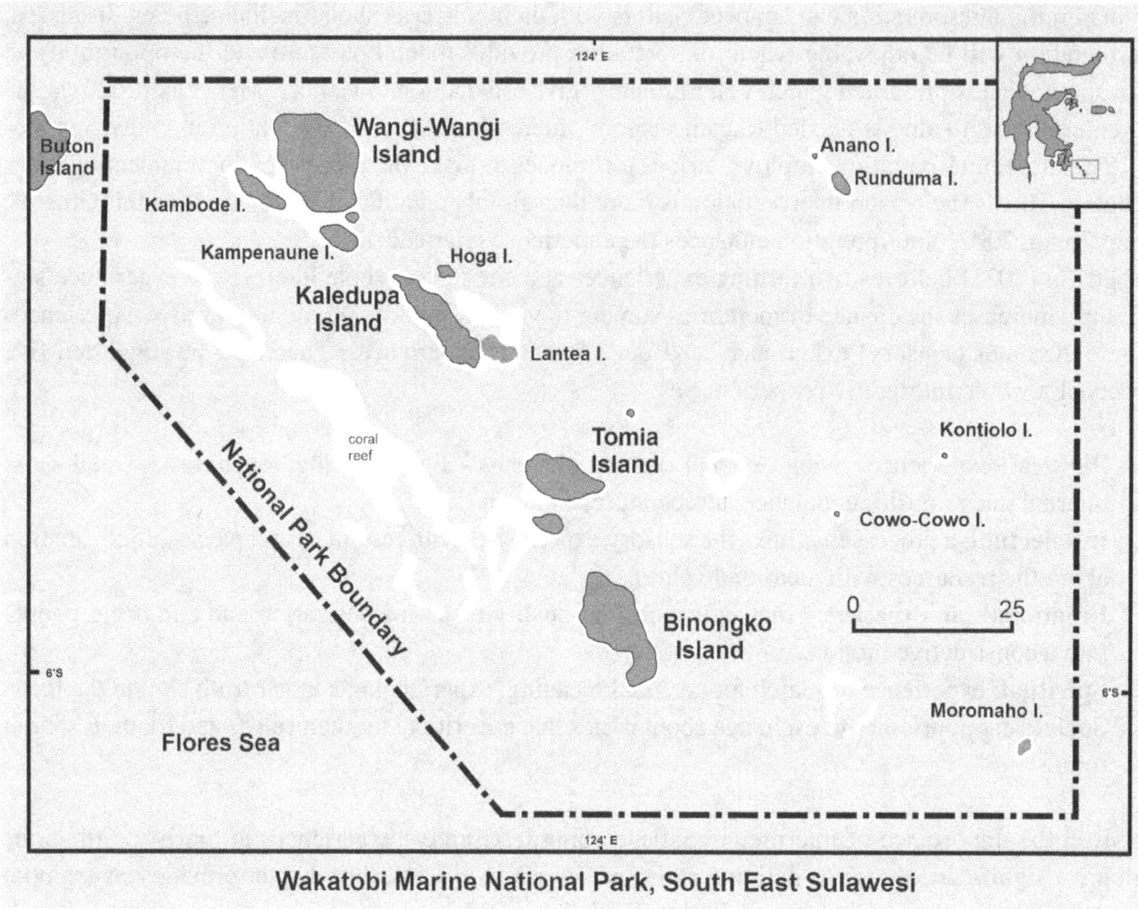

Wakatobi is one of the marine biodiversity hotspots in eastern Indonesia and the sea area makes up 97% of the park. Around 396 species of hard coral and 942 species of reef fish are found in the national park area. In addition to coral reef ecosystem biota, Wakatobi is also a habitat for megafauna species such as whales, dolphins and turtles (Firmansyah, et al., 2017). In 2005, the Government of Indonesia, through the Ministry of Forestry, proposed the national park as a UNESCO World Heritage site and it currently listed on the UNESCO Tentative List.

Recognizing the critical need to safeguard the region's marine and coastal resources, a multilateral cooperation called the Coral Triangle Initiative on Coral Reefs, Fisheries and Food Security (CTI-CFF) was formed in 2009 and members of the initiative include the governments of Indonesia, Malaysia, Papua New Guinea, the Philippines, Solomon Islands and Timor-Leste representing the custodians of the Coral Triangle area (Figure 2).

Figure 2. Wakatobi National Park in Coral Triangle Initiative on Coral Reefs, Fisheries and Food Security Implementation Area
Source: Coral Triangle Initiative (n.d.)

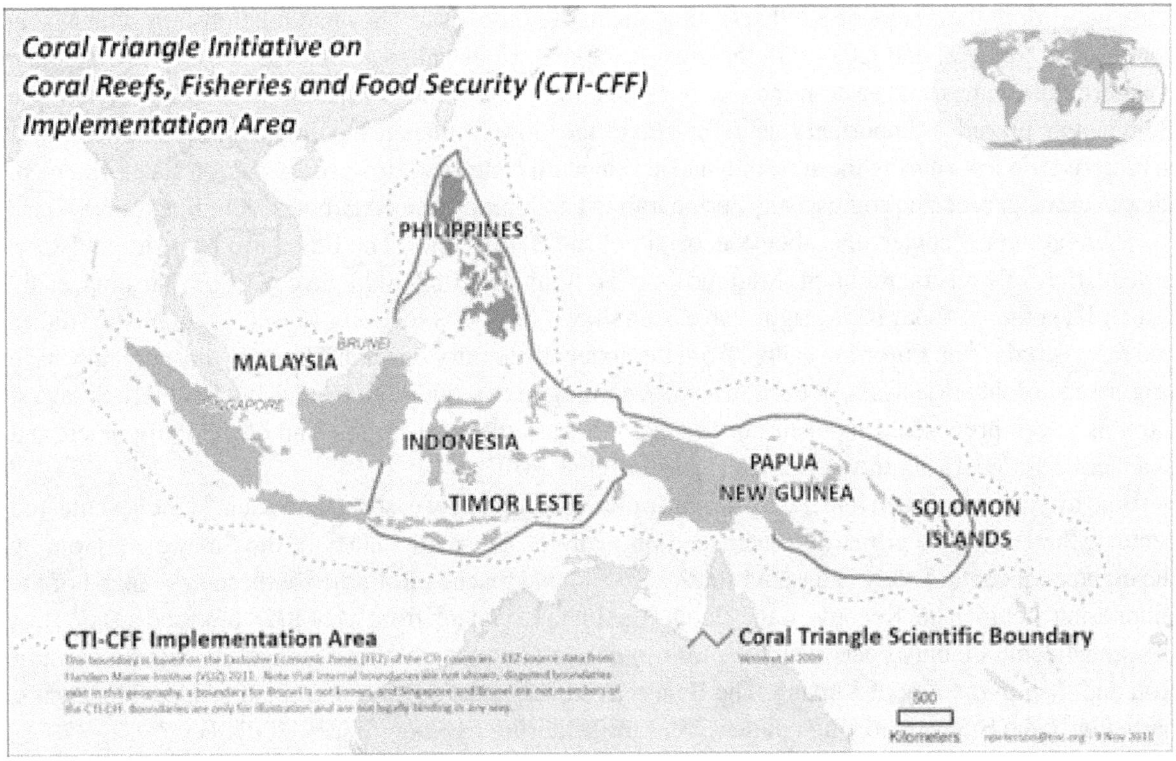

Wakatobi's marine resources are a source of livelihood for more than 50% of its inhabitants (Purwanto et al. in Firmansyah, et al. 2017) and play an important role in the marine biodiversity of the region (Turak in Firmansyah, 2017). Considering the importance of the preservation of marine biodiversity and the sustainable fulfilment of the needs of the population, the Indonesian government nominated it for the UNESCO Man and Biosphere Program, an intergovernmental scientific program that aims to establish a scientific basis for enhancing the relationship between people and their environments. In 2012, the park officially became one of 727 biosphere reserves. Today, Wakatobi is also designated as one of the ten national priority destinations to be developed. An analysis on the Wakatobi Regency population in 2016 showed that the Wakatobi ethnic group make up more than 90% of the Wakatobi population while Bajo or Bajau ethnic community makes up 8%.

Bajo Ethnic Group and Bajo Mola

The Bajo ethnic group is one of the two nomadic sea people remaining in Southeast Asia. The Orang Laut are found in the western waters of Indonesia, especially in the Riau Archipelago and around the Malay Peninsula, and the Bajo people in eastern Indonesia, Malaysia and in the Philippines. The Bajo are the largest group of nomadic sea dwellers (Chou, 2006; Stacey, 2007; Kusuma, P. et al.,2017). Places' names such as Labuhanbajau on We Island, one of the westernmost islands in Indonesia, and locations called Bajau or Bajo on Anambas Islands in the South China Sea indicate that Bajo seafarers had a wider sailing range in the past (Lapian, 2011). Sea nomads live entirely on boat and were not tied to a land territory (Chou, 2006; Kusuma, et al.,2017). Sea nomads might shape the initial maritime trade network in the waters of Southeast Asia, including Indonesia. They had relationships with Malay countries in the west and later, with the emergence of various sultanates on the coast of Kalimantan, southern Philippines, and eastern Indonesia (Sather 1995 and Sather 1997:329-33 in Chou, 2006). The Bajo people played an important role in the rise of the Gowa Sultanate's political and economic power in eastern Indonesia during the sixteenth and seventeenth centuries. However, as nation states emerged, the sea nomads became progressively peripheralized and impoverished (Chou, 2006; Stacey, 2007).

There are many conjectures about the origin of the Bajo people. The Bajo have no written history, instead, they rely on oral tradition (Muljono-Larue & Pranoto, 2002/2008; Stacey, 2007; Kusuma, et al., 2017). Hypotheses about their origins have been drawn from this folklore, as well as linguistic studies and rare records from European sailors from the sixteenth century onward. The Bajo diaspora may have originated in Johor, Malaysia, or even Arabia, according to oral tradition. Brunei and Southern Sulawesi have also been proposed. Linguistic surveys point toward the Sulu archipelago of the Philippines, and at an earlier stage, to Southeast Borneo (Kusuma, et al. 2017).

Bajo migration has been integral to the maritime development of eastern Indonesia. In the seventeenth century, the Bajo were primarily established on many of the small islands of the Sulawesi region; by the eighteenth century, they had sailed southward and had reached Roti, the southernmost island of the Indonesian archipelago. Records of the Dutch East India Company from May 1728 reported a Bajo fleet of some 40 small family boats searching for *trepang* or sea cucumber first along the southern coast of Roti and then in the Bay of Kupang. The Bajo thus sailed as the advanced scouts for other Macassans, particularly the Bugis, searching out new areas for gathering *trepang* (Stacey, 2007; ix).

However, a study on the genomic origin and dispersal of three Bajo communities - Bajo Kota Baru in southern Borneo, Bajo Derawan in northeastern Borneo, and Bajo Kendari in southeastern Sulawesi - in 2017 revealed that the three communities have an admixed profile with two major Asian components - Manyan Borneo and Igorot from the Philippines, as well a Papuan component, but in varying proportions. Bajo Kendari which is located in South East Sulawesi Province, near Bajo Mola, has more Papuan component than the two Borneo Bajo groups, consistent with their location further east (Kusuma, et al., 2017).

Sopher in Stacey (2007) indicated that the boat-dwelling life of Bajo declined after the nineteenth century. Some sources say that before the mid-twentieth century (Chou, 2006; Kusuma, P. et al.,2017) there were still many Bajo people living nomadically in boats. However, after the mid-twentieth century, the government's policy forced the Bajo people to settle in shallow waters which usually where their boat anchor points (Chou, 2006). It is now difficult to find Bajo communities who are not registered as residents of a village in Indonesia.

There are five Bajo settlements in Wakatobi National Park. The Bajo Mola is the only Bajo settlement in the waters of Wangi-Wangi Island. They have settled and are recorded as residents of a village. They still keep their life orientation toward the sea and some of them still go on fishing voyages for months. However, they will return to the village and their families in Bajo Mola.

Bajo people were permitted to live and build their stilted houses in Mola, the waters of Wangi-Wangi Island, by the Mandati ethnic group leader, the ruler of that particular area of the island in 1958. Since they live in an area that belongs to another ethnic group, they do not really have the freedom to develop the area. However, a reclamation project by the central government in 2013 created several areas where the traditional stilt houses have been remade in brick.

THE PROCESS

The development of interpretive stories of and experiences in Bajo Mola in Wakatobi National Park are part of the Mandiri Bersama Mandiri – Pariwisata Berkelanjutan (MBM-PB) program (in English: Self-sustained with Mandiri – Sustainable Tourism Program). This initiative was implemented in a number of locations between 2012-2015 as a joint program between the Bank Mandiri, a state-owned bank, and the British Council, which acted as the program implementor. The implementation of the MBM-PB in the Bajo Mola area took place in 2013-2015. The purpose of the MBM-PB Program was to:

- encourage economic development in a region by improving the community's ability to manage tourism businesses,
- increase income through the introduction of creative and productive activities in the tourism sector.

The Selection of the Location

The selection of locations in each area was carried out based on a number of criteria. First, the candidate village was required to be a recipient of Program Nasional Pemberdayaan Masyarakat Mandiri Pariwisata (PNPMMP) or National Program of Community Empowerment in Self-sustained Tourism funded by the Ministry of Tourism. By meeting this criteria, the MBM-PB Program was expected to support villages that had already started tourism initiatives. The Greater Mola area consists of five villages: Mola Bahari, Mola Utara, Mola Samaturu, Mola Selatan and Mola Nelayan Bakti. Only Mola Bahari and Mola Selatan, out of the five villages, had received the PNPMMP fund. However, because of the proximity of the five villages, the similarity in their physical and socio-cultural structures, Greater Mola was considered as one destination. Second, the candidate villages were in a destination proposed for a community-based tourism initiatives by the local government. One of the main reasons of proposing these villages because the destination is located on Wangi-Wangi Island, only ten minutes from the capital of Wakatobi Regency and thirty minutes from the airport and it is the main port for inter-island traditional boat transportation. Third, the village leaders of the candidate villages committed to supporting the program with basic logistics and policies necessary for community-based planning and implementation processes. While the previous MBM-PB program involved only one village, an agreement to become one destination and commitment from five village leaders in Greater Mola were obtained before the implementation of the program. Last, the candidate villages were assessed as having the potential resources and community support for a community-based tourism initiatives. The villages in the Greater Mola area have a distinct atmosphere of above-water settlement, with a unique maritime culture when compared with other villages on the island.

The MBM-PB Process in Bajo Mola

The MBM-PB program in the Greater Mola area was carried out over approximately twenty-three months. The process used for this program can be seen in Figure 3.

Figure 3. The Process of Mandiri Bersama Mandiri – Pariwisata Berkelanjutan Program
Source: Program Mandiri Bersama Mandiri Pariwisata Berkelanjutan

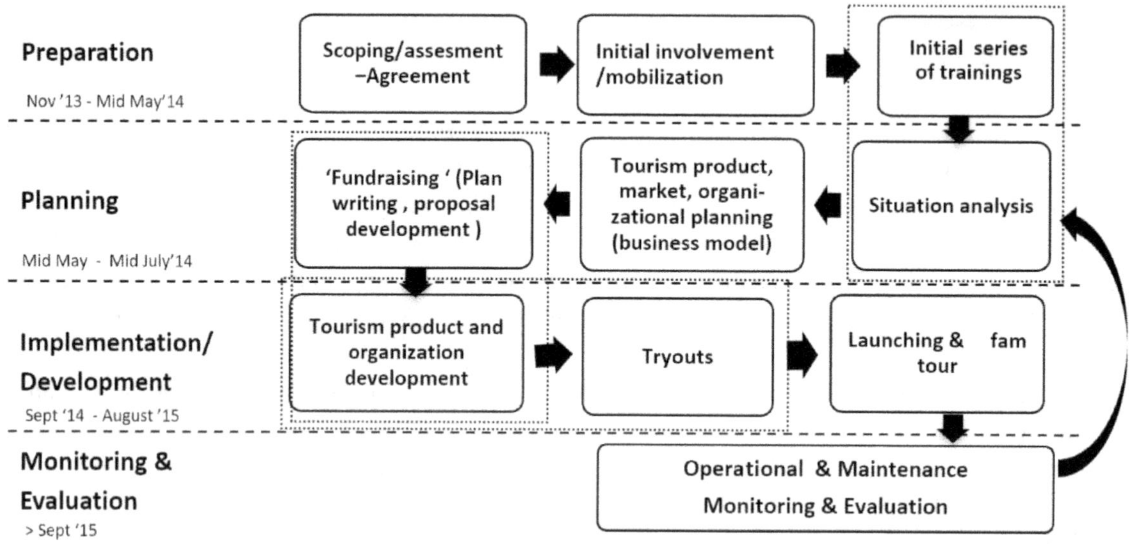

The process involved several steps directly related to the development of interpretive story and experience, namely:

1. Situation Analysis

In this step, the significant attributes of the Bajo Mola settlements were identified in a workshop with the village tourism planning team whose members were selected by each village leader. The significant attributes of the Bajo Mola community were as an ethnic group with a strong ability to 'read' nature, particularly nautical weather, wind, sea currents, marine animals. It is also the largest Bajo settlement in Indonesia or perhaps in the world. Bajo Mola fishermen supply tuna and reef fish in Wakatobi, Kendari, and even Bali. Southeast Sulawesi Province is in the fifth largest in Indonesian fish production. Therefore, Bajo fishermen contribute not only to the local economy but also to the regional and national economy. The community makes the most boats in the regency.

This stage recognized more than sixty potential points of interest within the Great Mola area and beyond which were then ranked based on their relevance to the significant attributes, scarcity, accessibility, attractiveness, constraint, resilience and several other criteria. The list includes places of:

- **Beautiful scenery**: sunset, sunrise, night sky view.
- **Wildlife appearance**: dolphins, birds, different kinds of fishes and sea creatures.
- **Fishery practices**: traditional and floating *keramba* or underwater fish cage made of net, mini plants, markets.
- **Talents**: traditional skills such as boat making and fishing equipment craftsmen as well as modern ones such as plastic waste recycling, dried shredded fish and snacks making.
- **Rituals or events**: Bajo life and their fishery activities, etc.

2. Tourism Product Planning

On several occasions, participants of various workshops in Bajo Mola said that they never thought that their dense and noisy settlement and their marine-focused way of life were interesting to tourists. They recognized their lack of understanding of tourism as one of their problems. As the case in many remote village communities, this is due to their limited opportunity to travel as tourists. Therefore, introducing them to different kinds of tourist experiences outside their world was an important step in the tourism product planning workshop.

With the Pine II and Gilmore's four experience realms (Figure 4) as a reference (Pine II and Gilmore, 1998), a community brainstorming session produced more than seventeen activities for tourists. These were then ranked based on several criteria such as the attractiveness or easiness of the activity to general tourists, potential competition, 'feasibility', and possible constraints. After further discussion, they were merged into ten activities. A consultation session with the village and religious leaders in the area resulted in dropping of three more activities from the list due to concerns about a plan for a night market by one village and religious concerns by others. After further analysis considering windows of opportunity to offer the activities throughout the year, the variety of experiences, the diversity of the target market served, and the complementary nature of each activity in terms of the time of delivery in a day, the community finally agreed to choose four tourist activities to create, namely, *Dolphin Watching, Bajo Cultural Walking Tour, Village Canoeing Tour,* and *Star-telling*.

Figure 4. Pine and Gilmore Model of Four Realms of an Experience
Source: Pine and Gilmore (1998)

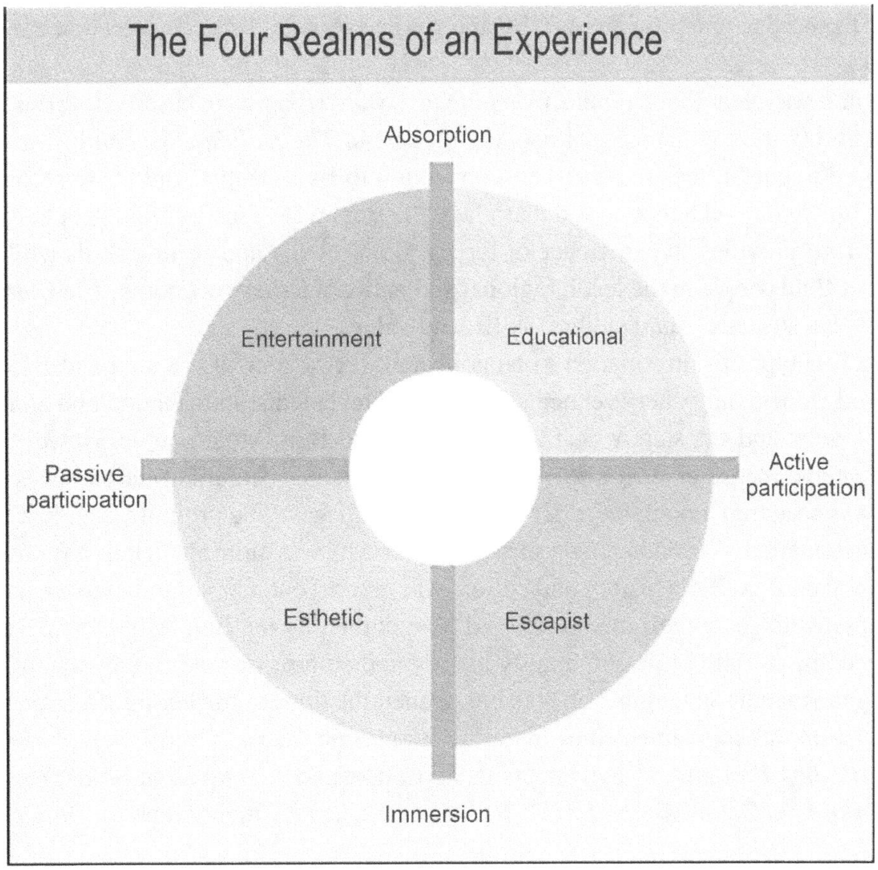

The Dolphin Watching Tour is based on the tradition that dolphins are a Bajo person's best friend. The boat tour begins before dawn and involves dawn stargazing, watching dolphin behavior, enjoying Bajo food as refreshment, and the experience of being on the open water at dawn. *The Bajo Cultural Walking Tour* involves experiencing Bajo culture and skills. Visitors take a walk through the village, visiting homes and seeing craftsmen and women at work then try Bajo food as refreshment. *The Canoeing Tour* provides an experience of typical Bajo rowing and fishing skills, visits some important and historic places in the village along the canals between houses; it also proposes to view the sunset and enjoy Bajo refreshments. *The Star-telling* activity takes place on the waterfront in the evening to enjoy the night sky, suggesting learning about how sky objects relate to Bajo people, and enjoying Bajo refreshments. These four products were aimed to complement the existing tourism products (snorkeling and diving) in Wakatobi National Park.

Each chosen activity then underwent a further process of defining the services necessary to make them enjoyable experiences for visitors. These services include interpretation, restoration, recreational services, safety and security, accessibility, and administration. Each service requires appropriate infrastructure and facilities, equipment, and skillful personnel. A list of basic required infrastructure and facilities, equipment, and training or workshop for personnel were developed for every experience planned along with an implementation plan and cost estimate.

One important activity which greatly contributed to refining the design of the experience was the study visit to Bali. The study visit by six key leaders and key people in charge of each tourism product allowed the villagers to learn not only how other communities provided their tourism products, but also about pricing, promotion, and other elements of marketing. The study visit was also a way for them to become tourists, to experience real tourism products and the life of a tourist. Therefore, they became more aware of the need to entertain guests with appropriate services and interpretation during their visit to Bajo Mola.

At this point, it was clear that *Dolphin Watching* and *Star-telling* were based on stories related to the Bajo people's ability to read nature. Tourists taking part in *The Dolphin Watching Tour* would listen to stories about dolphins in the area and their connection to Bajo people, and try to recognize dolphin behavior. *The Star-telling* audience would learn how the stars were used by Bajo people for navigation. *The Canoeing Tour* provides an experience of typical Bajo rowing and fishing skills while finding out about the role of Bajo people in the local, regional and national fishery economy. *The Cultural Walking Tour* would involve stories of Bajo culture, skills, and values.

There were five types of information to be collected: the history of the area and Bajo people; the Bajo culture and values; the fishery-related skills and achievements; information and stories about the dolphins in the area; and the stars which are meaningful to Bajo Mola people. However, Bajo Mola people do not keep a record of their history or cultural elements. The information necessary for story development was scattered among the elders' of the five villages. The program then had to find a way to collect the information. As conducting a survey or interview was quite challenging for the community to do, a workshop on Bajo Mola history and culture was carried out. Community leaders and traditional figures were invited to share their memories, and be recorded by the Bajo Mola team. This first elder and leaders meeting on Bajo Mola community history and culture was the first ever conducted. It was dominated by information on tangible objects but opened the door to further information seeking.

This information was supplemented by literature about Bajo such as a translation of "*Peuple nomade de la mer: Les Badjos d'Indonesie*" by Fancois-Robert Zacot and "*Orang Laut, Bajak, Laut, Raja Laut: Sejarah Kawasan Laut Sulawesi Abad XIX*" by Adrian Lapian. A monograph of "*Bugis Navigation*" by Gene Ammaral showing pictures of different stars and star clusters familiar to Bugis navigators was useful in giving examples and obtaining information from experienced fishermen during the workshop and further information gathering.

3. Tourism Product Development

The detailed design of each experience was further elaborated during the implementation of the plan or the development stage, particularly in a workshop on interpretation development. The workshop took about forty-four hours, spread over a 6-day workshop held between the afternoon to evening, the most suitable time for the participants. The workshop involved twenty interpreter candidates with half involved in the product planning stage, in charge of or interested in the development of the products. The rest were new faces who were either selected by their village leaders or interested in becoming an interpreter.

Aside from training and developing interpretation, some ten hours of this workshop were dedicated to learning practical astronomy, practicing the use of equipment, observing and identifying stars and other sky objects using national and international nomenclature. Some highly experienced fishermen took turns joining the three-night sky observation to assist the interpreters in finding out the sky objects. The whole process was facilitated by an astronomer.

The development of a story in the workshop used Ham's TORE approach and began with the establishment of the *Theme*. Since none of the participants had experience in preparing an interpretive theme, the process started with a brainstorming session. The session brought up many ideas and new elements for the stories and indicated the need for a further and deeper search for information. The brainstorming result on *Star-telling*, as an example, showed that not only stars, but planets, the moon, and meteors also had meaning for Bajo Mola people. Furthermore, sky objects were not only useful for navigation, but also for weather forecasting, estimating the time, determining the location of fish or collect particular fish, and even predicting marital events. These elements were included in the *Star-telling* story.

Based on the elements from the brainstorming session, the Bajo Mola interpreters formulated the initial themes for their stories. They were refined throughout the process of developing the story. The final theme statement of the Dolphin Watching Tour, for example, is "As (the descendant of) sea nomads, for a long time the Bajo people has used dolphins as indicators of dangers, safety, hope, and fortune at sea which led to the establishment of friendly relations between the Bajo people and these creatures". Therefore, the two and half hours *Dolphin Watching Tour* was enriched with information relevant to the development of that conclusion in the audience's minds.

The theme in every tour or experience includes at least one universal concept, an intangible or symbolic connection to notions that always have significance to humans everywhere, and for all time (Ham, 2013). The universal concept in the theme is expected to help visitors to identify with a larger truth behind the story. In the *Dolphin Watching Tour*, dangers, safety, hope, fortune, and friendship are the universal concepts. Since they are universal, most people, including tourists, can relate to them, regardless of their unfamiliarity with dolphins or the Bajo people. Hence, the relationship between the visitor and the resources was developed early in the theme development.

Before moving on to the next step or letter in the TORE approach, with the theme in their mind, interpreters responsible for each tourism product were encouraged to think of the intellectual and emotional objectives and even behavioral objectives of the interpretive experience. Intellectual objectives relate to knowledge interpreters would like the audience to take home. Emotional objectives were about the feelings interpreters would like the audience to feel during and right after the tour. Behavioral ones relate to the behavior of the audience which the interpreters would like to see, particularly during the tours or experiences. The objectives are a reference for the later steps when narrating the stories and selecting techniques that suit the objectives. The objectives were also useful as a basis for monitoring or observing and evaluating the quality of the visitors' experiences.

Once the initial theme and various objectives were settled, it was time to proceed to the *Organize* step in the TORE approach. The many ideas from the brainstorming session were used to create a sensible flow for the story. This involved drafting the outline of each story, breaking down the theme into sequential sub-themes which had to be harmoniously arranged, in a tour, with the points of interest to be visited. The interpreters responsible for the designed experiences managed to come up with a simple outline as can be seen for the *Village Canoeing Tour* in Table 1. This outline grew throughout the story development process, particularly during the writing of the narratives, field workshop, drillings, and tryouts as new things and ideas from interpreters and audiences arose. The work of Bajo Mola interpreters showed that when it comes to community, story-telling means story-sharing.

Table 1. Outline of story in the Village Canoeing Tour

Subthemes and subsubthemes	Stopping point
Staging and Introduction	Samaturu pier
Bajo Mola people and their boats a. Boats as home and mode of transportation for Bajo Mola people b. Bajo Mola skill in making the different types of boats	On the canoe on the way to floating *keramba*
Fishery in Bajo Mola Area a. Type of *keramba* in Bajo Mola b. A different function of *kerambas*	Floating *keramba*
Production of fish (tuna and reef fish) a. A different way to catch tuna and reef fish b. The handling process of fish caught by fishermen c. The quantity of fish production in Bajo Mola	The Mini-plant
The culture of catching/cultivating fish of the Bajo Mola community a. The role of the house in Bajo Mola fishermen's culture b. Unusual places to cultivate fish in Bajo Mola area	Fishermen house and stilt/traditional *keramba* Bahari Elementary School
The contribution of Bajo people to local economy	Fish market
Conclussion and Closing	Samaturu pier

Source: Bajo Mola Village Canoeing Storyline

With the outline available as a reference, the task of writing the stories became possible for Bajo Mola interpreters. A written story is probably not necessary for experienced interpreters. However, that is not the case for a community who has never undertaken interpretation before. Experience of drafting the narration of the story turned out to be helpful in recalling the details of the story when sharing it with visitors. The text of narration also serves as a tool for communication between the interpreters and the facilitator. It helped the facilitator to understand the interpreters' thoughts and to provide necessary assistance. It was also a useful tool for monitoring and evaluating the progress of interpreters in delivering the interpretive story and experience.

Writing the narration of the stories meant fleshing out the outline with necessary information beyond what had already been recorded in the workshop on Bajo Mola history and culture. Additional information was obtained from the community and published sources of information. The abovementioned literature about Bajo people was very informative about the values of the Bajo people compared to those of other ethnic groups. While reading these books was challenging for most interpreters, a couple of teachers who participated in the program and workshop found the book interesting and confirmed that about seventy percent of what was written by Zacot in his book about Bajo people in Torosiaje was also relevant to Bajo Mola. General information about Dolphins was collected from various online sources and specific information on Dolphins in the area was collected from the World Wildlife Fund Indonesia office in Wakatobi. Additional information about Wakatobi and the region fishery was also collected from online sources and the Fishery Office of Wakatobi Regency.

It was at this point that the interpreter needed to make the story *Relevant* to the audience following the TORE approach. Relevance means meaningful and personal. The interpreters were introduced to various ways to make the information in the story understandable and relate to something within the personality or experience of the audience (Ham, 2013).

The process of writing stories is the proper time to also think of a way to make the story *Enjoyable*, the fourth letter in TORE approach. There were various techniques to make interpretation enjoyable, engaging, and entertaining introduced to Bajo Mola interpreters. Using visual aids, involving the audience in activities, humor, and storytelling were just some of the examples.

Table 2 displays the techniques applied by interpreters of the *Cultural Walking Tour* to make the tour relevant and enjoyable. Various techniques were applied to ensure all aspects of a good interpretive experience were met. Activities included in the tour provided a direct physical experience. Stories, explanations, examples, and comparisons were ways to stimulate intellectual and emotional aspects of experiences. Universal concepts were expected to help visitors intellectually stimulate and find their personal meaning and experience a larger truth behind the selected facts conveyed by the interpreters. Questions were asked to create conversation or sharing opportunities. Interpretation is not merely storytelling. It includes story-sharing, story-showing, and experiencing the story.

While many techniques involving physical experiences were used to make the story enjoyable, only a few verbal communication techniques were used by Bajo Mola interpreters to make the story meaningful. The pattern was quite similar across all other interpretive experience outlines. Bajo Mola interpreters were not familiar in delivering formal, framed, or intellectual stories.

Table 2. Techniques Applied in the Cultural Walking Tour

Relevant		**Enjoyable**
Meaningful	**Personal**	
Explaining Giving examples Comparisons	Using universal concepts in theme and story narration	Body language Asking question Using five senses Using activity to try on/out Bajo's custom Telling short stories Using objects Using demonstration

Source: Bajo Mola Cultural Walking Tour Storyline

The first draft of the stories were rough due to the limited time and information available. It took some time and effort for the facilitator to review and enrich the stories with information collected from outside the Bajo Mola community and provide input for the detailed execution of the chosen techniques to make the stories and accompanying activities relevant and enjoyable. Once the enriched drafts were finished, they were then shared with the interpreters for them to finalize with more detailed information.

4. **Simulation and Tryouts**

Simulation and tryouts are pivotal in creating an experience. It is the moment when experiences designed on paper are transformed into real experiences (Figure 5 and 6). We tried to facilitate one simulated tour before the first tryout and at least three tryouts before a community could officially offer the experiences to the public. The simulation was internally conducted and whenever possible on-site, so the interpreters could get familiar with the location, the story, and its related activities without worrying about an audience. However, even without an audience, many interpreters had problem in speaking when friends, neighbours, and relatives watched them. During tryouts in Bajo Mola, we invited five to ten real guests, tourism stakeholders we knew in Wangi-Wangi Island, to provide a secure environment for the new interpreters and supporting team to practice. These guests include staff from the local tourism office, not-for-profit organizations, and tourism organizations from neighboring villages.

Since each experience takes about two to three hours, tryouts were usually carried out over two days. We left one up to two months between tryouts to allow the interpreter team to master the stories and activities as well as to give time for the completion of supporting facilities and infrastructure (tourism information center, pier for boarding for *Canoeing Tour* and *Dolphin Watching Tour* boats as well as *Star-telling* platform, signages) and various equipment (boats, visual aids, safety equipment, office equipment, tableware, etc.) for use. The third tryout was actually a rehearsal. Fortunately, there were some occasions when real tourists wanted to try some of the products even if they were not ready yet. Therefore, some interpreters had additional opportunities to do a tryout.

Tryouts provide opportunities for interpreters to monitor their progress in delivering stories and experiences. Significant improvement in the story and related activities delivered could be perceived in every Bajo Mola tryout. The interpreters became more relaxed in public with every tryout. Simulation and tryouts were very useful for fine-tuning the products as well. Additional information and activities suggested on-site by interpreters and guests were added. Ideas developed this way included canoe racing and singing a simple Bajo song about canoeing while enjoying the sunset on the way back to the pier. Trials also provide opportunities for interpreters to internalize the necessary attitudes and enforce the standard operational procedure they set.

Figure 5. Visitors learning about Bajo fishes and ways of fishing during the Village Canoeing Tour
Source: Lepa Mola Documentation

Figure 6. Visitor putting on Bajo sunscreen in the beginning of the Cultural Walking Tour
Source: Lepa Mola Documentation

The Bajo Mola Community launched their tourism products fifteen months after their tourism products planning workshop or ten months after the interpretation development workshop. It took quite some time because the MBM-PB Program developed not only these interpretive experiences, but also the main infrastructure and facilities to support the experiences, as well as carried out trainings and workshops (culinary, safety, hospitality, business model, administration, and finance, organization, etc.) necessary for providing required services. Some of the trainings or workshops were also intertwined with other activities such as Bajo Culture Campaign for elementary schools, marine-based skill competition, culinary competition and training to improve the community's sense of culture and pride.

A familiarization trip was carried out as part of the launch event. Seventeen participants including special interest tour operators, travel journalists, national TV station reporters, bloggers, and government officers representing local, provincial, and national tourism offices participated in the trip. All participants considered that the experience met their expectations and around half thought that the experience exceeded their expectations. The *Star-telling* and the *Dolphin Watching Tour* were scored 4.8 on a 5-point scale (where five means very enjoyable), while the *Village Canoeing Tour* and the *Cultural Walking Tour* scored 4.4 and 4.3, respectively. Participants liked listening about Bajo local wisdom heard during the tours and were surprised by the community's openness to visitors.

Feedback from the participants also demonstrated that they were willing to pay significant amounts for the experiences. The interpretive stories also provided a wider range of promotion content than the conventional resource-based material that is used by many community-based tourism organizers. During their first five months of operation, the Bajo Mola tourism organization, Lepa Mola, received around 109 visitors with many taking one of these tours. These new type of tourism products focused attention on Bajo Mola as central to Wakatobi tourism.

The Bajo Mola community still offers these tourism products despite experiencing various organizational, operational, and marketing difficulties. Some experiences were modified to accommodate recent development of the area. Some interpreters are now capable of delivering more than one experiences. The community entered the Indonesian Sustainable Tourism Award competition in 2018 and were one of the finalists.

During the COVID-19 pandemic, with the support of the Ministry of Tourism and Creative Economy, the Bajo Mola community was given a chance to adapt their on-site experiences into a live virtual one through a Zoom cloud meeting application. Delivering stories virtually was possible, however, delivering stories-related activities was quite challenging since a virtual tour relies on only two senses: sight and hearing. Therefore, while pictures, short movies, and real-time broadcast on-site situations were prepared, some new simple but relevant-to-story activities and games using everyday objects were introduced to better engage the audiences at home.

FUTURE RESEARCH DIRECTIONS

A community's culture and history are of great interest to visitors. They provide access to the unique and authentic elements of a place and people. Often however, a community's local wisdom, myths and legends, cultural practices and intangible cultural heritage are undervalued by the people of the community themselves. Stories are lost and history forgotten. In some countries there are research programs to capture the stories about a place. The UNESCO International Convention for the Safeguarding of Intangible Heritage (Aikawa, 2004) is one example of a such a program. Indonesia has a multiplicity of unique cultures across its archipelago and there remains much to be done in collecting its intangible cultural heritage. This chapter highlights how such knowledge can form the basis of valuable tourist experiences and create an income stream.

However, the chapter also notes that transforming community heritage into a tourist experience requires specialist skills applied over an extended period. It would be useful to develop materials to support community education about the value of cultural heritage and how to transform this into a tourism product. The approach applied in Bajo Mola had been replicated in several other communities in the country. Most of these community projects resulted in an introductory level of interpretation. Kohl (2020) terms this the second level of interpretive wholeness. In this type of interpretation, the guide provides non-controversial information which reinforces the audience's current perceptions of the site. In Bajo Mola, two of the interpretive stories that were developed serendipitously provide a non-consensus perspective on the community. These provide an advanced level of interpretation that may differ from the audience's understanding. Further research on factors necessary for a community to develop more advanced interpretive stories and their impact on the audience would be valuable. Research would also be useful on effective methods to train community members as tour guides and story tellers. Developing such skills can be difficult and challenging and training material in this area would be useful.

CONCLUSION

Experience creation is a crucial strategy to increase the quality of a destination's products. Tourists value the process of delivering the experience. A destination community that delivers the experience can also have an alternative source of income. The literature on tourist experience indicates that stories play an important role in experience creation and developing a context for a visit. One of the approaches to communicating the story to a visitor involves interpretation. Interpretation intertwines messages with activities and various techniques to provide an experience. The process of story and tourism product planning and development are intertwined and should not be done in isolation from each other.

Identifying and including community stories during the tourism planning process made the development of tourism products (along with the activities, supporting services, and necessary atmosphere) easier and better understood by the community. It also developed community awareness of the value of their stories and increased their involvement in the development process. In this case study, the community team member's sense of pride in their stories ignited their willingness to learn more about their history and to become the driving force in researching and investigating these stories. The process also brought the community closer. Local teachers in particular played an important role in this process including collating and checking information about Bajo culture obtained from external sources.

The interpretive story research and development process improved the community's understanding of their culture and of tourism, as well as their skills in communicating these to visitors. There was a lack of public speaking skills as villagers had little formal interaction with other communities in the island. Therefore, practice drills and tryouts were critical for success in delivering the stories and experiences. The program gave participants confidence enabling them to talk to foreign visitors, high-level officials, and the media. Interpretive storytelling embedded in Bajo Mola tourist experiences added value not only to their products, but also to the other aspects of the tourism marketing mix. The interpretive stories allowed the price of their tourism products to increase. Furthermore, the interpretive stories prepared during product development served as rich sources for preparing distinguished content of promotional materials.

ACKNOWLEDGMENT

This program was supported by the Bank Mandiri and the British Council.

REFERENCES

Aikawa, N. (2004). An historical overview of the preparation of the UNESCO international convention for the safeguarding of intangible heritage. *Museum International*, *56*(1-2), 137–149. doi:10.1111/j.1350-0775.2004.00468.x

Ammarell, G. (1999). *Bugis navigation*. Yale University Southeast Asia Studies.

Bassano, C., Barile, S., Piciocchi, P., Spohrer, J. C., Iandolo, F., & Fisk, R. (2019). Storytelling about places: Tourism marketing in the digital age. *Cities (London, England)*, *87*, 10–20. doi:10.1016/j.cities.2018.12.025

Binkhorst, E., & Den Dekker, T. (2009). Agenda for co-creation tourism experience research. *Journal of Hospitality & Leisure Marketing*, *18*(2–3), 311–327. doi:10.1080/19368620802594193

Chou, C. (2006). Research trends on Southeast Asian sea nomads. *Kyoto Review of Southeast Asia*, *7*, 1–11.

Coral Triangle Initiative. (n.d.). *History of CTI-CFF* [Infographic]. https://coraltriangleinitiative.org/about

Firmansyah, F., Mustofa, A., Estradivari, Damora, A., Handayani, C., Ahmadia, G., & Harris, J. (2017). *Satu dekade pengelolaan Taman Nasional Wakatobi: Keberhasilan dan tantangan konservasi laut* [A decade of Wakatobi National Park management: The achievements and challenges of sea conservation]. World Wide Fund for Nature-Indonesia.

Fisher, J. A., & Schoemann, S. (2019). Correction to: toward an ethics of interactive storytelling at dark tourism sites in virtual reality. In R. Rouse, H. Koenitz, & M. Haahr (Eds.), Lecture Notes in Computer Science: Vol. 11318. Interactive Storytelling. Springer., ICIDS 2018. doi:10.1007/978-3-030-04028-4_81

Gardiner, S., & Scott, N. (2017). Experience Design. In N. Scott, M. D. Martino, & M. V. Niekerkm (Eds.), *Knowledge transfer to and within tourism* (Vol. 8, pp. 13–25). Emerald Publishing Limited. doi:10.1108/S2042-144320170000008002

Gravili, S., Rosato, P., & Iazzi, A. (2017). Managing DMOs through storytelling: A model proposal for network and value co-creation in tourism. *International Business Research*, *10*(7), 8. doi:10.5539/ibr.v10n7p8

Gross, M., & Zimmerman, R. (2002). *Interpretive center: The history, design, and development of nature and visitor center*. UW-SP Foundation Press, Inc.

Ham, S. H. (1992). *Environmental interpretation: A practical guide for people with big ideas and small budgets*. Fulcrum Publishing.

Ham, S. H. (2013). *Interpretation: Making a difference on purpose*. Fulcrum Publishing.

Hartman, S., Parra, C., & de Roo, G. (2019). Framing strategic storytelling in the context of transition management to stimulate tourism destination development. *Tourism Management*, *75*(April), 90–98. doi:10.1016/j.tourman.2019.04.014

Kim, J. H., & Youn, H. (2017). How to design and deliver stories about tourism destinations. *Journal of Travel Research*, *56*(6), 808–820. doi:10.1177/0047287516666720

Kohl, J. (2020). Four levels of interpretive development based on Tilden's principle of wholeness. *Journal of Interpretation Research*, *25*(1), 51–57. doi:10.1177/1092587220963520

Kusuma, P., Brucato, N., Cox, M. P., Letellier, T., Manan, A., Nuraini, C., Grangé, P., Sudoyo, H., & Ricaut, F.-X. (2017). The last sea nomads of the Indonesian archipelago: Genomic origins and dispersal. *European Journal of Human Genetics*, *25*(8), 1004–1010. doi:10.1038/ejhg.2017.88 PMID:28513608

Lapian, A. B. (2011). *Orang laut bajak laut raja laut: Sejarah kawasan laut Sulawesi abad XIX* [Sea People Sea Pirate Sea King: The history of Sulawesi sea in 19th century]. Komunitas Bambu.

Li, Y. M. (2014). Effects of story marketing and travel involvement on tourist behavioral intention in the tourism industry. *Sustainability*, *6*(12), 9387–9397. doi:10.3390u6129387

Ludwig, T. (2021). How can interpetive experiences make people mindful? In *Proceedings of Interpret Europe Web Conference on Recreating Tourism through Heritage Interpretation* (pp. 22-44). Interpret Europe.

McCabe, S., & Foster, C. (2006). The role and function of narrative in tourist interaction. *Journal of Tourism and Cultural Change*, *4*(3), 194–215. doi:10.2167/jtcc071.0

Medlik, S., & Middleton, V. T. C. (1973). The Tourist Product and its Marketing Implications. *International Tourism Quarterly, 3.*

Mei, X. Y., Hågensen, A. M. S., & Kristiansen, H. S. (2020). Storytelling through experiencescape: Creating unique stories and extraordinary experiences in farm tourism. *Tourism and Hospitality Research*, *20*(1), 93–104. doi:10.1177/1467358418813410

Moscardo, G. (2020). Stories and design in tourism. *Annals of Tourism Research*, *83*, 102950. doi:10.1016/j.annals.2020.102950

Muljono-Larue, F., & Pranoto, I. B. (2008). *Orang Bajo Suku Pengembara Laut: Pengalaman Seorang Antropolog* [Nomadic people of the sea: The Badjos of Indonesia] (F. Zacot, Trans.). Kepustakaan Populer Gramedia. (Original work published 2002)

National Association for Interpretation. (2021). *What is interpretation?* https://www.interpnet.com/ nai/interp/About/What_is_Interpretation_/nai/_About/what_is_interp.aspx?hkey=b5ddeff3-03a8-4000bf73-433c37c8a7af

Pine, B. J., II, & Gilmore, J. H. (1998, July-August). Welcome to the experience economy. *Harvard Business Review*. https://hbr.org/1998/07/welcome-to-the-experience-economy

Prahalad, C. K., & Ramaswamy, V. (2004). Co-creating unique value with customers. *Strategy and Leadership*, *32*(3), 4–9. doi:10.1108/10878570410699249

Prebensen, N. K., Chen, J. S., & Uysal, M. S. (2017). Tourist experience creation An overview. *Co-Creation in Tourist Experiences,* 1–9.

Rasoolimanesh, S. M., Seyfi, S., Hall, C. M., & Hatamifar, P. (2021). Understanding memorable tourism experiences and behavioural intentions of heritage tourists. *Journal of Destination Marketing & Management*, *21*, 100621. doi:10.1016/j.jdmm.2021.100621

Scott, N., Gao, J., & Ma, J. (2017). Introduction: creating memorable experiences. In N. N. Scott, J. Gao, & J. Ma (Eds.), Visitor Experience Design (pp. 3–12). CABI Publishing.

Stacey, N. (2007). *Boats to Burn: Bajo fishing activity in the Australian fishing zone*. ANU E Press.

Tilden, F. (1977). *Interpreting our heritage* (3rd ed.). University of North Carolina Press.

UNESCO Man & the Biosphere Programme. (2012, September 30). *Wakatobi Biosphere Reserve, Indonesia*. https://en.unesco.org/biosphere/aspac/wakatobi

United Nation Education, Scientific and Cultural Organization. (2006). *Wakatobi National Park*. https://whc.unesco.org/en/tentativelists/2006/

United Nations World Tourism Organization (UNWTO). (2019, December 10). *Product Development*. https://www.unwto.org/tourism-development-products

Victoria University of Wellington. (n.d.). *Study sites* [Infographic]. https://www.wgtn.ac.nz/sbs/research/marine-biology-research/sponge-marine-ecology/research/study-sites

Wiyonoputri, W. T., Anindita, D., & Masita, A. (2021). *Pedoman Interpretasi Daya Tarik Wisata* [Guideline for Developing Interpretation in Tourist Attractions]. Kementerian Pariwisata dan Ekonomi Kreatif / Badan Pariwisata dan Ekonomi Kreatif Republik Indonesia.

ADDITIONAL READING

Beck, L., & Cable, T. T. (2011). *The gifts of interpretation: Fifteen guiding principles for interpreting nature and culture* (3rd ed.). Sagamore Publishing.

Brochu, L. (2015). *Personal interpretation: Connecting your audience to heritage resources*. InterPress.

Ellis, G. D., Freeman, P. A., Jamal, T., & Jiang, J. (2017). A theory of structured experience. *Annals of Leisure Research*, 1–22.

Gilson, J., & Kool, R. (2021). The place of inspiration in heritage interpretation: Views from the field. *Journal of Interpretation Research*, *26*(2), 49–72. doi:10.1177/10925872211063832

Regnier, K., Gross, M., & Zimmerman, R. (1994). *The interpreter's guidebook: Techniques for programs and presentations*. UW-SP Foundation Press, Inc.

Stern, M. J., & Powell, R. B. (2020). Field trips and the experiential learning cycle. *Journal of Interpretation Research*, *25*(1), 46–50. doi:10.1177/1092587220963530

Veverka, J. A. (1998). *Interpretive master planning: The essential planning guide for interpretive centers, parks, self-guided trails, historic sites, zoos, exhibits and programs*. Accorn Naturalist.

Ward, C. W., & Wilkinson, A. E. (2006). Conducting meaningful interpretation: A field guide for success. Fulcrum Publishing.

KEY TERMS AND DEFINITIONS

Bajo Mola: Bajo ethnic group living in Mola area, eastern waters of Wangi-wangi Island, Wakatobi National Park.

Interpretive Experience: Experience designed and delivered in accordance to the flow of story which is prepared by applying interpretive approach and using interpretive media – personal and or non-personal.

Story-Centric/-Based Visitor Experience: Visitor experience designed and delivered in accordance with the flow of previously prepared story.

Story-Sharing: The action of giving of a portion of own (community) story to jointly enjoyed with others.

Story-Showing: The action of showing something to illustrate or support the story.

Tourist Experience: Is a set of activities allowing tourists to build their travel experiences which satisfy their needs.

Tryouts: Series of trials to test the performance of the interpreter team to deliver interpretive experiences.

Chapter 5
From Story-Based to Story-Interactive Experiences:
Layers of Narrative Application in Theme Parks and Attractions

Carissa Baker
University of Central Florida, USA

ABSTRACT

Storytelling has been one of the defining attributes of the theme park industry. This chapter describes the strategic storytelling present in theme parks, one of the most popular visitor attraction sectors globally. It posits four types of story-connected layers present in theme parks: story-based experiences, story-adjacent experiences, story-driven experiences, and story-interactive experiences. Each of these layers provides narrative connections with visitors and enhances positive outcomes such as visitation, revenue, revisit intention, and brand loyalty. The chapter notes the value of utilizing strategic storytelling to stage impactful experiences and suggests narrative strategies that can be applied in multiple tourism contexts.

INTRODUCTION

The theme park industry in its contemporary iteration (1955-on, demarcated from the establishment of Disneyland in California) is one of the most narrative forms of tourism (Baker, 2018a; Hench, 2009; Lukas, 2012; Younger, 2016). The global theme park industry, which garners over 500 million guests with the top ten operators alone (Rubin, 2020), is rife with narrative strategies (Philips, 1999), and they appear at each stage of the visitor's journey. Storytelling is evident in marketing and staging experiences as well as through guests connecting with and recalling their experiences. Narrative is an impactful way to market destinations (Bassano et al., 2019; Kim & Youn, 2017), and theme parks frequently have storytelling-based promotion. Though guest experiences spawn powerful instances of narratives via word-of-mouth, travel stories, or social media output (Kang, Hong, & Hubbard, 2020; Lund, Cohen, & Scarles, 2018), this chapter focuses on theme park operators using strategic storytelling to stage memorable experiences, in other words, the narrative strategies employed on site. After describing some of

DOI: 10.4018/978-1-6684-3436-9.ch005

the concepts found in the background literature, this chapter clarifies four strata of narrative application and integration utilized by theme parks. These levels will be described by detailing attractions from the global industry that demonstrate the layers and their success in appealing to visitors and creating lasting memories. Finally, the chapter offers principles for practitioners and areas where future research can enrich knowledge of these practices. The chapter sets out to observe major ways in which narrative is integrated into theme parks, illustrating both its prominence in the success of these popular attractions and the significance of storytelling in human culture.

BACKGROUND

This entire volume tributes the power of storytelling, and it has particularly been essential for tourist destinations because of their connections to culture and place. Storytelling is a basic trait of human culture; indeed, people "think with narrative" (Abbot, 2012). Human culture is passed down through stories, and daily lives are organized through narratives. Experiences in destinations and attractions are likewise narrativized after the visit (Chronis, 2012). In tourism contexts, storytelling can coordinate an entire business, express the organization, and create a brand (Mossberg, 2008). Corporate storytelling can be powerful, both shaping the company culture and the perception of stakeholders (Gabriel, 2000).

With theme parks and other tourism sites, personal narratives are reaffirmed (and occasionally re-produced), making the spaces easier to connect to. Storytelling can be strategically deployed in most servicescapes, but it is inherent to the theme park. Storytelling, used interchangeably in this chapter with *narrative*, is understood in the narrative theory sense as the combination of *story* and *discourse*. Story is a sequence of events, often with characters and setting, in chronological order. The discourse, or the telling of the story, can be stated in any order and use any number of techniques. In the theme park, the *narrator* of stories can be humans using oral storytelling, media like film, robotic figures, or the environment itself. Like with other forms of storytelling, theme park narratives are closely tied to emotion and generational memories (Hover & Vugts, 2013).

Theme parks have antecedents in fairs, pleasure gardens, world expos, amusement parks, theatre, and cinema. Theme parks are distinguished from amusement parks by their attractions being organized through *themes* (underlying messages) and narratives (telling stories through various methods), rather than only rudimentary *motifs* (recurring images). In these parks, the theme becomes the "main part of the experience" rather than only used as decoration (Pikkemaat and Schukert, 2007). Early parks with clear themes include Knott's Berry Farm (themed Ghost Town section 1940s), holiday parks (e.g., Santa Claus Land, 1946, now called Holiday World), and storybook parks (e.g., Children's Fairyland, 1950, or Efteling, 1952). Seen as the "most important line of demarcation" (IAAPA, 2018) in the history of the industry, Disneyland (1955) is the theme park that sparked the contemporary industry and brought it to global prominence (Clavé, 2007; Cross & Walton, 2005; Jones & Wills, 2005; Price, 2004). Disneyland is associated with intentional themes, cohesive stories, symbolic architecture, a master-planned layout, scale, synergy with other business units (film, television, merchandise), and its influences on future themed endeavors and society itself. Perhaps because of this Disney influence, storytelling has become more prominent in the theme park medium to the point that it can be perceived as the "differentiating factor" (Younger, 2016).

Though initially less strategic, the utilization of storytelling has long been a part of the antecedents of the theme park. Some older themed entertainment had clear storylines including A Trip to the Moon

(1901), first seen at the 1901 Pan-American Exposition before it headlined the famous Luna Park in Coney Island, New York. This early dark ride involved riding the airship Luna to the moon and landing on its surface (made of papier-mâché). Though Hales Tours and Scenes of the World (1904), a simulated train ride, did not include a plot, it still had the basic trope of *you are in a different place* and became about the immersion into and experience of being in a location (Rabinovitz, 2006). This core concept would find its way into many traditional attractions even before Disneyland would make story and *you are in a fantasy world* a basic operating principle. Knott's Berry Farm's Ghost Town, considered one of the first themed lands, had authenticity in its real Ghost Town buildings but also created fictional spaces and characters, again illustrating that it is a different place. Universal Studios (1912, theme park 1964) began with real studio tours but morphed into a site with a ride through multiple artificial movie productions; this made the movie industry "more interactive and immersive" (Blackwood, 2018) and again employed placemaking for the purposes of telling a story.

Theme parks tell stories using several methods and techniques, with the most common attraction types being stage shows and *dark rides* (a generally narrative, enclosed ride with a sequence of scenes) (Baker, 2018a). Nonetheless, they have told stories in walkthroughs, simulators, roller coasters, films, immersive theatre, parades, projection shows, nighttime spectaculars (which usually have combinations of water effects, projections, fireworks, or live actors), augmented reality, virtual reality, and interactive quests. They tell stories using multiple narrative techniques including alluding to or symbolizing stories, containing a plot and characters, and displaying storytelling scenes within the attraction. Three narrative styles are present within theme parks. In *explicit* storytelling, an attraction has a story that is narrated through words, either written or an oral narrator (live human, recorded human, animated character, or animatronic). This is seen in most stage shows and in rides such as Universal's Cat in the Hat and the High in the Sky Seuss Trolley Train Ride, both of which use the words of Dr. Seuss books (both written and oral) to tell the story. In *implicit* storytelling, a particular story is meant to be told, but it is done so through environmental elements (architecture, landscaping, music, props, lighting, static or animated figures, etc.). This is a common method in traditional theme park design, with famous dark rides like Pirates of the Caribbean and Haunted Mansion having some dialogue but primarily being told through scenery, effects, and character figures. Attractions may have no dialogue whatsoever, making them rely on implicit storytelling. One other, less common, category exists, and this is *interpretive* storytelling. This would be an attraction where a designer has not endorsed a particular story through the design; it is the guest in this case who determines the meaning of the experience and it may be different based on the individual. An example of this is Efteling's ride Droomvlucht/Dreamflight, where designer Ton van de Ven encouraged riders to "'invent' their own story in the attraction" (as cited in Younger, 2016, p. 152).

Different media types have distinct properties that make them appealing to audiences. Oral storytelling has human interaction, literature has great verbal depth, film has visual detail, and video games have interactive and exploratory features. The theme park's key property is its presence in physical space. Theme park stories are effective because they are a form of spatial transmedia storytelling (Williams, 2020). They rely on *environmental* storytelling (Jenkins, 2004; Wolf, 2012), a dimensional method of storytelling that allows environmental features (architecture, lighting, sound, props, etc.) to carry a narrative message. Visitors are able to walk and ride through the park's elaborate storyworlds. The Wizarding World of Harry Potter at several Universal Studios parks, which includes many immersive elements (i.e., architecture, music, rides, restaurants, shops, and shows) has particularly been commented on for its use of environmental storytelling to bring guests into the popular stories (Baker, 2018c; Godwin, 2017; Waysdorf & Reijnders, 2018).

In Pine and Gilmore's (2011) well-known concept of the experience economy, theme parks are posited as the epitomes of staged experience. This evolution of economies goes beyond conventional products or services and closer to transformations. Comparing enterprises to theatre, Pine and Gilmore elaborate on the ways in which theme parks use multi-sensory design to construct settings that develop into complex, memorable experiences. Their four realms of experience (entertainment, education, esthetic, and escapist) are explored in much literature, with theme parks capable of functioning in the middle spot between all realms (Lee, Jeong, & Qu, 2020). Similarly, Schmitt's (1999) experiential dimensions (sense, feel, think, act, relate) have been applied to the theme park context, with both emotional engagement and "feeling a part of the story" (a feature of cognitive involvement) important features (Tasci & Milman, 2019). Finally, several factors are considered predictors of attitudes and behaviors related to theme parks including sociodemographic characteristics, past visit experiences, and theme park visit quality; both tangible and intangible aspects contribute to satisfaction and loyalty (Godovykh, Milman, & Tasci, 2019). Immersion in storyworlds represents both tangible (architecture, rides, merchandise) and intangible parts (culture, emotion, memory) of the experience.

Narrative immersion in a theme park is a process that is initiated by experience stagers and gone through by visitors. Organizations contribute to *storybuilding* (creating a story for a place), *storytelling* (telling the story to visitors), and *storylistening* (listening to visitor feedback about the story) (Bassano et al., 2019). Destinations create atmospheric immersion with a sequence of techniques: *gripping* (bringing a guest into an interesting setting), *immersing* (with a detailed setting and strong narratives), *unfolding* (where guests become participants in the narrative), *connecting* (guests relating the experience to their lives), and *fulfilling* (where the narrative becomes part of their lives) (Kerz, 2016). Looking at other narrative contexts including written texts, Green and Brock (2000) covered the concept of *narrative transportation*, wherein a narrative can allow one to feel as if they have traveled somewhere because of its immersive qualities. Needed for transportation are structure, unresolved questions, imagery, involvement in a story, and emotional impact. Theme parks physically transport people and emphasize narratives that complete a journey. Looking at heritage sites, Chronis (2012) observed that narrative familiarity assists visitors engaging in staged experiences. Once at the site, visitors go through a narrative construction process: *narrative enrichment* (identifying the story), *narrative imagining* (adding to the story through imagination), and *narrative closure* (connecting with the story). Wolf (2012) found types of immersion to be *physical*, *sensual*, *cognitive*, and *emotional*. Fu et al. (2022) refined these categories after studying how narrative immersion happens in a theme park; the resulting categories were *spatio-temporal immersion*, *sensory immersion*, *conceptual-imaginative immersion*, and *emotional immersion*. The strategic storytelling of theme parks leads to success in each of these categories. In this model, the immersive continuum itself impacts positive outcomes including meaning-making, satisfaction, and revisit intention.

STRATA OF NARRATIVE INTEGRATION

The theme park industry uses several layers of narrative application to stage experiences. They strategically use narrative to create space, immerse guests in a world, and in some cases as a design process (Rohde, 2007). Within a theme park, there can be examples of non-story attraction implementation, where story was not considered in its design, presentation, or experiential qualities. An example of this would be a roller coaster in a parking lot that is not connected to any narrative or intellectual property (legally protected source material such as from recent books or films, henceforth IP). Increasingly, however, sites use strategic storytelling to draw guests in, keep them there longer, and create a desire to return.

One of the core concepts of theme park design is the idea of the environment being a multi-layered story space. The late Walt Disney Imagineering legend John Hench (2009) once explained that the Disneyland original ride Peter Pan's Flight (a suspended dark ride) was so successful because it had "three-dimensional staging," meaning the dimensions of representing the film, sensory information in seeing London from the perspective of the characters (with familiar music and images), and the symbolic layer of flying. Within attraction experiences themselves, in other words, there are multiple layers. Parks will purposely embed stories into space and then layer additional stories on top of spaces. This can be called "concentricity," with concentric layers that surround a guest (Rohde, 2007; Younger, 2016). From the idea of the physical berm created at Disneyland to keep out the real world to elaborate detail as the "inner berm" (Rohde, 2017) that saturates guests to help them believe they are in a storyworld, to the actual manifestation of story in plot and story experiences, the theme park contains narrative strata. In what Naversen (2017) called the "sliced onion technique," layers continue to be designed for visitors.

When discussing augmented reality (AR), Sandvik (2010) described it as a palimpsest, an overlaying of actual space with a fictional universe that becomes a "story space." Theme parks are mixed reality sites, offering virtual reality (VR) ride packages, AR mobile applications, interactive quests overlaid on physical locations, and entire lore created in park details to be discovered (such as Disney's multi-park Society of Explorers and Adventurers/S.E.A. storyline) or in other media types that relate back to the park (comic books, television shows, novels, etc.). This foray into "locative interactive narrative" (Eddy et al., 2020) results in sometimes exceptionally detailed storyworlds. As an example, *Star Wars*: Galaxy's Edge (at Disneyland in California and Disney's Hollywood Studios in Florida) has novels about it that can be read before the visit. During the visit, there is an intricate world to be traversed with several attractions including rides, dining, and retail. The rides have comprehensive queues (the waiting line) before the ride experience that initiate visitors into plotlines and visual references. In addition, the *Star Wars*: Datapad app adds a layer where guests can interact with the space and *Star Wars* characters through completing specific tasks within the land, highlighting the location-based nature of this kind of entertainment. Those guests staying at the hyper-immersive *Star Wars*: Galactic Starcruiser hotel are given even more additional layers and tasks as part of the role-playing activities inherent in the experience.

A land or even a single attraction can include one or several strata of narrative integration. Described in order from least to most intensity within the experience, these are: *story-based experience, story-adjacent experience, story-driven experience,* and *story-interactive experience*. Details of these types are expressed in figure 1. While some theme park attractions may only employ one level, others utilize several or all layers. For example, several roller coasters have used comic book-based IPs with extensive storyworlds (Superman, Batman, etc.) on attraction names, signage, colors, and occasionally the queue, but this may be the extent of story connection, making them *story-based*. On the other hand, Ghost Town Alive! at Knott's Berry Farm is based on an original story, created because of its placement in the Ghost Town section of the park, performed as an elaborate exercise in an unfolding and branching story, and designed with heavy visitor participation in mind, thus employing all four layers.

Figure 1. Layers of story-connected experiences in attractions
(Source: Author's own)

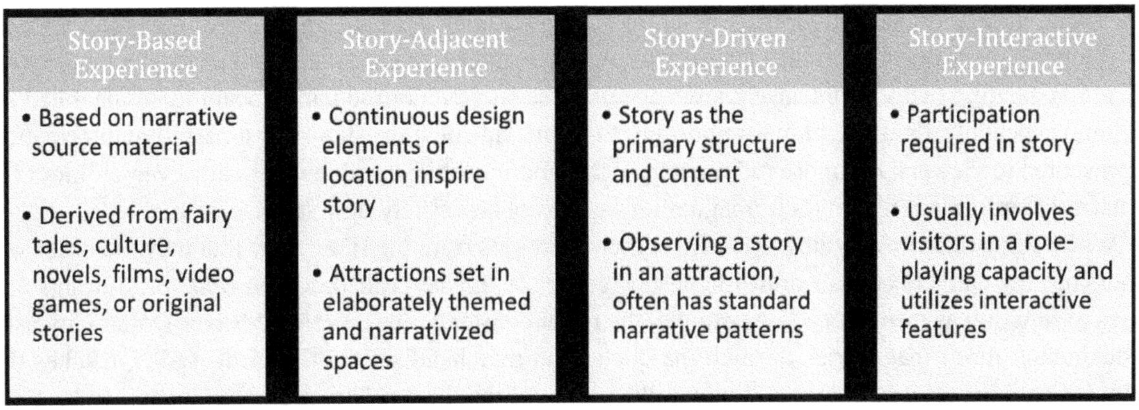

Story-Based Experience

Story-based experience indicates that the foundation of the attraction is a story, whether a well-known cultural tale, IP, or original tale. Being story-based does not imply how story-centric the attraction itself is, however. For instance, the Disneyland rides Alice in Wonderland and Mad Tea Party both use the source material of the 1951 Disney animated film *Alice in Wonderland*. The Alice in Wonderland dark ride is structured like a story and has narrative content on the ride including the characters and scenes from the book and movie. Mad Tea Party, on the other hand, is a simple spinner (or "flat ride") with no storyline, but it uses the symbolic icons of the film's music, color palette, and fanciful decorations to reference the original tea party with the Mad Hatter.

The presence of particular source material is one of the major discussions surrounding the success of theme park narratives (e.g., Hobbs, 2015; Hover and Vugts, 2013; Philips, 1999). The most well-known theme park operators, Disney and Universal, have had attractions based on their movies for decades. Their popularity in the film industry was conducive to the success of their theme parks to begin with. Over the years, the most common source materials in theme parks in all world regions have been IPs (especially novels and films) and cultural stories such as fairy tales or historical legends (Baker, 2018a). These stories are generally familiar to a majority of guests and are a kind of shorthand to the overall storyworld, allowing for less detailed exposition. They may instantly resonate emotionally with visitors and connect to generational storytelling. In the last few decades, there has been an emergence of lands based on single IPs (for example, *Arthur and the Invisibles*, *Avatar*, *Batman*, *Cars*, *Harry Potter*, *Jurassic Park*, *Pirates of the Caribbean*, *Sesame Street*, and *Star Wars*) or lands that are a combination of several properties from one studio (i.e., Marvel superheroes at several parks, DC superheroes at some parks, Motiongate Dubai's DreamWorks area, or Warner Bros. World Abu Dhabi lands Cartoon Junction and Dynamite Gulch).

Entire parks such as Efteling in the Netherlands (with its Sprookjesbos/fairy tale forest) and storybook-based "kiddie" parks (Children's Fairyland, CA; Storybookland, NJ; Story Land, NH, et al.) were founded on the concept of popular fairy tales and nursery rhymes for children. Based on the *Märchenwald* (fairy tale forest in German) concept, narrative vignettes are framed by landscaping and greenery. Particularly important in these parks are *symbolic architecture* and *dramatic play* (Baker, 2022). Many of these parks tell stories through static or kinetic scenes or dioramas. Architecture is often used to symbolize a story, for instance a caterpillar on a mushroom to allude to *Alice in Wonderland* or a giant whale for *Pinocchio*. One scene may be exchanged for the whole, for example Rapunzel standing on the tower with long hair or trying the slipper on Cinderella. These examples of *synecdoche* substitute the part for the whole with the understanding that visitors have knowledge of these stories. For the parks meant to be inhabited by children especially, dramatic play is important. In a dark ride or stage show, most narrative information is presented to viewers. For more rudimentary or symbolic architectural features, however, children fill in narrative information with their imaginations or create an entirely new story.

While some theme park rides or lands get their identities from the IP or story that they are based on, others use the name to act as a draw to guests ("come see the new ride based on *Batman*!") or in some cases to function as metaphors. For example, the roller coaster Time Traveler at Silver Dollar City uses a steampunk motif that carries through the queue and merchandise; the time travel idea matches the spinning roller coaster type. A well-known illustration of the thrill ride as metaphor technique is at Six Flags parks that have Superman-based "flying" roller coasters. As the guest rides prone and under the track, it feels like flying and thus connects back to the idea of the superhero. These rides may include signs about the superheroes and villains, a Superman logo, and Superman colors, but they are not story-driven experiences the way that the Justice League interactive dark ride is at several Six Flags parks. The roller coaster is selected for type and thrill factor; the Superman motif and adventure theme are applied later rather than the approach of creating the story first and then picking the ride type that will most easily express the story. Nonetheless, even most of the rides with light themes and motifs may include a gift shop that sells related products (e.g., Superman capes or figures), which reinforces the power of story brands.

The effect of narrative familiarity with intellectual properties is perhaps best understood through their use in ancillary features within a park like restaurants, retail shops, or character meet and greets. Magic Kingdom (at Walt Disney World in Florida, USA), the most visited theme park in the world (Rubin, 2020), has features such as character meet and greets from several films, a *Beauty and the Beast*-based restaurant, a *Dumbo*-based store, and a *Pirates of the Caribbean*-based interactive quest. In addition, Magic Kingdom has a rest area and bathroom complex designed to simulate the environs of the *Tangled* (2010) film. The set of restrooms is based on specific locations in the film (Rapunzel's tower and the Snuggly Duckling bar) and plays the soundtrack. The park uses environmental storytelling to bring visitors into the story, even adding subtle hoof patterns in the ground related to one of the story characters. Guests' narrative familiarity allows them to quickly identify and connect with a space (Chronis, 2012), making it an effective strategy even for something as mundane as bathrooms.

Story-Adjacent Experience

Close to the story-based category would be story-adjacent attractions that draw their inspirations from being placed in a design area next to another story. For instance, experiences in Magic Kingdom's Adventureland (Tortuga Tavern restaurant, Pirates Bazaar shop, A Pirate's Adventure interactive quest, etc.) developed organically to match the existing story of the Pirates of the Caribbean ride and enrich its storyworld. A few theme park operators in China (e.g., Songcheng parks or Sunac Land sites) utilize in-park stories that relate to historical and environmental surroundings. Nanchang Sunac Land is in the Jiangxi province, an area with a historical connection to pottery. Thus, the park is full of pottery references including a Pottery Village area and pottery-based rides, restaurants, historical signs, and an entire pottery motif mall. Representing local stories and physical features has lent to perceived authenticity and satisfaction in historical and cultural parks, so this strategy makes sense (Cheng, Fang, & Chen, 2016; Ludwig & Wang, 2020). The parks become more popular because of their story adjacency, which may contribute to additional offerings that continue to complement the approach. Comparably, ancillary features including food and beverage or merchandise offerings will be purposely connected to the adjacent area for increased authenticity or connection, which in turn leads to more satisfaction and revenue.

When determining how to utilize space, theme parks may employ the strategies of *popular design*, where IPs and popular tales are brought in and determine new experiences, or *aspirational design*, where attractions under development are inspired by current park themes (Younger, 2016). An example of the former would be the single-IP lands mentioned earlier, while the latter is exemplified by the Ghost Town Alive! event at Knott's Berry Farm, which draws its themes, storyline, and characters from the classic, existing land. Regardless, both forms of design create story-adjacent attractions. Kang and Kodos' Twirl 'n' Hurl, a spinner ride at Universal Studios Florida, uses characters from *The Simpsons* because it was placed in the land using that IP. Conversely, Kings Island's recent Orion roller coaster was a continuation of the paranormal theme already in the Area 72 section of the park. Had it been placed in a different area, it might have carried a different theme or storyline. Something similar has happened to many of the "Madhouse" rides around the world. This mild thrill ride that gives the illusion of flipping upside down tends to be very narrative with obvious plots and environmental storytelling. While some are designed based on a storyline that designers wanted to create, others are connected to adjacent tales. For instance, Hex – The Legend of the Towers relates to local lore near Alton Towers, Cassandra's Curse holds a mythology narrative because of its placement in the Greece section of Europa-Park, and Feng Ju Palace similarly carries a Chinese plot and architecture because of its inclusion in Phantasialand's Chinese area.

Story-adjacent attractions would also consist of "vast narratives" (Harrigan & Wardrip-Fruin, 2009) that may be included in a theme park. For example, there are multiple attractions in Disney parks around the world that link to the Society of Explorers and Adventurers (S.E.A.) storyline (Mystic Manor at Hong Kong Disneyland, Tower of Terror and Fortress Explorations at Tokyo DisneySea, Big Thunder Mountain Railroad at Magic Kingdom, Miss Adventure Falls at Typhoon Lagoon, the Oceaneer Lab on Disney Cruise Line, specific restaurants, etc.). These attractions have their own stories they are based on but connect to the larger storyworld. This allows "forensic fans" (Mittell, 2013) the ability to identify and analyze connections between stories, which Williams (2020) calls "retrospective transmedia," as visitors assemble the pieces of the narrative to complete the whole. Rides or restaurants are purposefully tied into the vast narrative, and their adjacency to this fan-embraced storyline make them more appealing.

Story-Driven Experience

Just as storytelling can act as the framework for a business (Mossberg, 2008) or a corporate culture (Gabriel, 2000), narrative can be the primary structure of experience creation in theme park attraction development (Rohde, 2007) or the most significant aspect of a particular brand (Hover & Vugts, 2013). In the context of this chapter, story-driven refers to experience staging strategies with storytelling at the center. If the primary content of an attraction is narrative, and the visitor goes through an event or space that has the structure of a narrative, it is story-driven. Though these experiences may be occasionally passive, guests are more often given roles to play in a story in contemporary attractions. This category will focus on those narrative attractions where the delivery itself may not be overtly interactive, but they have explicit or implicit connections to an often-detailed narrative.

Story-driven attractions may choose to emphasize story experience above thrill or other factors. For example, one of the most well-regarded, cutting-edge rides in the world is *Star Wars*: Rise of the Resistance at Disney's Hollywood Studios and Disneyland. This experience brings guests through multiple story scenes and an evident multi-act dramatic structure (Quesenberry & Coolsen, 2019) but mostly avoids thrills. The attraction is a hybrid experience with live actors, animatronics, and media screens acting as narrators. It contains both practical and virtual effects and has walking sections and three separate ride systems in use. Rise of the Resistance begins with the design of the area and implicit and environmental storytelling related to hiding underground in the queue and continues with explicit storytelling including a call to action with the character Rey recruiting guests into the Resistance and a shuttle mission to a secret base plot. It moves to two pre-show areas where angry officers interact live with guests and the First Order leaders General Hux and Kylo Ren on screen to interrogate guests. The ride then ends in a trackless dark ride with several design techniques utilized including callbacks to the characters, visual details, and a soundtrack from the films. If using the *Star Wars*: Datapad app, guests can play the Resistance Encampment mission, which directly relates to the ride content. It is even possible to have pre-existing knowledge of the events prior to the ride, with the transmedia novel *Star Wars Galaxy's Edge: Black Spire* by Delilah Dawson. All of these layers make for an exciting, story-driven experience that people wait several hours for.

Story-driven experiences can also be shows, quests, and even thrill rides. Some companies have opted to give story structure, characters, a plotline, and a soundtrack to roller coasters instead of only a light motif. These include rides like Baron 1898 at Efteling, Cobra's Curse at Busch Gardens Tampa, Mystery Mine at Dollywood, Mystic Timbers at Kings Island, Nemesis at Alton Towers, Revenge of the Mummy at Universal Studios, TRON Lightcycle Power Run at Shanghai Disneyland, and many others. While the story may not be the primary part of the experience in all cases (thrills are still paramount), it adds to a more complete experience. In some cases, story elements are limited to the queue and station areas of rides when budgets are constrained. For example, Flying Aces at Ferrari World Abu Dhabi, Xpress: Platform 13 at Walibi Holland, and Star Trek: Operation Enterprise at Movie Park Germany have extensive uses of theming and story before the ride begins. However, once the ride starts, the clear focus is on the traditional roller coaster experience.

Despite the many possibilities for story-driven experience, the theme park's best examples of the strategy are dark rides. With their more than a century-old history, dark rides have long told stories whether through purposeful narrative or the *you are in a different place* trope. With the modern refinement of operators such as Disney (with Peter Pan's Flight and hits such as Pirates of the Caribbean and Haunted Mansion) and the emergence of companies that create dark rides for many parks like Sally Dark Rides (famous for their Ghost Blasters model in addition to other work), dark rides have become a staple of the industry. While traditionally track-based, now boat rides, enhanced motion vehicles, slot cars, and trackless vehicles have all become means to conduct guests through stories. One roller coaster model, the suspended powered coaster from Mack, was purposely designed to mimic the controlled directional changes of a classic dark ride, with Arthur: The Ride at Europa-Park and Dragon Gliders at Motiongate Dubai examples of the story-driven experience possible with this kind of ride system. Dark rides have been studied as significant forms in immersive media (Zika, 2018), drivers of immersion and loyalty (Langhof & Güldenberg, 2019), conduits of emotion (Cornelis, Strijbosch, & Corsius, 2022), and even children's literature (Rahn, 2010). Many dark rides utilize a standard dramatic structure, environmental design, characters, soundtracks, and interactive or role-playing elements. They will continue to help define this entertainment genre and inextricably link the parks to the art of storytelling.

Story-Interactive Experience

The final layer involves a higher level of intensity for guests in relation to their participation. People's tendencies of media consumption have shifted over the years with transmedia storytelling (storyworlds distributed across media platforms such as books, movies, video games, and theme parks), the convergence of media types, and the rise of participatory culture (Jenkins, 2006). This desire to participate has increased the amount of interactive, role-playing features of most entertainment forms. In tourism and other experiential fields, there is also a concept of co-creation, with storytelling a powerful conduit for co-creating brands and experiences (Pera, 2017). In theme parks, there can be mild interactive features such as can be found on Spaceship Earth at EPCOT, where the guest must input choices about future desires into a touchscreen interface or a medium level such as Millennium Falcon: Smugglers Run at Disney's Hollywood Studios and Disneyland, where guests control the steering and blasters on the ship and characters on screen comment to the guests (but the ship will still fly if no one does anything). However, theme park iterations of interactivity can also be more akin to the *storyliving* concept (Maschio, 2017), wherein the story demands visitor engagement to be considered interactive (Brigante & Elger, 2020). These can be experiences overlaid on places to engage the curious visitor like the former Sorcerers of the Magic Kingdom and current installations like A Pirate's Adventure or Harry Potter Wands (Baker, 2016) and other locative augmented reality games (Eddy et al., 2020).

While theme parks have not been as inventive as some other entertainment (electronic literature, video games, immersive theatre), they have begun to bring in the experimental forms of other mediums. It is not uncommon now in theme park storytelling to have a *bi-narrative structure* wherein two stories exist: the third-person *observable story* where riders watch characters go through events and the second-person *experiential story* where riders themselves go through a story (Kay, 2013). With story-interactive attractions, however, the observable story is less important or occasionally non-existent. In some cases, the guest is the primary agent in an attraction, for instance in Avatar Flight of Passage at Disney's Animal Kingdom, where the rider's journey through Pandora and its beauties and challenges are the primary plotline. However, story-interactive experiences include the emerging immersive theatre installations found in Ghost Town Alive! at Knott's Berry Farm or Forbidden Frontier at Cedar Point, which have guests as characters, guest agency, and sometimes branching story paths. These go even further to creating experiences with the most amount of interaction, as there would be no experience without the engagement of guests. For instance, Magic Kingdom's Enchanted Tales with Belle tells a story with the guests playing the parts from the *Beauty and the Beast* film. If there were not enough guests in the show, it would not make sense or function as a narrative.

Lastly, the story-interactive category includes the group of parks where the entirety of the experience is about engaging with a story and role-playing in a storyworld (i.e., Evermore Park in the US, Puy du Fou in France, and Songcheng in China). These types of parks have few or no rides, with immersive experience being the primary driver of the visit. The practice of walking through the parks and going through events and meeting characters is the primary purpose of the spaces. Puy du Fou emphasizes shows but also immersive experience. Songcheng (especially the Hangzhou location) also has shows but focuses on reproducing a historical but idealized Song Dynasty; they emphasize being a part of the environmental scenes and cosplaying in Song attire. Evermore, seen as a different kind of interactive park, allows for joining guilds, doing specialized story missions, and immersing in the environment, is story-interactive to the point that an individual who prefers passive entertainment is far less likely to enjoy the experience. The key to reaching a diverse group of guests with story-interactive experiences is to have levels of participation available so that guests can choose how deeply to engage.

STRATEGIC STORYTELLING SOLUTIONS

Theme parks have developed a set of narrative strategies that have been shown to be effective in attracting visitation, revenue, and brand loyalty. Repeat visitation, considered one of the core aspects of managing a successful park (Clavé, 2007), is often prompted by the appealing and multi-layered narrative environment. The more visits undertaken by guests, the better a park performs financially, which is then reinvested in projects that repeat the cycle. The more appealing the spaces, the more likely guests are to initiate this cycle and make the attraction competitive. While destination theme parks are known for having enormous amounts of capital to spend on staging experiences, great story application can be found in regional, local, and niche parks as well. It has also become more common to apply narrative strategies found in theme parks in other types of visitor attractions like museums, heritage sites, and science centers (Baker 2018b).

The following principles can be applicable to practitioners building and managing their spaces. The recommendations for usage are accompanied by descriptions of successful examples from both theme parks and other sectors.

Source Material

Base attractions on *popular stories*, whether cultural tales or IP. This helps guests digest the information and allows for pre-existing familiarity with these narratives. Using appealing *characters* is important whether using known stories or not, as it personalizes events and makes the experience more memorable. A majority of theme park attractions relate to existing stories such as fairy tales, cultural legends, or intellectual properties including novels and films. Those that have original stories often draw on genre traditions (adventure, romance, horror, etc.), common tropes, and new but relatable characters that drive the plot. For instance, Disney's Mystic Manor dark ride (Hong Kong Disneyland) invents a character (Lord Henry Mystic) who expresses the explorer trope and a cute sidekick character (Albert the monkey) while also utilizing the Monkey King figure important to Chinese culture in addition to visual allusions to popular Disney attractions like Haunted Mansion and Enchanted Tiki Room. Museums have been known to add characters that a visitor can follow throughout a space (e.g., exhibits at York, England's Castle Museum) or to empathize with in dark tourism sites (such as Holocaust-based museums). Likewise, several national parks in the United States have mascot animals one can adopt, read a story about, and buy related merchandise of.

Structure

Employ a standard *story structure* and the *techniques of drama*. Whatever the attraction type, it is in some ways a staged experience that could benefit from purposeful composition. Visitors appreciate clear structures with a beginning, middle, and end. They prefer obvious exposition and pointers as to the direction of content scenes. Using techniques of the stage including narration, dialogue, lighting and sound effects, and musical scores will also help with immersion into the space. In addition to the stage shows and immersive theatre installations at parks, many theme park rides are elaborate theatre. While not all of them go through an apparent plot and story structure, many successful ones do. Scooby Doo: The Museum of Mysteries at Warner Bros. World Abu Dhabi is a great example of a clear story structure. The mystery is set up in the queue and in the opening scene of the dark ride. The middle section of the ride includes tension with a haunted museum and a plan amongst the main characters to uncover the culprit. It ends with a tidy conclusion and the mystery solved. Environmental scenes, dialogue, favorite characters, and music all add to the feeling that the ride is just another story in the classic television series. A particularly effective example of these principles in a museum context is in Het Spoorwegmuseum, the national train museum of The Netherlands. In The Great Discovery, visitors are asked to listen via headphones and take a walking tour of a theatrical set. In a tour, the guest hears the voice of John Middlemiss, a British driver who narrates the story of the first Dutch engine, *De Arend*. When one approaches the physical train at the end of the narrative, it has meaning and is not simply another train in the museum's collection with typical descriptive signage.

Space

Utilize the *space* that is already there. Theme parks and museums have both been known to repurpose spaces when needed. Both have also found that existing environments or other constraints may limit story options. Nonetheless, think of ways to incorporate current space or to overlay new narratives on top of existing space. An example of the latter is Legoland Florida's use of props and figures to enhance the remaining gardens of Florida's original Cypress Gardens theme park. In this way, it drives demand from new audiences while also staying true to the history of the park, sparking repeat visitation and potential brand loyalty. The Jorvik Viking Centre in England is an example of the more permanent utilization of space. Sitting atop a real Viking site with a moderate footprint, the museum has traditional display cases, a glass floor over the archaeological dig, and a dark ride (with "time capsules") that take visitors through a tour of history complete with sights, sounds like ancient Norse, and even unpleasant smells. Though a major flood damaged the attraction, it reopened a few years later, showing the resilience and sustainability of the concept.

Environmental Design

Use *environmental design,* which theme parks have proven is powerful. Today's visitors are accustomed to elaborately designed spaces, as these have become more common in retail, restaurants, visitor attractions, and public spaces. Developing multi-sensory places with combinations of architecture, landscaping, sounds, music, props, lighting effects, and other aspects contributes to a sense of authenticity and immersion. While theme parks are known for the immersive lands developed around very popular intellectual properties like *Jurassic Park* or *Pirates of the Caribbean*, Phantasialand in Germany instead opted for lands of ornate details based on original stories. This includes their most recent two lands, the rustic old Norse, basalt canyon-based Klugheim or the steampunk, new age of flying Rookburgh. Other visitor attractions have been more likely to utilize these techniques now, but the World of Beatrix Potter in England is an example of environmental design as a way to immerse visitors in reproductions of her famous storyscapes. Studio Ghibli's theme park, under construction in Japan, will make extensive use of environmental design to replicate the whimsical worlds of Hayao Miyazaki. Several zoos over the years have used environmental design to simulate the habitats and cultures related to where animals live in the wild.

Use the Right Attraction for the Purpose

Ensure that the type of new attraction being built *meets the purpose* of the space and the *current needs of the site*. One of the key concepts of amusement and theme park design is *attraction mix*, meaning the variety of attractions at the site is both diverse and fits the needs of the markets that visit. For some parks, a children's dark ride makes more sense than a thrilling roller coaster and vice versa. The challenges of limited money, space, and available labor are constants. Thus, there is not a single type of attraction that will always work. Organizations building a new attraction or exhibit within an existing attraction should make sure it fits with market demand and available resources. Some theme parks challenge conventional wisdom that rides are needed to be successful. Several children's story parks have thrived despite a lack of thrills or adult-oriented attractions. Florida's Kennedy Space Center has chosen to utilize several types of attractions including walkthroughs, immersive shows, films, and a simulator ride based on what the exhibition is meant to accomplish. Similarly, world expositions over the years have relied on different presentation types. For instance, Expo 2020 Dubai utilized techniques from stage shows or projection mapping to immersive films, walkthrough exhibits to environmental design.

Participation

Give guests a *role in the story*. There has been a transition in entertainment from more passive forms to a participatory culture with opportunities for interaction and co-creation. In theme parks, everything from shows to rides have become more participatory. Interactive quests, scavenger hunts, and immersive theatre are all about the capabilities of role-playing to engage audiences. Parades, including Universal's Mardi Gras or the Cedar Fair company's parades during Grand Carnivale are about soliciting audience participation in the parade. Attractions like KidZania are purpose-built for the role-playing of children. Some of the most elaborate of the participatory attractions are the interactive quests associated with the company Creative Kingdoms. Over the years, they have built complex interactive adventures that give guests a role in a narrative. These have included attractions built for science centers (e.g., CosmicQuest at Kennedy Space Center, Dino Quest at Discovery Cube), water parks (SplashQuest, such as the former PearlMaster Quest at Yas Waterworld), and their entries at hotels (the well-known MagiQuest at Great Wolf Lodge resorts or Adventure Trek: Quest of the Guardians at Six Flags Great Escape Lodge).

Keep it Updated

Keep the site and the presentation *up to date*. The late Imagineering great Marty Sklar (2015) argued that keeping it up and maintaining the space was as important as the strategic storytelling and design in theme parks. This is absolutely true, as an ill-kept space communicates to guests a lack of care, a form of bad service. In the theme park industry, however, this can also mean making an experience fresh by *updating it*, changing it, or "plussing" it as Walt Disney called it. There are several reasons why this can happen including technological or cultural change. Guests' tastes, sensibilities, and expectations change over time. While an attraction may be completely removed because of this, others may be altered to create a better experience. Classic dark rides may be updated to stay more in line with technological or cultural expectations, for instance when Six Flags Over Georgia's dark ride Monster Plantation was refurbished as Monster Mansion. An attraction offensive in its depiction of race, Monsieur Cannibale at Efteling, was reimagined as part of The World of Sindbad. Similarly, museums change exhibits to reflect new trends in exhibition, as when the previously discussed Castle Museum updated their Victorian Street to be more immersive. It now has character stories that can be followed along with and costumed actors that lend to the atmosphere.

Make it Enjoyable

Give the people an *enjoyable experience*. It is often said that the amusement and theme park industry is the "business of fun." While educational content can be found at museums, heritage sites, zoos, national parks, world expos, and others, the edutainment model has proven effective. A park like Disney's EPCOT includes museum exhibits and educational rides such as Spaceship Earth or Living with the Land, but over the years they have used environmental design, ride systems, music, clear narrative, and other techniques to make the content more engaging. Heritage sites like Colonial Williamsburg or the Jamestown Settlement increase enjoyment through physical sets to inhabit and live actors playing roles of people during the time. Museums like the American Revolution Museum in Yorktown, Virginia use contemporary techniques including environmental vignettes, interactive exhibits, and multi-sensory films to provide enjoyment for visitors rather than using only a more traditional, dry presentation.

FUTURE RESEARCH DIRECTIONS

The future of theme park narratives will likely include further reliance on technologies, increased customization of the product, continued interactivity and role-playing features, becoming a more focused part of transmedia storytelling webs, and an emphasis on more and more layers of experience. As theme parks compete with several other types of attractions and develop within emerging markets, they need to continually innovate with technologies and strategic storytelling. Story-based, story-adjacent, story-driven, and story-interactive layers will continue to immerse guests in many strata of narrative experience to ensure their engagement. From role-playing and immersion to environmental design and co-creative opportunities, theme parks provide compelling encounters with narrative at the very core of its experiential model. There are opportunities to study narrative structures more deeply within the space, the diversity of attractions for telling stories, and which layers are the most successful in engaging guests to facilitate satisfaction, loyalty, and revisit intention.

Theme parks will continue to be a context explored by scholars because of their commercial success. Their narrative capabilities will also be something significant to study in even more depth, as it is crucial to their ability to get millions of visitors. It would be useful to investigate how theme parks have impacted other visitor attractions such as immersive art like Meow Wolf, how they have reshaped the concept of destination, or their lasting impact on urban design. The reciprocal influence of theatre forms or video games would also be useful to discern. The effect of the theme park paradigm on fan cultures is a continuing topic of discussion in fan studies but influences on greater communities will also be valuable to discover. Differences between theme park narratives and other forms of entertainment or visitor attractions remain evident, but it remains to be seen whether theme parks will try the more experimental narratives now found in some mediums. Guest expectations and satisfaction criteria for attractions including theme parks will remain a common inquiry with both scholars and practitioners.

CONCLUSION

Pine and Gilmore (2011) found theme parks to be the paragon of their experience economy principle and found theming strategy to be about the "unfolding story." Some of the current success of the industry is due to its complex fabric of narrative that parks weave for their visitors. In this chapter, four narrative layers were described that parks utilize to enhance guest experience and involve visitors in storyworlds that lead to guest satisfaction and loyalty. As noted, narrative content improves attendance, revenue, and repeat visitation. Though many organizations lack the capital of destination theme parks, many of their strategies can and have been leveraged in other spaces. Storytelling is one of the significant aspects of human culture. It organizes patterns in culture (Abbott, 2012) and in theme parks, it organizes space in a way that humanizes environments and provides multiple forms of immersion including physical, sensual, cognitive, and emotional (Wolf, 2012). Scholars and practitioners alike can benefit from understanding ways in which narrative is incorporated in theme park attractions and how integral narrative has become to this form of tourism. By using this knowledge, destinations can apply strategic storytelling to create meaningful experiences for visitors that simultaneously tap into their humanity.

REFERENCES

Abbott, H. P. (2012). *The Cambridge introduction to narrative*. Cambridge University Press.

Baker, C. (2016). Creative choices and fan practices in the transformation of theme park space. *Transformative Works and Cultures, 22*.

Baker, C. (2018a). Summary of exploring a three-dimensional narrative medium: The theme park as "de Sprookjessprokkelaar," the gatherer and teller of stories. *Journal of Themed Experience and Attractions Studies*, *1*, 1–9.

Baker, C. (2018b). The use of themed entertainment design in museums and heritage sites. *The Museum Review, 3*(1).

Baker, C. (2018c). Universal's Wizarding World of Harry Potter: A primer in contemporary media concepts. In A. Firestone & L. A. Clark (Eds.), *Harry Potter and convergence culture: Essays on fandom and the expanding Potterverse* (pp. 55–66). McFarland.

Baker, C. (2022). Magic keys and magic mushrooms: Storylands and early dramatic play. In K. M. Jackson & M. I. West (Eds.), *Storybook worlds made real: Essays on the places inspired by children's narratives* (pp. 125–146). McFarland.

Bassano, C., Barile, S., Piciocchi, P., Spohrer, J. C., Iandolo, F., & Fisk, R. (2019). Storytelling about places: Tourism marketing in the digital age. *Cities (London, England)*, *87*, 10–20. doi:10.1016/j.cities.2018.12.025

Blackwood, G. (2018). Dream factory tours: The Universal Pictures movie tour attraction in the 1960s. *Historical Journal of Film, Radio and Television, 38*(3), 516–535. doi:10.1080/01439685.2017.1369238

Brigante, R., & Elger, S. A. S. (2020). *2020 Immersive Entertainment Industry Annual Report*. HERE Institute.

Cheng, Q., Fang, L., & Chen, H. (2016). Visitors' brand loyalty to a historical and cultural theme park: A case study of Hangzhou Songcheng, China. *Current Issues in Tourism, 19*(9), 861–868. doi:10.1080/13683500.2015.1006589

Chronis, A. (2012b). Tourists as story-builders: Narrative construction at a heritage museum. *Journal of Travel & Tourism Marketing, 29*(5), 444–459. doi:10.1080/10548408.2012.691395

Clavé, S. A. (2007). *The global theme park industry*. CABI. doi:10.1079/9781845932084.0000

Cornelis, P. C. M., Strijbosch, W., & Corsius, P. (2022). Conceptualizing the dark ride experience through the dark ride cube: Evidence from the EMEA region. *Journal of Hospitality & Tourism Research (Washington, D.C.)*, 1–26. doi:10.1177/10963480221080331

Cross, G., & Walton, J. (2005). *The playful crowd: Pleasure places in the twentieth century*. Columbia University Press. doi:10.7312/cros12724

Eddy, R., Baker, C., Macy, R., Murray, J., & Salter, A. (2020). Hacking droids and casting spells: Locative augmented reality games and the reimagining of the theme park. *Proceedings of the 31st ACM Conference on Hypertext and Social Media*, 37-46. 10.1145/3372923.3404801

Fu, X., Baker, C., Zhang, W., & Zhang, R. (2022). Theme park storytelling: Deconstructing immersion in Chinese theme parks. *Journal of Travel Research*. Advance online publication. doi:10.1177/00472875221098933

Gabriel, Y. (2000). *Storytelling in organizations: Facts, fictions, and fantasies*. Oxford University Press. doi:10.1093/acprof:oso/9780198290957.001.0001

Godovykh, M., Milman, A., & Tasci, A. D. A. (2019). Theme park experience: Factors explaining amount of pleasure from a visit, time allocation for activities, perceived value, queuing quality, satisfaction, and loyalty. *Journal of Tourism and Leisure Studies (Champaign, Ill.)*, *4*(2), 1–21. doi:10.18848/2470-9336/CGP/v04i02/1-21

Godwin, V. (2017). Theme park as interface to the Wizarding (story) World of Harry Potter. *Transformative Works and Cultures, 25*.

Green, M. C., & Brock, T. C. (2000). The role of transportation in the persuasiveness of public narratives. *Journal of Personality and Social Psychology, 79*(5), 701–721. doi:10.1037/0022-3514.79.5.701 PMID:11079236

Harrigan, P., & Wardrip-Fruin, N. (2009). Introduction. In P. Harrigan & N. Wardrip-Fruin (Eds.), *Third person: Authoring and exploring vast narratives* (pp. 1–9). MIT Press.

Hench, J. (2009). *Designing Disney: Imagineering and the art of the show*. Disney Editions.

Hobbs, P. (2015). *Walt's utopia: Disneyland and American mythmaking*. McFarland.

Hover, M., & Vugts, O. (2013). Efteling and the Fairytale Tree: Storytelling between tradition and innovation. *Trendrapporttoerisme, Recreatie en Vrijetijd*, 272-275.

International Association of Amusement Parks and Attractions (IAAPA). (2018). *IAAPA history*. IAAPA.org.

Jenkins, H. (2004). Game design as narrative architecture. In N. Wardrip-Fruin & P. Harrigan (Eds.), *First person: New media as story, performance, and game* (pp. 118–130). MIT Press.

Jenkins, H. (2006). *Convergence culture: Where old and new media collide*. New York University Press.

Jones, K., & Wills, J. (2005). The invention of the park: Recreational landscapes from the Garden of Eden to Disney's Magic Kingdom. *Polity*.

Kang, J.-A., Hong, S., & Hubbard, G. T. (2020). The role of storytelling in advertising: Consumer emotion, narrative engagement level, and word-of-mouth intention. *Journal of Consumer Behaviour*, *19*(1), 47–56. doi:10.1002/cb.1793

Kay, I. (2013, September 17). Bi-narrative structure. *Pure Imagineering: Theme Park Narratology*. https://pureimagineering.blogspot.com/2013/09/experiential-and-presentational-stories.html

Kerz, C. (2016). Atmosphere, immersion, and authenticity in Colonial Williamsburg. In S. Lukas (Ed.), *A reader in themed and immersive spaces* (pp. 195–205). ETC Press.

Kim, J.-H., & Youn, H. (2017). How to design and deliver stories about tourism destinations. *Journal of Travel Research*, *56*(6), 808–820. doi:10.1177/0047287516666720

Langhof, J. G., & Güldenberg, S. (2019). Pirates, ghosts and customer loyalty: Reviewing the dark ride experience. *Tourism Management Perspectives*, *31*, 398–420. doi:10.1016/j.tmp.2019.06.004

Lee, S., Jeong, E., & Qu, K. (2020). Exploring theme park visitors' experience on satisfaction and revisit intention: A utilization of experience economy model. *Journal of Quality Assurance in Hospitality & Tourism*, *21*(4), 474–497. doi:10.1080/1528008X.2019.1691702

Ludwig, C., & Wang, Y.-W. (2020). Open-air museums and historical theme parks in the UK and China. In C. Lugwig, L. Walton, & Y.-W. Wang (Eds.), *The heritage turn in China: The reinvention, dissemination, and consumption of heritage* (pp. 131–168). Amsterdam University Press.

Lukas, S. (2012). *The immersive worlds handbook*. Focal Press. doi:10.4324/9780240820989

Lund, N. F., Cohen, S. A., & Scarles, C. (2018). The power of social media storytelling in destination branding. *Journal of Destination Marketing & Management*, *8*, 271–280. doi:10.1016/j.jdmm.2017.05.003

Maschio, T. (2017). *Storyliving: An ethnographic study of how audiences experience VR and what that means for journalists*. Google News Lab.

Mittell, J. (2013). Forensic fandom and the drillable text. *Spreadable Media*. https://spreadablemedia.org/essays/mittell/index.html#.X0bapshKhnI

Mossberg, L. (2008). Extraordinary experiences through storytelling. *Scandinavian Journal of Hospitality and Tourism*, *8*(3), 195–210. doi:10.1080/15022250802532443

Naversen, N. (2017). *The sliced onion technique: How to theme a space*. Themed Attraction.

Pera, R. (2017). Empowering the new traveller: Storytelling as a co-creative behavior. *Current Issues in Tourism*, *20*(4), 331–338. doi:10.1080/13683500.2014.982520

Philips, D. (1999). Narrativised spaces: The functions of stories in theme parks. In D. Crouch (Ed.), *Leisure/Tourism geographies: Practices and geographical knowledge* (pp. 92–108). Routledge.

Pikkemaat, B., & Schukert, M. (2007). Success factors of theme parks: An exploratory study. *Tourism (Zagreb)*, *55*(2), 197–208.

Pine, B. J., & Gilmore, J. H. (2011). *The experience economy* (2nd ed.). Harvard Business School Publishing.

Pollock, S., Quesenberry, K. A., & Coolsen, M. K. (2019). Drama goes viral: Effects of story development on shares and views of online advertising videos. *Journal of Interactive Marketing*, *48*, 1–16. doi:10.1016/j.intmar.2019.05.001

Price, H. (2004). *Walt's revolution! By the numbers*. Ripley Entertainment.

Rabinovitz, L. (2006). From Hale's Tours to Star Tours: Virtual voyage, travel ride films, and the delirium of the hyper-real. In J. Ruoff (Ed.), *Virtual voyages: Cinema and travel* (pp. 42–60). Duke University Press. doi:10.2307/j.ctv1131b73.6

Rahn, S. (2010). The dark ride of Snow White: Narrative strategies at Disneyland. In K. M. Jackson & M. I. West (Eds.), *Disneyland and culture: Essays on the parks and their influence* (pp. 87–100). McFarland.

Rohde, J. (2007). From myth to mountain: Insights into virtual placemaking. *Computer Graphics*, *41*(3), 1. doi:10.1145/1281324.1281325

Rohde, J. (2017, May 5). Details, the inner berm. *Instagram*. Available at: https://www.instagram.com/p/BTugAK_AoZN/

Rubin, J. (Ed.). (2020). *Global attractions attendance report 2019*. TEA/AECOM.

Sandvik, K. (2010). *Crime scenes as augmented reality: Models for enhancing places emotionally by means of narratives, fictions and virtual reality*. Channel View Publications.

Schmitt, B. (1999). *Experiential marketing: How to get customers to sense, feel, think, act, and relate to your company and brands*. Free Press.

Sklar, M. (2015). *One little spark!: Mickey's ten commandments and the road to Imagineering*. Disney Editions.

Tasci, A., & Milman, A. (2019). Exploring experiential consumption dimensions in the theme park context. *Current Issues in Tourism*, *22*(7), 853–876. doi:10.1080/13683500.2017.1321623

Waysdorf, A., & Reijnders, S. (2018). Immersion, authenticity and the theme park as social space: Experiencing the Wizarding World of Harry Potter. *International Journal of Cultural Studies*, *21*(2), 173–188. doi:10.1177/1367877916674751

Williams, R. (2020). *Theme park fandom: Spatial transmedia, materiality, and participatory cultures*. Amsterdam University Press.

Wolf, M. J. P. (2012). *Building imaginary worlds: The theory and history of subcreation*. Routledge.

Younger, D. (2016). *Theme park design and the art of themed entertainment*. Inklingwood Press.

Zika, J. (2018). Dark rides and the evolution of immersive media. *Journal of Themed Experience and Attractions Studies*, *1*, 54–60.

ADDITIONAL READING

Berger, A. (2013). *Every guest is a hero: Disney's theme parks and the magic of mythic storytelling*. BCA Press.

Blackwood, G. (2018). Dream factory tours: The Universal Pictures movie tour attraction in the 1960s. *Historical Journal of Film, Radio and Television*, *38*(3), 516–535. doi:10.1080/01439685.2017.1369238

Clavé, S. A. (2007). *The global theme park industry*. CABI. doi:10.1079/9781845932084.0000

Hench, J. (2009). *Designing Disney: Imagineering and the art of the show*. Disney Editions.

Hill, B. R. (2020). *Imagineering an American dreamscape: Genesis, evolution, and redemption of the regional theme park*. Rivershore Press.

King, M. (2002). The theme park: Aspects of experience in a four-dimensional landscape. *Material Culture*, *34*, 1–15.

Lukas, S. (2012). *The immersive worlds handbook*. Focal Press. doi:10.4324/9780240820989

Marling, K. A. (1998). *Designing Disney's theme parks: The architecture of reassurance*. Flammarion.

Pierce, T. J. (2016). *Three years in wonderland: The Disney brothers, C.V. Wood, and the making of the great American theme park*. University Press of Mississippi.

Rohde, J. (2007). From myth to mountain: Insights into virtual placemaking. *Computer Graphics*, *41*(3), 1. doi:10.1145/1281324.1281325

KEY TERMS AND DEFINITIONS

Attraction: Within a theme park, the primary draws for visitors. This can include roller coasters, dark rides, water rides, thrill rides, walkthrough exhibits, stage shows, projection shows, parades, fireworks, and occasionally restaurants or shops.

Dark Ride: An enclosed ride with multiple show scenes. Generally found in amusement and theme parks, these rides can be in many forms (busbar, enhanced motion vehicle, trackless, boats in canals, and occasionally roller coasters).

Explicit Storytelling: A story told through written or oral means.

Implicit Storytelling: A story told through environmental means (sound, music, lighting, props, architecture, landscaping, etc.).

Interpretive Storytelling: A story that is open to interpretation from visitors rather than a specific story being told (explicit) or a specific story that is meant to be discovered (implicit).

Narrative/Storytelling: The way in which a story is told; it can be in any order (i.e., it can include flashbacks, removed sections, etc.). In a theme park, this can be explicit, implicit, or interpretive; it can be in many formats including dark rides, stage shows, or nighttime spectaculars.

Story: A sequence of events in chronological order; usually involves characters, events, and settings.

Story-Adjacent Experience: An attraction that is influenced by its location in a theme park to be part of a structured narrative.

Story-Based Experience: An attraction that has narrative source material (usually books, films, video games, culture, and fairy tales). The attraction may or may not use the story in its presentation.

Story-Driven Experience: An attraction that has the structure and content of a narrative. The presentation reinforces its narrative.

Story-Interactive Experience: An attraction that involves participation from visitors to fully experience a narrative.

Theme Park: Visitor site with an underlying theme(s), filled with a variety of attractions. Several categories exist including destination parks (e.g., Disney, Universal), regional parks, local parks, or niche parks. Antecedents include amusement parks, world expos, theatre, and cinema.

Chapter 6
The Influence of Storytelling Strategies on Attitude and Purchase Intention

Luís José Andrade
Universidade Europeia, Portugal

Maria João Ferreira
Universidade Europeia, Portugal

ABSTRACT

This study seeks to examine the influence of storytelling strategies on individuals' attitudes and purchase intentions. Seven storytelling strategies were classified as dimensions, which served as a starting point for storytelling design and studying its impact on individuals. In this research, it is assumed that storytelling strategies positively influence attitude and purchase intention. Methodologically, the research adopts a causal-comparative approach. Accordingly, two hypotheses were tested through design and comparison of two scenarios with and without storytelling strategies. Analyses confirmed the study's hypotheses.

INTRODUCTION

A story is a discourse that narrates interrelated actions and consequences in chronological order (Gilliam & Flaherty, 2015). According to Baskin (2004), storytelling is the most important and most human thing that our species does. Stories have been told since the beginning of time, be it war stories or heroic deeds.

This study focuses on analysing the influence of storytelling strategies on individuals' attitude (ATT) and purchase intention (PI), examining how storytelling may or may not influence the positive weighing of an individual regarding his/her ATT and PI. According to Sole and Wilson (2002), storytelling is defined as the sharing of knowledge and experiences through narratives to communicate lessons, complex ideas, concepts, and casual connections. The power of storytelling has been widely recognized in various disciplines, including Sociology, Psychology, Management, and Marketing (Gilliam & Flaherty, 2015),

DOI: 10.4018/978-1-6684-3436-9.ch006

thus demonstrating its importance and usefulness in different fields of study and business areas, as in the case of tourism and destination marketing.

The theoretical model adopted in this study was developed by Denning (2005) in his book *The Leader's Guide To Storytelling*, which seeks to describe the importance of storytelling, not as a cultural phenomenon, but as a set of tools that can be worked and shaped according to certain objectives. This study will serve as a basis for businesses to develop and apply storytelling strategies leading to individual ATT and PI. Thus Denning's model (2005) will serve as a starting point and a theoretical reference of what a strategic storytelling approach may be. Authors' application of this approach will be replicated from a digital perspective, using storytelling in a blog narrative format as a means of exposure.

By itself, however, storytelling may not be enough to cause direct influence on PI. Woodside (2010) refers to the ability of a story to affect the way individuals perceive the world. This ability can also affect how individuals act, react, or face certain common situations. Belk (1975) suggests that the individual is susceptible to change in his behaviour through situational variables. These variables may originate in various formats and/or occasions provoked or unrelated to the individuals themselves. In the present study, ATT is taken as a variable of connection between storytelling and PI, i.e. the applied storytelling strategies expectedly will create or change the individual's ATT. In turn, this ATT may or may not generate PI.

Storytelling may also be one of the most effective communication tools for persuasion and relationship building (Mcgregor & Holmes, 1999). Therefore, brands tend to be interested and invest in the means and ends, which allow them to take advantage of this communication tool. This brand-consumer proximity may result in an emotional connection and reinforce the effectiveness of the intended marketing message (Budtz, Fog & Yakaboylu, 2005), facilitating not only the brand's strengths and values but also the individuals' primary choice. for example, in the tourism sector and destination management, storytelling is a narrative adequate to communicate the value proposition of tourism service systems and enhance the value co-creation process of all stakeholders involved (Bassano, Barile, Piciocchi, Spohrer, Iandolo & Fisk, 2019).

This chapter presents a conceptual model based on studies concerning storytelling, strategies, ATT, and PI. The main objective is to develop an understanding of strategies, ATT, and PI and relationships between them. After the literature review, the research model and data collection and analysis methods are described. Further on, results obtained are also presented. Lastly, the discussion section discusses the results, the study's limitations and implications, and suggestions for future studies. As a final reflection, this chapter's expectation is to contribute both at an academic and managerial level to a better perception and use of storytelling strategies. This study's findings help companies to choose the strategy that best fits their objectives (e.g. improving the individual's perception of the brand value) and to design the combination of strategies according to the results businesses want to achieve.

BACKGROUND

Definition of History, Story and Storytelling

Story or history? When and how should we use each term? One online source states that the word 'history' is the narration of real facts, such as social and political facts, memories of the past, or a bibliography (https://www.infopedia.pt/dicionarios/lingua-portuguesa/história, Accessed 11 February, 2019).

The same source defines 'story' as a story of a fictional or popular character, such as fairy tales (https://www.infopedia.pt/dicionarios/lingua-portuguesa/estória, Accessed 11 February, 2019). However, over the years, the word 'history' has been discontinued from common use. Thus 'story' began to represent both meanings, the telling of real and fictional facts. That said, in this chapter, 'story' will be used to represent the two possible facets: history and story.

David (1991) elucidates that the content of a story is critical in classifying it as a story. Nevertheless, people do not tell narratives, they tell stories. Someone saying "Let me tell you a narrative" would sound strange (Frank, 2000). According to the author, narrative is the structure that underpins the story and narrative analysis underpins the structures that storytellers rely on to tell stories. This structure consists of two important elements: chronology (temporal dimension of the story) and causalities (relationships between story elements) (Delgadillo & Escalas, 2004). In this way, narrative is understood as an integrated part of the story, which complements and supports it and helps in understanding and sharing it.

A story can be defined as the narration of a chain of events that are told to pass on knowledge. Stories are a powerful tool to make complex and multifaceted content explicit (Escalfoni, Braganholo, & Borges, 2009). For these authors, the chain of events in a story should follow an order that facilitates knowledge sharing and conveying more meaning to facts. This approach is objective since, according to Woodside (2010), the story can trigger behaviours among the audience, help raise awareness and provide emotional understanding. However, these factors may affect the preconceived order consisting of the description of the chain of events (Escalfoni et al., 2009). McGregor and Holmes (1999) draw attention to the great distortion and invention of facts to turn a story into a good story. In general, researchers opt for defining the story and storytelling, not as concepts, but how they should be constructed, conceived and transmitted: after all, what is a good story and good storytelling? This approach will be further developed in the following section.

Good Stories and Good Storytelling

Morgan and Dennehy (1995) outline five sequential steps for constructing a good story. firstly, the setting, secondly, the crescendo (the unfolding of the problem), thirdly, the crisis or climax, fourthly, the learning, and fifthly, the new behaviour or awareness. authors explain that scenario is composed of time, space, context, and characters in more detail. in this way, it is understood that a good contextualization helps to attract the listeners' attention. they also highlight the climax as the highest point in the story, the key event when something is expected to happen, and the story takes a turn. furthermore, as the final step, they identify the new behaviour or consequence, in which the story's moral is addressed. finally, stories relate to experiences of other people, which can fill in the gaps in our own experience (Wilkins, 1984), by assimilation of the consequences of behaviours to be avoided in the future.

Morgan and Dennehy (1997) state that for a story to be powerful and to remain in the consumer's memory, it must be concrete, have some common sense connotation, be credible to the point of creating an impact on the listener, have a connection with social dogmas and be unique to single the storyteller out from the rest. Wilkins (1984) and Martin, Feldman, Hatch, and Sitkin (1983) initially developed these assumptions before they were addressed and extended by Morgan and Dennehy (1997). Denning (2004) states that a storyteller must be involved with the story so that the audience is drawn into the world of the narrative. Frank (2000) presents storytellers as agents who do not intend their stories to be analysed, but rather people who want their stories to become a stage for sharing experiences, discovering common aspects, and building relationships. Both approaches define the storyteller agent as fundamental to the story's unfolding and sharing, but primarily highlight the importance of the storytelling agent to meet the listeners' expectations and perceptions.

McDonald (2009) argues that there is no correct formula for storytelling construction. However, MacDonald highlights three principles that the storytelling technique should include: conflict, authenticity, and entertainment. Conflict is the stage at which the characters' goals or desires collide, something that motivates action in all good stories. Authenticity is described as the sense of believability created when characters display genuine emotions and react believably, this principle promotes empathy between the audience and the characters. Finally, entertainment refers to the ability of the story to hold the audience's attention/interest. In contrast, Gilliam, Flaherty, and Rayburn (2014) state that there is not enough evidence to indicate the most important dimensions for storytelling. Gilliam et al. (2014) developed a qualitative study to determine the dimensions of storytelling important for consumers. According to Gilliam et al. (2014), four dimensions of storytelling are demonstrated to stand out in the consumer's attention: the relevance of the story to the consumer, the humour component that the story incorporates, the personal connotation (of the storyteller), and plausibility. Inherent to this set of dimensions presented is the degree of subjectivity, resulting from audience diversity.

Pera and Viglia (2016) approach storytelling as a relationship established between production and consumption of a story. storytelling uses production and consumption components to represent how an individual consumption experience can be transformed into a collective relationship experience (Pera & Viglia, 2016). according to these authors, a story is considered authentic when it appears to be 'original' or referring to 'something real.' they also argue that stories are perceived through two primary receptors: firstly, empathy, which allows the individual to distance him/herself from the 'real' world; and secondly, mental images that help him/her following the story plot.

The Leader's Guide to Storytelling by Denning (2005)

This section covers a survey on relevant information from the study that by Denning (2005). Details such as sounds, smells, and images confer sensorial and visual cues to the narrative. Denning (2005) argues that the story can be perceived as external or internal. The external perspective has the advantage of providing stability and clarity to those who perceive it, but has the disadvantage of being outside the experience of the story itself. An internal perspective, experienced as a participant, is immediate and intense but can be elusive, inconstant, and vulnerable to abuse. This perception of the story stimulates the listener's imagination and creates states of tension and release. Therefore, the listener is classified as a partner with an active state of thought rather than a passive receiver of information. Denning (2005) adds that a story can be conveyed through words, pictures, video, or mimicry. While recognizing the advantageous suitability of verbal language in communicating narrative, studying it in its non-verbal manifestations without requiring verbal narration is possible (Denning, 2005). He also argues verbal language is not an essential and indispensable factor for storytelling. A well-told and compelling story is effective regardless of the purpose of the story (Denning, 2005).

A story is not a tool, but a set of various strategies that can help achieve several purposes, such as: induce the audience to action, communicate aspects that demonstrate who the speaker is (be it an individual or a collective person), convey specific values on a given topic, share knowledge, 'tame the vine' (a popular expression for taming rumours), and, finally, lead people into the future (audience) for the future (Denning, 2005). Table 1 details the description of storytelling purposes, which explains how to use the various types of storytelling according to the intended purpose. Subsequently, the approach of strategies as storytelling techniques is analysed, followed by an evaluation of their impact on individuals.

Table 1.

If your objective is:	You will need a story:	In telling it, you will need to:
Sparking action	Describes how a successful change was implemented in the past, but allows listeners to imagine how it might work in their situation	Avoid excessive detail that will take the audience's mind off its own challenge
Communicate "who you are"	Provides audience-engaging drama and reveals some strength or vulnerability from your past	Provide meaningful details but also make sure the audience has the time and inclination to hear your story
Transmitting values	Feels familiar to the audience and will prompt discussion about the issues raised by the value being promoted	Use believable (though perhaps hypothetical) characters and situations, and never forget that the story must be consistent with your own actions
Fostering collaboration	Movingly recounts a situation that listeners have also experienced and that prompts them to share their own stories about the topic	Ensure that a set agenda doesn't squelch this swapping of stories – and that you have an action plan ready to tap the energy unleashed by this narrative chain reaction
Tame Taming the grapevine (rumors)	Highlights, often through the use of gentle humor, some aspect of a rumor that reveals it to be untrue or unreasonable	Avoid the temptation to be mean-spirited – and be sure that the rumor is indeed false!
Sharing knowledge	Focuses on mistakes made and shows, in some detail, how they were corrected, with an explanation of why the solution worked	Solicit alternative – and possibly better – solutions
Lead people into the future	Evokes the future you want to create without providing excessive detail that will only turn out to be wrong	Be sure of your storytelling skills. (Otherwise, use a story in which the past can serve as a springboard to the future.)

Using the strategies identified in Table 1 is seen as beneficial because it is possible to avoid making some of the most frequent mistakes. In Denning's work (2005), the importance of stories in the business arena is also addressed and questioned, due to the great scepticism towards the topic. "What is the ROI of storytelling?" is often the first question a business audience asks when discussing storytelling (Denning, 2005). Denning (2005) argues the question is a pretext for non-action or a way of making a negative statement. In this way, the measurement of the value of storytelling should start from a genuine attempt to understand what abilities are possessed.

The collection of ideas presented in Denning (2005) meets the theme discussed in this chapter, as it addresses concepts, techniques, and perceptions of storytelling in a transversal way. The most exciting key factor in this research is related to the fact that, even recognising the main components and the objectives of storytelling narrative, it suggests it is possible to identify which may influence the individuals' attention.

Storytelling in the Digital Age

The roots of digital storytelling (DST) date back to the 1980s, when Dana Atchley first used the term when experimenting with the use of multimedia elements in narrative performances (Rossiter & Garcia, 2010). however, the digital technologies emerging in the 1990s offer new tools for expression and 'fertile ground' for experimentation with DST (https://www.storycenter.org). DST emerged from the integration of multimedia and traditional storytelling to meet the needs of individuals (Sarica & Usluel, 2016). Culmination of these experiences, 'traditional' storytelling and multimedia, assists individuals in meeting their communication, self-expression, and teaching performance needs, which consequently improves their skills (Sarica & Usluel, 2016).

For Rossiter & Garcia (2010), DST combines the art of storytelling with multimedia objects, as images, audio, and video. However, creation of DST involves particular skills: the ability to conceptualize, write, execute, select, create images, integrate, and produce meaningful content (Benmayor, 2008). In the same vein, Clarke and Adam (2010) argue that DST encompasses all forms and processes of narrative produced and shared digitally and involve a range of formats: the narrative, the image-only story, the radio, or the multimedia narrative integrated with images sounds and/or text.

According to Lundby (2008, P. 3), DST is intrinsically linked to digital content sharing platforms. Bryan (2011) conceptualizes DST through practical examples that put into practice Lundby's theory (2008). Practical examples from Bryan's work (2011) include an alien invasion fraudulently using transmedia through multiple Twitter accounts, a blog novel about the journey of two teenagers through political and personal landscapes of America in 1968, and a life story of a Holocaust victim retold through Facebook. Similarly, Couldry (2008) believes that DST allows multiple possibilities of transmission, retransmission, and transformation and that these are the characteristics that distinguish it from traditional storytelling. DST enables also the distribution of a scarce resource (such as representing the world around us) using a shared infrastructure, such as social networks. As suggested by Couldry (2008) and Lundby (2008), DST is the link between storytelling and multimedia, but also the representation of how the storytelling concept has been evolving, thus following the evolution of digital platforms.

According to Lambert (2013), DST is composed of seven components: 1) point of view; 2) a dramatic issue; 3) emotional content; 4) the voice gift; 5) the soundtrack; 6) the economy; and, 7) the rhythm. For the Centre for Digital Storytelling (CDS), these 7 components define DST. Lambert (2013) discusses the larger world of options when storytelling connects to media. Although there is a line of thought for the creation of DST, there are even more possibilities of creation when combining with media and advertising.

According to Ohler (2006), the most important tools that should be used in the creation of DST are writing scripts (digital skills in composing a story) and story treatments (type of writing). Dreon, Kerper, and Landis (2011) build on Ohler's (2006) approach to argue that although the seven elements, advocated by Lambert (2013) and Robin (2008), can describe how storytelling is effective, DST creation process involves leveraging a wide variety of skills, such as researching topics, writing scripts, creating storyboards and also final editing using story building software. Robin (2008) adds that there has been a constant and significant evolution in the way storytelling is produced and suggests that the cause of this evolution lies in the creation and constant improvement of new storytelling development software.

Meehan (1977) and Lebowitz (1984) discussed the theme of developing programs and software that, based on the collection of specific data, develop and construct a story, culminating in two software packages: UNIVERSE and TALE-SPIN. UNIVERSE, the first phase in the development of a storytelling program, aims to understand the cognitive processes that human authors use in the creation of a story and subsequently explore the practical perspectives of computer programs that can influence the formation of a story for educational or entertainment purposes (Lebowitz, 1984). TALE-SPIN, studied by Meehan (1977), consists of simulating and mapping the behaviour of individuals to construct a story. Therefore, Smeda, Dakich, and Sharda (2014) argued that there is no single definition for DST, but that most authors emphasize the technique using multimedia tools. Consequentially, based on references above, it may be said that DST is primarily defined by the format through which it is transmitted and the program used to developed it.

We now live in a world of stories, images, brands, and relationships are developed through many media platforms. Transmedia storytelling (TST) represents the integration of entertainment experiences using the diverse digital platforms (Jenkins, 2010). Currently, engagement with and among audiences has become a need. Engaging with the audience is possible through the transmission, development, and sharing of storytelling simultaneously in the diverse digital platforms. For Long (2007), the combination of storytelling and the types of media simultaneously results in transmedia storytelling. TST occurs on numerous media platforms, and each platform contributes to communicating a story in a distinct and valuable way. From Scolari's (2009) perspective, TST is not only defined by its transmission through various platforms (cinema, comics, television, video games, among others), but also through the different types of language available, verbal, iconic, such as emojis and gifs. The different media and languages participate and contribute to the construction of the TST (Scolari, 2009). Digital evolution and technological disruptions led to TST and also to 'cross-media,' 'multiple platforms,' 'hybrid media,' 'intertextual commodity,' 'transmedial worlds,' 'transmedial interactions,' 'multimodality,' and 'intermedia'. According to Clarke and Adam (2010), DST has been developed in various ways, often shaped by advances in technology. In the last decade, DST has presented itself as a tool for communicating, teaching, researching, and applying personal reflection (Clarke & Adam, 2010).

Attitude

Maze (1973) defines Attitudes (ATT) as rationalizations of individuals' hidden or inconsistent impulses. On the other hand, Ajzen and Fishbein (1977) argue that ATTs are held towards aspects of the individual's world, such as another individual, a physical object, or a behaviour. Although many definitions of 'ATT' have been proposed, most authors agree that an individual's ATT represents the evaluation of an entity of some sort (Ajzen & Fishbein, 1977). However, and even though definitions vary, personality traits and ATTs are typically defined as relatively enduring dispositions that exert a pervasive influence on a wide range of behaviours (Ajzen, 1987).

For Robinson, Stimpson, Huefner, and Hunt (1991), there are two fundamental approaches to the nature of ATT. One approach holds that ATT is a one-dimensional construct and is only represented by the affective reaction. The other is known as the tripartite model, which states the three dimensions of ATTs are affect, cognition, and conation. Olson and Zanna (1993) agree with this approach. Petty and Wegener (1996) argue that in the 1970s, studies on ATTs were quite common, but suffered from conceptual inconsistency. That said, two vital models on ATTs were developed: the Elaboration Likelihood Model (ELM) and the Heuristic-Systematic Model (HSM) (Petty & Wegener, 1996). The ELM, developed by Richard Petty, John Cacioppo, and other researchers, is an example of a 'dual process' approach to the study of persuasion (O'Keefe, 2008). According to Bhattacherjee and Sanford (2006), this model classifies individuals' influence mechanisms from the perspective of the information they process, examining the circumstances by which an individual's ATT is influenced and evaluating the long-term effects of each influence mechanism. HSM is comprised of three foundations for explaining why individuals perform information processes: accuracy, defence, and impression motivation (Zhang, Zhao, Cheung, & Lee, 2014). Thus, Zhang et al. (2014) explain that individuals cognitively consider which of the information they receive is the most relevant, elaborate/evaluate that information, and subsequently make a judgment based on the evaluation they have previously made.

According to Ajzen and Fishbein (2000), ATT is based on relevant and accessible information in the memory of individuals. However, Ajzen (2001) adds that individuals do not form only a single ATT toward an object or situation and that the studies conducted to date are a simplistic conception. Ajzen suggests that when there is a change in ATT, a new ATT rises and consequently overrides the previous one. Eagly and Chaiken (2007) propose a comprehensive definition for ATT change and discuss three significant characteristics, evaluation, object of ATT, and tendency, that together assess an individual's propensity to evaluate an entity (positively or negatively). appraisal refers to all classes of evaluative responses, whether overt or covert, cognitive, affective, or behavioural (Eagly & Chaiken, 2007).

Although effect plays a role in ATT affirmation, the role of cognition has also been widely recognized, the relationship between cognition/beliefs and evaluative ATT is thus integrated into the most popular model of ATT formation and structure, the expectancy-value model (Ajzen & Fishbein, 2000). According to Cohen, Fishbein, and Ahtola (1972), the expectancy-value model analyses an individual's ATT toward a particular object based on the assumption of a cognitive structure composed of beliefs about the potential of that object to achieve or block the achievement of an evaluative state. For Gawronski and Bodenhausen (2006), the studies contain a conceptual weakness, arguing that there are two types of ATTs (explicit and implicit) and should be studied as individual variables. Explicit ATTs consist of the deliberate formation of an ATT. In turn, implicit ATTs are considered automatic ATTs that individuals produce. In other words, models of ATT change do not distinguish between implicit and explicit ATTs and most models that distinguish them consider implicit ATTs to be stable evaluative representations arising from long-term socialization experiences.

Purchase Intention

Researchers use purchase intention (PI) as a proxy measure for purchase behaviour (Morwitz, Steckel, & Gupta, 2007). However, most behavioural and quantitative studies on PI focus on the occurrence of the purchase (the when) and on-brand choice (the what) (Wansink, Kent, & Hoch, 1998). Spears and Singh (2004) argue that PI refers to action tendencies related to the brand. Thus, a concise definition of PI can be as follows: PI is an individual's conscious plan to attempt to purchase a brand, product, or service (Spears & Singh, 2004). Based on the above theories, PI is linked to the claim that consumer behaviour (CB) can be influenced by situational variables, i.e. it can be determined on specific conditions and situations, individual and situational factors (Belk, 1975).

Fishbein and Ajzen (1975) developed the theory of reasoned action (TRA) to better understand and perceive ATTs and behaviours and, consequently, ATT towards a given situation. For Vallerand, Deshaies, Cuerrier, Pelletier, and Mongeau (1992), the TRA describes an individual's behaviour as the behavioural intention that emanates a 'real' behaviour. For the authors, behavioural intention is divided into two main factors, personal and social. The link between individuals' ATT and behaviour is confirmed. Sheppard, Hartwick, and Warshaw (1988) state that studies on situation understanding and prediction are not consistent with the TRA framework developed by Fishbein and Ajzen in the 1970s. Sheppard et al. (1988) argue that there are limitations regarding behaviours of individuals in the TRA. Posteriorly, in the 1990s, Ajzen developed the theory of planned behaviour (TPB). The central factor in TPB is the intention of the individual to perform a particular behaviour. Intentions capture the motivational factors that influence individuals' behaviour (Ajzen, 1991). Madden, Ellen, and Ajzen (1992) add that TBP extends the condition to limiting pure volitional control, specified by TRA. This change is possible because there is inclusion of beliefs about resource ownership and opportunities needed to perform a particular behaviour (Madden et al., 1992). The environmental and marketing stimuli, present in the model of stimulus (S) and response (R), penetrate in the individual's consciousness and, combined with a set of psychological factors that hold certain characteristics of the consumer, lead to decision processes and ultimately to the purchase decision (Keller & Kotler, 2012 p. 172). The psychological factors that influence the consumer to make a purchase decision are motivation, perception, learning, and memory.

For Bettman and Park (1980), the two main aspects in the consumer choice environment are: 1) the large amount of information available about the product; 2) consumers prior experience with the product (Bettman & Park, 1980). Chang and Wildt (1994) consider that pi is formed on a transaction hypothesis and that consequently it is an indicator of the 'real' purchase - the decision making. Keller and Kotler (2012) also highlight the stages of the purchase decision process as problem recognition, demand for information, purchasing decisions, and post-purchase behaviour. Based on Kevin and Kotler's (2013) study, although the purchase decision process involves certain stages, individuals' behaviour is not always linear. According to Hsin and Wen (2008), in the online world, the buyer cannot interact with the company and the products, and so perceived risk is higher for online purchases. In their study, Brorn, Pope, and Voges (2003) compare online shopping with catalogue shopping. Catalogue shopping suggests that convenience is the main reason why consumers use the Internet. However, these claims are unsound, and little empirical evidence is available to support them (Brorn et al., 2003). There are different patterns of information, new ways of evaluating alternatives, and a varied selection of products in online shopping.

Hausman and Siekpe (2009) further note that in addition to convenience, consumers use the internet for the browsing experience, organization and/or structure of the site, ease of task, and entertainment. According to Yoon (2002), most studies on online buying behaviour focus on purchase motives. based on previous studies, Yoon (2002) highlights the reasons why individuals buy online: convenience (88%), price (58%), ability to buy goods unavailable elsewhere (48%), breadth of selection (42%), and sellers (38%) (Yoon, 2002). However, Van der Heijden, Verhagen, and Creemers (2003) argue that online CB can be investigated using frameworks already developed for studying offline/traditional consumer behaviour. This claim is based on the stage model, according to which individuals go through stages until they reach the purchase decision (the purchase decision process). These stages are relatively abstract and do not consider the purchasing environment. Thus, they can be applied to the online environment also (Van der Heijden et al., 2003). However, there are two main differences between online and offline shopping: 1) online consumers need to interact with technology to purchase the goods and services they need, i.e., the physical store environment is replaced by an online shopping environment; 2) a greater degree of trust is required in an online shopping environment than in a physical store (Van der Heijden et al., 2003).

Lu et al. (2016) report that the development of social commerce allows the perceived risks of online shopping to decrease. The decrease is made possible by the development of new features. Social media-based design and Web 2.0 technologies (including recommendation lists, ratings, comments, social proof, and reciprocity applications) increase individuals' participation and allow them to harvest socially information-rich resources, which ultimately result in a more trustworthy and sociable online transaction environment (Lu et al., 2016). In turn, Yoon (2002) argues that certain factors increase consumer satisfaction when they are making an online purchase, thus, four factors used in evaluating a website are crucial in forming trust: transactional security, website properties, search functionality, and personal variables (Yoon, 2002).

Having this said, there was also the need to develop a theory that studied consumer behaviour in an online environment. Thus, Davis (1980) developed the Technology Acceptance Model (TAM). TAM is an adaptation of the TRA used to assess computer acceptance by the user. acceptance is measured by intention and influence of ATT, perceived usefulness, perceived ease of use about the intention to use (Lim, Osman, Salahuddin, Romle, & Abdullah, 2016). In short, PI is still a topic researchers have widely addressed. The phenomenon of PI stems from the social and technological development that has allowed consumers to experience new ways of buying. Therefore, it is necessary to monitor how individuals behave and react to the different stimuli that new ways of buying provide them.

Relationship between Storytelling, ATT and PI

Storytelling helps create a type of atmosphere that provides continuous improvement of individuals (Hensel & Rasco, 1992). For Hensel and Rasco (1992), stories help individuals acquire values and ATTs. Delgadillo and Escalas (2004) believe that there are advantages to using narrative (stories). This advantage is that narrative has exceptional similarities with a 'natural' information structure. This information is acquired naturally through experiences lived by individuals. Word of mouth (WOM) and customer referral marketing are areas where stories can be particularly relevant, given the possibility that they can be retold (Gilliam et al., 2014).

Storytelling positively influences the overall brand equity of the brand, consumers' positive word-of-mouth regarding the brand, and consumers' affective engagement with the brand (Sousa, 2019). Therefore, narrative advertising may be the best way to communicate a brand or service image (Delgadillo & Escalas, 2004). The emotional connection that the brand persona creates for its audience is based on the ability to address deeply felt needs and ATTs, though often poorly articulated and implicit (Herskovitz & Crystal, 2010). Telling a story with personas allows reaching individuals more deeply and, consequently, may influence ATTs and emotions. For Zebregs, Putte, Neijen, and Graaf (2014), storytelling allows individuals to experience moments as if they were real. The affective responses elicited affect intention and, depending on their valence, can lead to a state of action or inaction. According to Hsiao, Lu, and Lan (2013), there is also a positive relationship between ATT and behavioural intention; however, the increased knowledge that individuals currently have about products may affect the development of consumer intention. Laroche, Kim, and Zhou (1996) suggest that consumers' PI is not only determined by an ATT (toward something), but rather by a set of ATTs. A consumer's intention to purchase a specific brand is determined by ATT toward that brand and ATTs toward other competing brands (Laroche et al.,

1996). Some researchers argue that there is a conceptual distinction between ATT and intention, with intentions determined by ATT (Spears & Singh, 2004).

Cardoso (2017) developed a study on the influence of storytelling (as a narrative structure) on consumers' ATTs, perceptions, and behaviours. Cardoso (2017) concluded that the narrative structure positively influences the consumer's connection with the brand and that, in turn, this connection positively influences PI. consumers increase their propensity to purchase a specific brand by establishing and reinforcing their connection with it (Cardoso, 2017). Sousa (2019) found that DST influences consumer behaviour, concluding that: "[...] By influencing the value consumers perceive of the brand, their engagement with the brand and the way they talk about it (the brand) to others, brand storytelling campaigns are promoting a relationship with the brand that will influence purchase intent of consumers" (Sousa, 2019, p. 39).

DEVELOPMENT OF HYPHOTESES AND METHODS

Current research mostly examines storytelling from the perspective of narrative structure or construction technique with inherent characteristics. There are however exceptions to this. Gilliam et al. (2014) researched which characteristics individuals claim to appreciate in a storytelling context in a physical store. They concluded that story's relevance and humour are the main characteristics arousing more interest in individuals. McDonald (2009) argues that there are three principles that the storytelling technique should include, conflict, authenticity, and entertainment. These characteristics are found in storytelling strategies, such as the inducement to action dimension, which has credibility as a characteristic (above in Table 1).

In this study, a new approach to storytelling is taken, consisting in considering storytelling interpretation as a set of strategies, namely, inducing action, communicating 'who you are,' transmitting values, stimulating collaboration, taming rumours, sharing knowledge, and leading people to the future. Unlike other researchers, such as Delgadillo and Escalas (2004) and Morgan and Dennehy (1995), this research, in addition to user characteristics and/or stages in the development of storytelling, integrates techniques that aim to achieve certain objectives using the strategies mentioned above. These objectives are outlined before the strategy selection.

Extant studies show that storytelling strategies positively influence individuals' ATTs and increase their propensity to buy. In addition to evaluating the various storytelling strategies, this comparative study also aims examine if there is a positive relation between these strategies, ATT, and PI. Therefore, the following research hypotheses are proposed:

H1. If individuals are submitted to a situation with storytelling strategies, their ATT towards a certain social object will be more favourable than when submitted to a situation with narrative structure.

H2. If individuals are submitted to a situation with storytelling strategies, their PI towards a certain social object will be more favourable than when submitted to a situation with narrative structure.

Therefore, and bearing in mind that the study has a comparative character, it was necessary to create two scenarios (and questionnaires too): the scenario composed of seven storytelling strategies and the scenario constituted by a narrative structure of events and facts. The two scenarios were developed and presented in text format. Events described were the same for both scenarios.

The sample used in this study is non-probabilistic, with no participant selection process and geometric spread. According to Spreen (1992), this is called the snowball sampling method. It consisted of 130 participants, 66 of whom answered the questionnaire without storytelling strategies, and 64 answered with storytelling strategies. Both questionnaires were created and answered through Google Forms platform, and individuals accessed them through online dissemination on social networks (*Linkedin, Facebook*, and *Instagram*) and also private dissemination, through email contact. Before starting the participation in the study, participants chose one of the presented questionnaires. Questionnaires had both identical label, i.e. before opening did not show any difference, so as not to influence choice or the intended answers. Following data collection, a descriptive analysis was carried out. Descriptive analysis of the storytelling strategies was conducted and subsequently the comparative analysis of two scenarios was performed. The two scenarios allowed evaluation of ATT and PI when used (or not) the storytelling strategies (narrative structure). Thus, it was possible to understand if the storytelling strategies positively influenced participants' ATT and PI (Figure 1). In addition to a new academic approach, this study will also help in the use of storytelling in different business areas, such as destination marketing, allowing more correct use of each of the strategies, depending on the intended purpose.

Figure 1.

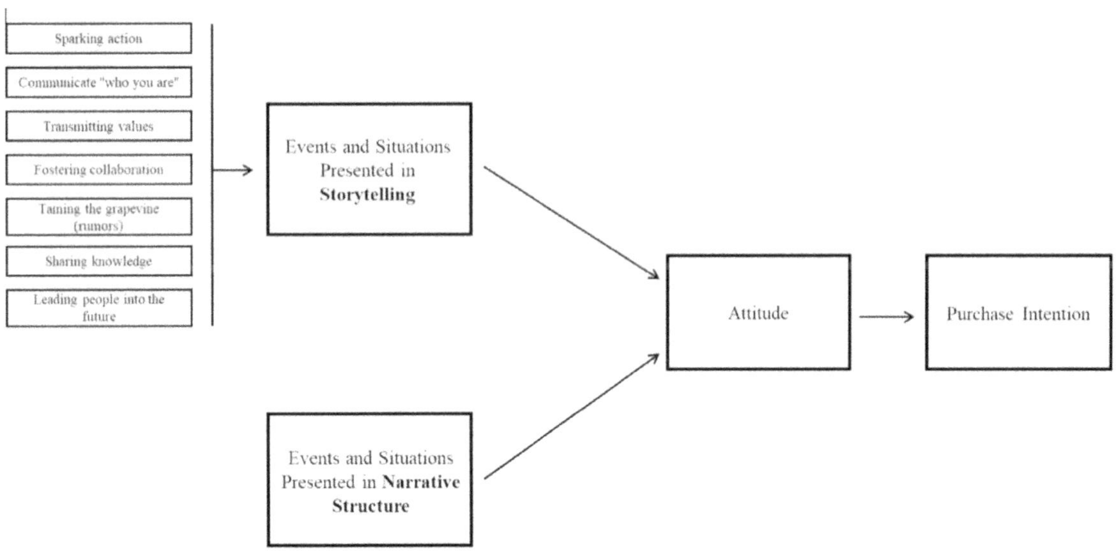

RESULTS

The variables under study and the storytelling strategies obtained positive values of reliability: communicating 'who you are' (0,88), encouraging/fostering collaboration (0,71), taming rumours (0,83), sharing knowledge (0,66) - except the dimensions sparking action (0,56), transmitting values (0,59), and leading people into the future (0,59). However, these variables were considered essential, taking into account the context and nature of the study. Reliability given to ATT (0,88) and PI (0,82) is also acceptable according to Peterson (1994), according to which reliability is unacceptable when it is < 0.6, low from 0.6 to 0.7, moderate to high from 0.8 to 0.9 and high when it is > 0.9. Variables and dimensions applied show the overall sensitivity of the scales. The validity analysis was not developed considering the unidimensionality of the variables and the small sample for the storytelling variable. However, and despite the low significance of the results obtained, the study shows that storytelling strategies influence individuals. Further, it shows which strategies are more important, particularly strategies intended to transmit values, spark action, and lead people into the future.

A GIC Matrix was created to better visualize results on the importance of the storytelling dimensions, which evaluates the degree of importance and agreement of the strategies by comparing the mean values and the standard deviation values obtained. This matrix has four quadrants. Quadrant A represents a high degree of agreement and importance. Quadrant B, a low degree of agreement and high degree of importance. Quadrant C, high degree of agreement and low degree of importance. Quadrant D, a low degree of agreement and importance. The dimensions of storytelling are identified by the initials of the name: SA (sparking action), CW (communicating 'who you are'), TV (transmitting values), EC (encouraging/fostering collaboration), TR (taming rumours), SK (sharing knowledge), and LPF (leading people into the future).

Figure 2.

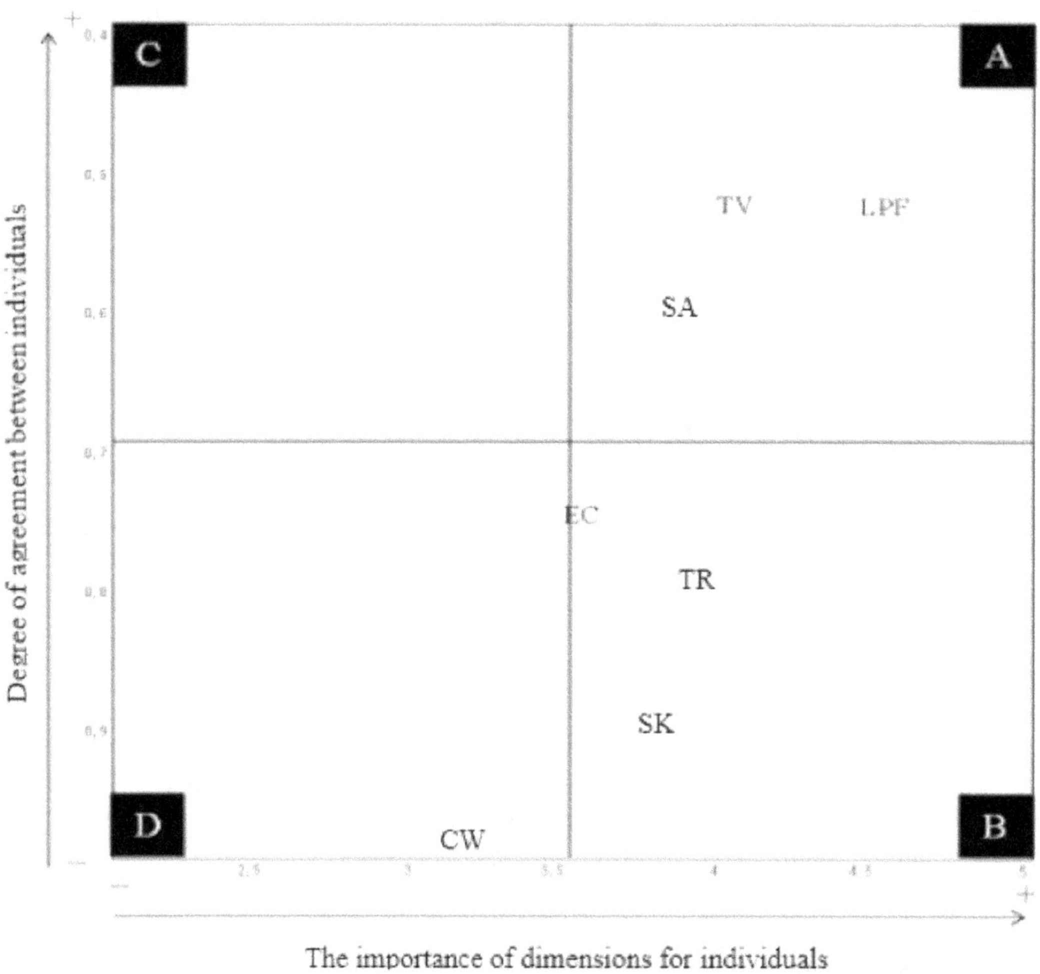

A comparative analysis was conducted to analyse the OUT/ST scenario (without storytelling strategies) and the W/ST scenario (with storytelling strategies). Two graphs were created, one for ATT and one for PI, comparing the means obtained in each item. In general, data show an overall tendency of the averages of the W/ST scenario to be higher than the averages of the OUT/ST scenario (Figure 3 and Figure 4).

Figure 3.

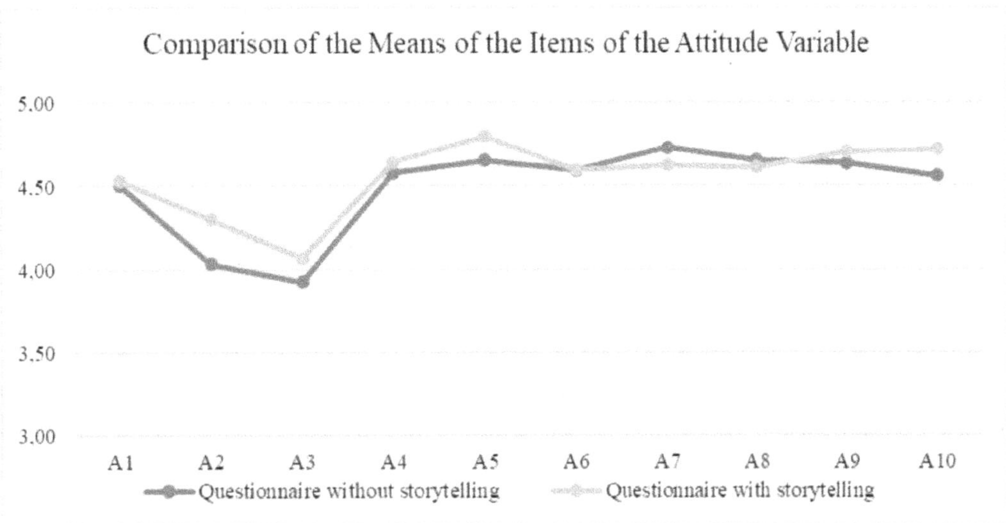

Figure 4.

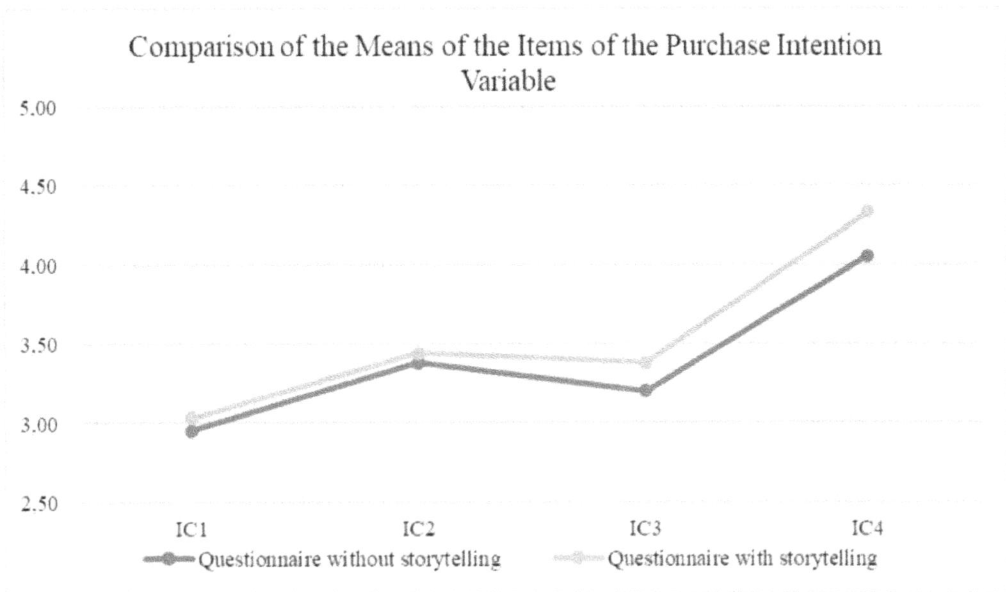

Although results support the study's hypotheses, future iterations of this study may obtain significant values. Thus, it is recommended that future studies consider a larger sample size, as the size of this study's sample made it impossible to carry validity analysis which subsequently limited the reliability analysis.

Authors recommend that questionnaire length to be reduced. Questionnaire length may have complicated the interpretation and retention of the information, and making it more difficult for participants to answer the questions. On the other hand, the scenario with storytelling had a high number of questions. These two characteristics may have contributed to a reduced sample size.

FUTURE RESEARCH DIRECTIONS

It is suggested examination of storytelling influence on an individual's ATT in relation to previously created memories. Ajzen and Fishbein (2000) argued that ATT is based on relevant and accessible information in the memory of individuals. Memory plays an important role in the ATT that individuals may or may not build or modify about something. Likewise, adding the variable 'context' to a future study would contribute to further explore the influence played by context in which storytelling is transmitted.

As storytelling presents a subjective facet, a qualitative analysis of storytelling strategies would be pertinent to explore potential meanings according to participants' own voices. Through this analysis, it would be possible to provide a better evaluation of the stimuli triggered by the storytelling strategies. Accordingly, after reading the texts, individuals should perform the task of replicating in text the story they have read in two different moments in time to understand which events of the story remain in (1) the short-term and (2) in the long term. This experimentation allows comparison of strategies effect in short and long term memory, thus providing a more detailed analysis of the inherent characteristics of the strategies.

The scenario where the storytelling strategies were employed comprised strategies used simultaneously. The individual use of each strategy would allow to understand if obtained results when they are jointly applied together are in line with the results obtained when they are applied individually. Comparison of the same storytelling in various formats, such as video and text, would enable to determine which of the two formats has a greater influence on ATT and/or PI. This suggestion derives from the assumption that stimuli received through video are different from the stimuli received through text.

Another avenue for future research would be to understand the influence of a pre-existing brand/product connection on results in a similar study. Before applying the storytelling strategies, it is important to understand if individuals already have a memory of the brand or product. Even when there is little or no memory bias towards the story events, its essence stems from the effect on subsequent ATTs about the characters and/or brands in the story (Delgadillo & Escalas, 2004). In what concerns PI, it is observed that when storytelling strategies are used, the propensity to purchase increases considerably. This phenomenon may derive from the connection that is created with the brand. The connection to the brand positively influences the PI and, therefore, consumers increase their propensity to buy a certain brand when they establish and strengthen their connection with it (Cardoso, 2017). It would be important to understand if this pre-existing connection with the brand/product maintains the same influence on consumers when storytelling is applied.

CONCLUSION

The present study aimed to examine the influence of storytelling strategies on ATT and PI. Indirectly, it was also analysed the importance each storytelling strategy has for individuals. Therefore, two scenarios were developed, one with the storytelling strategies and another, which presents a narrative structure of events. The techniques developed by Denning (2005 p.18) in his book *The Leader's Guide to Storytelling* were used as a strategic basis for the construction of storytelling. Specifically, in this study *sparking action*, *communicating 'who you are'*, *transmitting values*, *encouraging/fostering collaboration*, *taming rumours*, *sharing knowledge*, and *leading people into the future* were the strategies analysed. A blog format was used to present both scenarios.

Results from each scenario were studied through comparative analysis of variables and respective items. Through analyses, it was possible to understand which of the two scenarios obtained the best results in terms of having the greatest influence on individuals. Subsequently, a descriptive analysis was developed about the storytelling strategies, with the purpose of understanding which ones were more important to this study's participants.

Results supported both hypotheses. Hypothesis 1 suggesting storytelling strategies have a positive influence on ATT was confirmed. Of the two scenarios studied, the scenario where storytelling strategies were applied showed a greater influence on participants' opinion. It is concluded that storytelling strategies positively influence ATT. Results obtained are comparable to those in other studies. Hsiao et al. (2013) conducted a study on storytelling with a sample of 368 individuals and concluded that storytelling communicated in blogs indirectly affects ATTs through development of empathy. Although with a different objective, Herskovitz and Crystal (2010) found that when a person is used in the unfolding of a story (a technique equivalent to the inducement to action strategy used in scenario building with storytelling), deep processes are unconsciously triggered, which influence ATT and emotions. Zebregs et al. (2014) investigated the effects that narrative and statistical evidence have on beliefs, ATTs, and intentions. While the results were marginally significant, they concluded that ATT is influenced by narrative. This study, despite not using storytelling as a variable, is additional evidence that results of this research are reliable. Hypothesis 2 posited the influence of storytelling on PI. Storytelling strategies developed and applied have a greater influence on individuals than narrative structure. Therefore, it is concluded that individuals create greater involvement with the product/service when storytelling strategies are applied in communication.

This study is not the first focusing attention on evaluating individuals' buying behaviour when storytelling is applied. Zebregs et al. (2014) also concluded that storytelling has a great impact on intention. Cardoso (2017) also found that there is a positive influence of storytelling on PI, stating that narrative positively influences brand attachment, memorability, and affective involvement. Brand attachment positively affects PI. Similarly, Gilliam et al. (2014) examined the influence that storytelling may have on individuals' intention to buy. Although the study was developed in a physical context (physical store), it provides evidence that storytelling strategies influence PI. Results in Gilliam et al. (2014) suggest it would be valuable to undertake a comparison between dimensions developed in their study and the strategies examined in this chapter. Most studies exploring relationships between storytelling, ATT and PI corroborate the existence of a positive effect. In the case of ATT, the hypothesis that ATT can be changed or even positively created when storytelling strategies are applied is thus confirmed.

REFERENCES

Ajzen, I. (1987). Attitudes, traits, and actions: Dispositional prediction of behavior in personality and social psychology. In L. Berkowitz (Ed.), Advances in Experimental Social Psychology (Vol. 20, pp. 1–63). Academic Press. doi:10.1016/S0065-2601(08)60411-6

Ajzen, I. (1991). The theory of planned behavior. *Organizational Behavior and Human Decision Processes*, *50*(2), 179–221. doi:10.1016/0749-5978(91)90020-T

Ajzen, I. (2001). Nature and operation of attitudes. *Annual Review of Psychology*, *52*(1), 27–58. doi:10.1146/annurev.psych.52.1.27 PMID:11148298

Ajzen, I., & Fishbein, M. (1977). Attitude-behavior relations: A theoretical analysis and review of empirical research. *Psychological Bulletin*, *84*(5), 888–918. doi:10.1037/0033-2909.84.5.888

Ajzen, I., & Fishbein, M. (2000). Attitudes and the attitude-behavior relation: Reasoned and automatic processes. *European Review of Social Psychology*, *11*(1), 1–33. doi:10.1080/14792779943000116

Baskin, K. (2004). Complexity, stories, and knowing. *International Workshop on Complexity and Philosophy*.

Bassano, C., Barile, S., Piciocchi, P., Spohrer, J. C., Iandolo, F., & Fisk, R. (2019). Storytelling about places: Tourism marketing in the digital age. *Cities (London, England)*, *87*, 10–20. doi:10.1016/j.cities.2018.12.025

Belk, R. W. (1975). Situational variables and consumer behaviour. *The Journal of Consumer Research*, *2*(3), 157–164. doi:10.1086/208627

Benmayor, R. (2008). DST as a signature pedagogy for news humanities. *Arts and Humanities in Higher Education*, *7*(2), 188–204. doi:10.1177/1474022208088648

Bettman, J. R., & Park, C. W. (1980). Effects of prior knowledge and experience and phase of the choice process on consumer decision processes: A protocol analysis. *The Journal of Consumer Research*, *7*(3), 234–248. doi:10.1086/208812

Bhattacherjee, A., & Sanford, C. (2006). Influence processes for information technology acceptance: An elaboration likelihood model. *Management Information Systems Quarterly*, *30*(4), 805–825. doi:10.2307/25148755

Brorn, M., Pope, N., & Voges, K. (2003). Buying or browsing? An exploration of shopping orientations and online PI. *European Journal of Marketing*, *37*(11/12), 1666–1684. doi:10.1108/03090560310495401

Bryan, A. (2011). *The new digital storytelling: Creating narratives with new media*. Praeger.

Budtz, C., Fog, K., & Yakaboylu, B. (2005). *Storytelling: Branding in practice*. Springer.

Cardoso, R. J. M. (2017). *A influência do storytelling (estrutura da narrativa) nas perceções, atitudes e comportamento dos consumidores* [Doctoral dessertation]. Instituto Politécnico de Leiria.

Clarke, R., & Adam, A. (2010). DST in Australia: Academic perspectives and reflections. *Arts and Humanities in Higher Education*, *11*(1–2), 157–176. doi:10.1177/1474022210374223

Cohen, J. B., Fishbein, M., & Ahtola, O. T. (1972). The nature and uses of expectancy-value models in consumer ATT research. *JMR, Journal of Marketing Research*, *9*(4), 456–460. doi:10.1177/002224377200900420

Couldry, N. (2008). Mediatization or mediation? Alternative understandings of the emergent space of digital storytelling. *New Media & Society*, *10*(3), 373–391. doi:10.1177/1461444808089414

David, M. B. (1991). Consulting and change in the storytelling organisation. *Journal of Organizational Change Management*, *4*(3), 7–17. doi:10.1108/EUM0000000001193

Davis, F. (1980). *A technology acceptance model for empirically testing new end-user information systems: Theory and results* [Doctoral dissertation]. Massachusetts Institute of Technology]

Delgadillo, Y., & Escalas, J. E. (2004). Narrative word-of-mouth communication: Exploring memory and ATT effects of consumer storytelling. *Advances in Consumer Research. Association for Consumer Research (U. S.)*, *31*, 186–192.

Denning, S. (2004). Telling tales. *Harvard Business Review*, *82*(5), 122–129. PMID:15146741

Denning, S. (2005). *The leader's guide to storytelling*. Wiley.

Dreon, O., Kerper, R. M., & Landis, J. (2011). Digital storytelling: A tool for teaching and learning in the YouTube generation. *Middle School Journal*, *42*(5), 4–10. doi:10.1080/00940771.2011.11461777

Eagly, A. H., & Chaiken, S. (2007). The advantages of an inclusive definition of ATT. *Social Cognition*, *25*(5), 582–602. doi:10.1521oco.2007.25.5.582

Escalfoni, R., Braganholo, V., & Borges, M. R. (2009). Applying group storytelling to capture innovation features. *Proceedings of the 13th International Conference on Computer Supported Cooperative Work in Design*, 209-214. 10.1109/CSCWD.2009.4968060

Fishbein, M., & Ajzen, I. (1975). *Belief, ATT, intention, and behaviour: An introduction to theory and research*. Addison-Wesley.

Frank, A. W. (2000). The standpoint of storytellers. *Qualitative Health Research*, *10*(3), 354–365. doi:10.1177/104973200129118499 PMID:10947481

Gawronski, B., & Bodenhausen, G. V. (2006). Associative and propositional processes in evaluation: An integrative review of implicit and explicit ATT change. *Psychological Bulletin*, *132*(5), 692–731. doi:10.1037/0033-2909.132.5.692 PMID:16910748

Gilliam, D. A., & Flaherty, K. E. (2015). Storytelling by the sales force and its effect on buyer – seller exchange. *Industrial Marketing Management*, *46*, 132–142. doi:10.1016/j.indmarman.2015.01.013

Gilliam, D. A., Flaherty, K. E., & Rayburn, S. W. (2014). The dimensions of storytelling by retail salespeople. *International Review of Retail, Distribution and Consumer Research*, *24*(2), 231–241. doi:10.1080/09593969.2013.849611

Hausman, A. V., & Siekpe, J. S. (2009). The effect of web interface features on consumer online PIs. *Journal of Business Research*, *62*(1), 5–13. doi:10.1016/j.jbusres.2008.01.018

Hensel, W. A., & Rasco, T. L. (1992). Storytelling as a method for teaching values and ATTs. *Academic Medicine, 67*(8), 500–504. doi:10.1097/00001888-199208000-00003 PMID:1497776

Herskovitz, S., & Crystal, M. (2010). The essential brand persona: Storytelling and branding. *The Journal of Business Strategy, 31*(3), 21–28. doi:10.1108/02756661011036673

Hsiao, K. L., Lu, H. P., & Lan, W. C. (2013). The influence of the components of storytelling blogs on readers' travel intentions. *Internet Research, 23*(2), 160–182. doi:10.1108/10662241311313303

Hsin Chang, H., & Wen Chen, S. (2008). The impact of online store environment cues on PI: Trust and perceived risk as a mediator. *Online Information Review, 32*(6), 818–841. doi:10.1108/14684520810923953

Jenkins, H. (2010). Transmedia storytelling and entertainment: An annotated syllabus. *Continuum: Journal of Media & Cultural Studies, 24*(6), 943–958. doi:10.1080/10304312.2010.510599

Keller, K., & Kotler, P. (2012). *Marketing Management* (14th ed.). Pearson.

Lambert, J. (2013). *Digital storytelling: Capturing lives, creating community*. Routledge. doi:10.4324/9780203102329

Laroche, M., Kim, C., & Zhou, L. (1996). Brand familiarity and confidence as determinants of PI: An empirical test in a multiple brand context. *Journal of Business Research, 37*(2), 115–120. doi:10.1016/0148-2963(96)00056-2

Lebowitz, M. (1984). Creating characters in a story-telling universe. *Poetics, 13*(3), 171–194. doi:10.1016/0304-422X(84)90001-9

Lim, Y. J., Osman, A., Salahuddin, S. N., Romle, A. R., & Abdullah, S. (2016). Factors influencing online shopping behaviour: The mediating role of PI. *Procedia Economics and Finance, 35*(1), 401–410. doi:10.1016/S2212-5671(16)00050-2

Long, G. A. (2007). *Transmedia storytelling: Business, aesthetics, and production at the Jim Henson Company* [Doctoral dissertation]. Massachusetts Institute of Technology.

Lu, B., Fan, W., & Zhou, M. (2016). Social presence, trust, and social commerce PI: An empirical research. *Computers in Human Behavior, 56*(1), 225–237. doi:10.1016/j.chb.2015.11.057

Lundby, K. (Ed.). (2008). *Digital storytelling, mediatized stories: Self-representations in new media*. Peter Lang.

Madden, T. J., Ellen, P. S., & Ajzen, I. (1992). A comparison of the theory of planned behaviour and the theory of reasoned action. *Personality and Social Psychology Bulletin, 18*(1), 3–9. doi:10.1177/0146167292181001

Martin, J., Feldman, M. S., Hatch, M. J., & Sitkin, S. B. (1983). The uniqueness paradox in organizational stories. *Administrative Science Quarterly, 28*(3), 438–453. doi:10.2307/2392251

Maze, J. R. (1973). The concept of ATT. *Inquiry, 16*(1-4), 168–205. doi:10.1080/00201747308601684

Mcdonald, J. K. (2009). Imaginative instruction: What master storytellers can teach instructional designers instructional designers. *Educational Media International, 46*(2), 111–222. doi:10.1080/09523980902933318

Mcgregor, I., & Holmes, J. G. (1999). How storytelling shapes memory and impressions of relationship events over time. *Journal of Personality and Social Psychology*, *76*(3), 403–419. doi:10.1037/0022-3514.76.3.403

Meehan, J. R. (1977). TALE-SPLIN: An interactive program that writes stories. *Proceedings of the Fifth International Joint Conference on Artificial Intelligence*, *77*, 91-98.

Morgan, S. W., & Dennehy, R. F. (1995). Organizational storytelling: Telling tales in the business classroom. *Developments in Business Simulation & Experiential Exercises, 22*, 160-165.

Morgan, S. W., & Dennehy, R. F. (1997). The power of organizational storytelling: A management. *Journal of Management Development*, *16*(7), 494–501. doi:10.1108/02621719710169585

Morwitz, V. G., Steckel, J. H., & Gupta, A. (2007). When do PIpredict sales? *International Journal of Forecasting*, *23*(3), 347–364. doi:10.1016/j.ijforecast.2007.05.015

Ohler, J. (2006). The world of digital storytelling. *Educational Leadership*, *63*(4), 44–47.

Olson, J. M., & Zanna, M. (1993). ATTs and ATT change. *Annual Review of Psychology*, *44*(1), 117–154. doi:10.1146/annurev.ps.44.020193.001001

Pera, R., & Viglia, G. (2016). Exploring how video DST builds relationship. *Psychology and Marketing*, *33*(12), 1142–1150. doi:10.1002/mar.20951

Peterson, R. A. (1994). Meta-analysis of Alpha Cronbach's Coefficient. *The Journal of Consumer Research*, *21*(2), 381–391. doi:10.1086/209405

Petty, R. E., & Wegener, D. T. (1996). ATT change: Multiple roles for persuasion variables. In D. Gilbert, S. Fiske, & G. Lindzey (Eds.), *The handbook of social psychology* (Vol. 1, pp. 323–390). McGraw-Hill.

Robin, B. R. (2008). Digital storytelling: A powerful technology tool for the 21st century classroom. *Theory into Practice*, *47*(3), 220–228. doi:10.1080/00405840802153916

Robinson, P. B., Stimpson, D. V., Huefner, J. C., & Hunt, H. K. (1991). An ATT approach to the prediction of entrepreneurship. *Entrepreneurship Theory and Practice*, *15*(4), 13–32. doi:10.1177/104225879101500405

Rossiter, M., & Garcia, P. A. (2010). Digital storytelling: A new player on the narrative field. *New Directions for Adult and Continuing Education*, *126*(126), 37–48. doi:10.1002/ace.370

Sarica, H. Ç., & Usluel, Y. K. (2016). The effect of DST on visual memory and writing skills. *Computers & Education*, *94*, 298–309. doi:10.1016/j.compedu.2015.11.016

Scolari, C. A. (2009). Transmedia storytelling: Implicit consumers, narrative worlds, and branding in contemporary media production. *International Journal of Communication*, *3*, 586–606.

Sheppard, B. H., Hartwick, J., & Warshaw, P. R. (1988). The theory of reasoned action: A meta-analysis of past research with recommendations for modifications and future research. *The Journal of Consumer Research*, *15*(3), 325–343. doi:10.1086/209170

Smeda, N., Dakich, E., & Sharda, N. (2014). The effectiveness of DST in the classrooms: A comprehensive study. *Smart Leaning Environments*, *1*(6), 1–21. doi:10.1109/ICALT.2013.157

Sole, D., & Wilson, D. G. (2002). Storytelling in organizations: The power and traps of using stories to share knowledge in organizations. *Training & Development*, *53*, 44–52.

Sousa, C. S. M. (2019). *A influência do DST no comportamento do consumidor: O caso da Lemon Jelly* [Doctoral dissertation]. Instituto Politécnico De Leiria.

Spears, N., & Singh, S. (2004). Measuring ATT toward the brand and PIs. *Journal of Current Issues and Research in Advertising*, *26*(2), 53–66. doi:10.1080/10641734.2004.10505164

Spreen, M. (1992). Rare populations, hidden populations, and link-tracing designs: What and why? Bulletin of Sociological Methodology. *Bulletin de Methodologie Sociologique*, *36*(1), 34–58. doi:10.1177/075910639203600103

Vallerand, R. J., Deshaies, P., Cuerrier, J.-P., Pelletier, L. G., & Mongeau, C. (1992). Ajzen and Fishbein's theory of reasoned action as applied to moral behaviour: A confirmatory analysis. *Journal of Personality and Social Psychology*, *62*(1), 98–109. doi:10.1037/0022-3514.62.1.98

Van Der Heijden, H., Verhagen, T., & Creemers, M. (2003). Understanding online PIs: Contributions from technology and trust perspectives. *European Journal of Information Systems*, *12*(1), 41–48. doi:10.1057/palgrave.ejis.3000445

Wansink, B., Kent, R. J., & Hoch, S. J. (1998). An anchoring and adjustment model of purchase quantity decisions. *JMR, Journal of Marketing Research*, *35*(1), 71–81. doi:10.1177/002224379803500108

Wilkins, A. L. (1984). The creation of company cultures: The role of stories and human resource systems. *Human Resource Management*, *23*(1), 41–60. doi:10.1002/hrm.3930230105

Woodside, A. G. (2010). Brand-consumer storytelling theory and research: Introduction to a psychology & marketing special issue. *Psychology and Marketing*, *27*(6), 531–540. doi:10.1002/mar.20342

Yoon, S. J. (2002). The antecedents and consequences of trust in online-purchase decisions. *Journal of Interactive Marketing*, *16*(2), 47–63. doi:10.1002/dir.10008

Zebregs, S., Putte, B., Neijen, P., & Graaf, A. (2014). The differential impact of statistical and narrative evidence on beliefs, ATT, and intention: A meta-analysis. *Health Communication*, *30*(3), 282–289. doi:10.1080/10410236.2013.842528 PMID:24836931

Zhang, K. Z. K., Zhao, S. J., Cheung, C. M. K., & Lee, M. K. O. (2014). Examining the influence of online reviews on consumers' decision-making: A heuristic-systematic model. *Decision Support Systems*, *67*, 78–89. doi:10.1016/j.dss.2014.08.005

ADDITIONAL READING

Hamelin, N., Thaichon, P., Abraham, C., Driver, N., Lipscombe, J., & Pillai, J. (2020). Storytelling, the scale of persuasion and retention: A neuromarketing approach. *Journal of Retailing and Consumer Services*, *55*, 102099. doi:10.1016/j.jretconser.2020.102099

Luth, M. (2018). *The Best Story Wins: How to Leverage Hollywood Storytelling in Business and Beyond.* Morgan James Publishing.

McErlean, K. (2018). *Interactive narratives and transmedia storytelling: Creating immersive stories across new media platforms.* Routledge. doi:10.4324/9781315637570

Tsai, M. C. (2020). Storytelling Advertising Investment Profits in Marketing: From the Perspective of Consumers' PI. *Mathematics*, *8*(10), 1704. doi:10.3390/math8101704

KEY TERMS AND DEFINITIONS

Attitude: Is an act or behavior of a person in relation to another person, objects, or specific situations.

Causal-Comparative Methodological Approach: Confirmatory study adopting a methodological approach and comparison between two or more scenarios.

History: The method by which a series of events is told, directly connected with sharing the evolution of humanity, with the critical and detailed narration of social, political, and economic facts.

Narrative Structure: The detailed account of a fact or sequence of events/events.

Purchase Intention: The desire/willingness of a person to acquire/buy a product or service.

Storytelling: The way a story is told with the aim of creating impact or triggering an attitude, be it emotional, social, or behavioral.

Storytelling Strategies: The methods used to tell a story.

Section 2
Storytelling and Destination Marketing and Management

Chapter 7
Promoting Tourism Destinations Through Storytelling

Emmanuel Nii-Ayi Solomon
Accra Technical University, Ghana

Lorna Kankor Adu-Debrah
University of Ghana, Ghana

Stephen Mahama Braimah
University of Ghana, Ghana

ABSTRACT

Tourism destinations must be encouraged to tell their stories and share them with visitors. The objective of this study is to, firstly, increase the understanding of the role of storytelling in tourism destination promotion and, secondly, to ascertain the impact of storytelling on tourism destination promotion. In this chapter, tourism destinations are approached through the intriguing stories that are told about them. Stories have the potential to enhance and promote the qualities of a destination; hence, the chapter discusses the potential of storytelling as a strategic tool in "wrapping" or promoting destinations. It also provides the platform for discussion between various stakeholders at the destination in finding creative ways of promoting the destination. This chapter also discusses place storytelling and emphasises its relevance in promoting tourism destinations, and how stakeholders such as tour guides are employed as storytellers to promote key attributes of the destination.

INTRODUCTION

Tourism destinations compete to attract tourists to remain relevant and competitive (Mossberg et al., 2010). On the other hand, tourists that visit these destinations hope to have an unforgettable, unique, and superior experience which is largely viewed as very relevant to the strategic goal of tourism expansion and development, particularly concerning destination attractiveness and sustainability (Perovic et

DOI: 10.4018/978-1-6684-3436-9.ch007

al., 2018). Generally, destinations are seen as "a well-defined geographical area, such as a country, an island or a town" (Hall, 2000; Davidson & Maitland, 1997). However, it is increasingly recognized that a destination can be considered a perceptual concept, interpreted subjectively by consumers, depending on their travel itinerary, cultural background, purpose of visit, educational level, past experience (Buhalis, 2000), and to a large extent the stories that are told about it.

Most destinations are characterized by iconic structures and buildings like the Elmina castle and Gamla Stan, natural attractions, such as the Ngorongoro Crater and Tsavo East National Park, and amusement parks, like Legoland and Gold Reef City, all serving as a tool in destination development and marketing (Mossberg et al., 2010). In the area of destination promotion, it appears that most places lack major attractions that can become a strong marketing tool for destination promotion; hence there is the need to identify new creative and innovative ways to market the attractiveness of the destination, and to that end the storytelling approach is here analysed.

Stories have become an integral part and a centrepiece in destination marketing and tourism business promotion (Moscardo, 2020). According to Ben Youssef et al. (2019), this is a result of the increased usage of new technologies leading to a situation where travellers' decision is influenced by narratives and stories about the destination they seek to visit. Several studies have shown that whether a story about a destination is real or fictive, it elevates the uniqueness of the destination and gives tourists a more meaningful and exciting experience (Chronis, 2005; Mossberg, 2008). Destinations can therefore be recognized as *storyscapes*, i.e. "commercial environments where narratives are negotiated, shaped and transformed through the interaction of producers and consumers" (Chronis, 2005, p. 389).

Stories have been featured heavily in many different campaigns for tourism destinations. According to Moscardo (2020, p. 2), in 2016, "Abu Dhabi invited viewers to create their own story in the 'Your Extraordinary Story' campaign, in 2018 Booking.com told us 'These are our stories. What will yours be? Book your next story', the Korean Tourist Organization invites visitors to 'Let your story begin', and Airbnb encourages tourists to tell their stories using the hashtag #airbnb and then selects these guest stories to present in their own social media, adding to the stories from hosts in their website 'Stories from the Airbnb Community'. Tourism Australia has storytelling as one of nine 'philausophies' in its 2019–2020 promotional campaign, which includes selling a book of stories from famous Australians. These examples are focused on stories as a promotional tool and generally include destination stories and sometimes tourist stories".

According to Choi (2016, p. 1), storytelling is a "combined action of delivering a destination's story by appealing to a vivid story persuasively through using senses such as vision, hearing, and smell". Ho and Mi (2008) support the view that destination or place storytelling has a deeper meaning which is embedded not only in the verbal narrative category and folktales but also through nonverbal means as a form of communication which includes movies, dance, dramas relatable to the destination's offerings. This becomes a significant symbol shared by tourists and locals in creating a unique and memorable tourism experience that can be shared by all. Stories can be employed as a promotional tool for tourism destinations; but in some cases, the co-production or creation process must be considered. The co-creation process is very relevant because the intention of most tourism practitioners is focused on ascertaining and telling stories to tourists rather than helping tourists to create their own stories (Moscardo, 2020).

Research has shown that storytelling is enhancing the value of destinations by stimulating tourism activities (Järv 2010; Akgün et al., 2015). It is not surprising that stories have been given special attention in tourism practice, particularly as a promotional tool for boosting destination image and tourists experience (Moscardo, 2018). The surprise, therefore, is that stories have only recently been examined in more detail in tourism research. According to Moscardo (2020, p. 1), "the lack of attention to stories in tourism research may mirror the influence of twentieth-century modernist, positivist approaches to research as tourism researchers strive to establish some sense of legitimacy within traditional academic institutions". Polletta et.al., (2011, p. 110) argue that some academics and institutions still see stories as "trivial, ...entertaining but unserious", hence the less attention given to it. It is again not surprising that stories have been employed in functional areas such as public education promotions (Rhodes, Toole, & Arpan, 2016; Steinemann et al., 2017), staff training (Gill, 2015), and advertising (Yu & Chang, 2013). However, systematic research aiming to analyse the storytelling methods of different tourism destinations on a theoretical level is scarce (Zavattaro et al., 2015). This raises the question of how destination marketers evaluate the success of their respective tourism destinations.

The objective of this chapter is to: first, increase the understanding of the role of storytelling in tourism promotion and second, ascertain the impact of storytelling on tourism destination promotion. In this chapter, tourism destinations are approached through the intriguing stories that are told about it. Stories have the potential to enhance and promote the qualities of a destination; accordingly, the chapter discusses the potential of storytelling as a strategic tool in "wrapping" or promoting tourism destinations. It also provides the platform for discussion between various stakeholders at the destination in finding creative ways of promoting the destination.

STORYTELLING AND ITS IMPACT ON DESTINATION PROMOTION

It is widely appreciated that the tourism industry these days is gradually striving to differentiate itself to gain a competitive edge because of the increasing activities of tourists at tourism destinations which, to an extent, has created competition between destinations and those that develop tour packages (Woodside and Martin, 2015). As a result, most destinations are beginning to employ storytelling as a promotional strategy to deliver brand value and enhance the value of their respective destinations (King, 2002).

Many authors (including Chronis, 2012; Kim et. al, 2016) have identified the link between storytelling and its relevance in the promotion of places and tourist experiences. In promoting tourism destinations, Destination Marketing Organizations (DMOs) seek to create and communicate interesting stories to attract people to visit their destination and to promote tourism services (Stevenson, 2019). They are encouraged to develop rich stories for a wide range of visitors that stimulate positive emotions and enhance the perceived authenticity of the destination (Kim et. al, 2016). These stories are told in a culturally unique way to appeal to the senses of tourists (Rojek & Urry, 1997) combining images, words, tastes, dance, music, dramatic expressions, smells, and creating emotions that are engaging to their imagination and emotion. For example, dark or morbid tourists may want to visit historical sites associated with death and tragedy due to the stories they have heard about them. In 2009, President Obama and his family visited the monstrous slave castle in Cape Coast, Ghana, to trace his ancestral root and listen to the infamous stories of the Trans-Atlantic slave trade. Ghana as a tourism destination promotes heritage tourism and prides itself to be the destination having the oldest and surviving European structure (Elmina Castle) in sub-Saharan Africa. The Elmina castle was built by the Portuguese in 1482 as **Castelo de São Jorge da Mina** (*St. George of the Mine Castle*) as the first trading post built on the Gulf of Guinea.

In promoting the attractiveness of such tourism destinations, destination managers have adopted the use of storytelling as a key approach to marketing their destinations (Chronis, 2012; Lund, Cohen, & Scarles, 2017). According to Tobin (2006), stories and telling them can help define human existence.

Sole and Wilson (2002, p.73) assert that "stories including myths, legends, and folktales pass on wisdom, knowledge, and culture for thousands of years". Storytelling activities are therefore seen to be pervasive throughout human history, not excluding day-to-day dealings, at meals, while traveling, in songs, in captured images about the destination by photographers, and even while dreaming. "Storytelling begins with the very history of mankind, there has never existed anywhere a people without stories" (Barthes et al., 1969). Hence, stories are employed as a device aimed at stimulating tourists' interest and preserving memories, or sometimes to make sense of ongoing change, interpreting and reinterpreting the present by looking at the past and the future (Dawson & Sykes, 2018).

Stories have therefore been seen as a central part of people and have been discussed quite extensively in marketing and advertising literature as they impact consumers' experiences through the emotions they exhibit and elicit (Chronis, 2008). On the other hand, storytelling has been used by destination managers and marketers as a promotional tool to communicate a destination's attributes, engage tourists' imagination, and influence their expectations and attitudes about the destination (Escalas, 2007; Tussyadiah and Fesenmaier 2008). According to Mossberg (2008), destinations do not market themselves, stories market destinations hence stories about tourism destinations can enhance the image of the destination thereby making it competitive and can make tourists' visits more meaningful. Tourism destinations are actively using storytelling strategies to deliver the brand value of destinations and enhance their competitiveness as the role played by storytellers cannot be underestimated (Bassano et al., 2019).

Most of these destinations achieve this aim by employing tour guides as destination storytellers. Burner (2005) supports this assertion by claiming that tour guides are powerful stakeholders that bring to life the history and culture of people and places even to the extent that they compete with professional historians with their historical narrations. The role played by these tour guides at tourist destinations is crucial as they are seen as individuals who construct historical narratives and direct tourists' attention and gaze towards the images they wish to create about a destination's buildings or memorials (Hanna et al., 2004). Iles (2008) is of the view that "most tourists, especially visitors at dark tourists sites, are not there just to sightsee but also to identify and empathize and it is the job of tour guides to capture their client's emotional engagement with the area in addition to providing them with comprehensive accounts of the history" (Iles, 2008, p. 151). According to Reed (2002), the skill of storytelling is very important for tour guides as they must be able to paint the real image of a destination through their narratives. Salazar (2005, p. 642) also reiterated this point by saying that tour guides must be focused on being "agents of 'glocalization' in 'the way they (re)present and actively (re)construct local culture for a diversified global audience". In using stories to elicit feelings and emotions, it is imperative to consider the character of tour guides and their ability to create historical empathy among tourists by retelling the past in selective and emotionally evocative ways (Bryon, & Bryon, 2012).

In recent years, studies on storytelling have not been given the needed attention, instead, there is more focus on consumption and effects of destination brand storytelling (Ben Youssef et al., 2019). According to Pike (2007), there has been a significant increase in investments into destination branding since the 1990s but there seems to be relatively limited evidence on how destination marketers are promoting their destinations using storytelling (Hanna & Rowley, 2013), particularly with tour guides as the storytellers. The tourist destination which is an object of sightseeing has to a large extent been treated as branded merchandise, carrying cognitive value for the tourists (Hee, 2001). It is therefore worthy to note that what then makes one destination different from the others is the story that is told about it. Bassano et al. (2019) posit that, across centuries, people have always told each other stories about the places they are from, the places they live in, and the places they have visited. People like to share their stories and, in the same vein, they enjoy listening to stories. People who live in a specific place have considerable experience in that place and often develop a deep love for that place which is often carried out through storytelling. Increased travel and competitiveness between destinations have led destination marketers to use place storytelling as a promotional tool in marketing their destination (Bassano et al., 2019). To communicate the experiential value of the destination, storytelling is therefore given the greatest consideration.

For these reasons, this chapter of the book examines the methodological approaches of destination marketers towards storytelling. Moreover, this chapter also has the objective of identifying and examining the thematic contents of storytelling, its strategic purpose, its use as a promotional tool, and its impacts. Indeed, how do destination marketing practitioners conceptualize storytelling in promoting a destination? How does storytelling contribute to destination sustainability and promotion? And what are the gains expected from promoting destinations through storytelling?

In response to these questions, recent literature calls for better integration of place branding theory into practice by analyzing theoretically how the concepts become apparent or not (see Hanna & Rowley, 2013; Pike, 2007) in terms of destination promotion. Tourism destinations are characterized by an amalgam of stakeholders and the concept of developing storytelling necessitates close contact or relationship with different stakeholders such as Destination Management Organizations (DMOs), public administration at the municipality or regional level, public and private attractions, various types of tourism-related service providers and actual storytellers (Mossberg et al., 2010). Bringing together these interest groups and ensuring their commitment to the storytelling concept seems to be an important step toward promoting tourism destinations (Therkelsen & Halkier, 2010), just as providing the stakeholders with a means of mutual communication.

Generally, destination marketing activities, in the long run, must lead to the optimization of tourism impacts and the achievement of the overall objectives set by the stakeholders involved. Nevertheless, to have a strong impact in employing storytelling, it needs to be used effectively not only in the promotion of tourism destinations as a unique product but also in the design and creation of the product and how it is delivered. This means that for the destination to achieve its objectives, the messages embedded in the story told by tour guides or other storytellers at the destination must be presented in such a way to communicate the key attributes of the destination to the tourists by actively involving them in participating in the interpretation and creation of the destination experience.

A study conducted by Lichrou et. al. (2010, p. 135) found that "stories are a useful conceptual tool for the diverse interpretations of place meanings, which are important to marketers, consumers, and other social actors involved in a place". Aesthetic aspects that are sometimes presented in storytelling such as still pictures, videos, and audio increase the story's appeal to the listeners (Hsiao et al., 2013). This helps tourists to make sense of what they see and feel at the tourism destination. Moreover, the perceived aesthetics, story structure, and self-reference are key elements of stories that influence the tourist experiences, mostly because these elements evoke positive emotions and empathy (Akgün et al., 2015). Asides from the physical attractions like natural beautiful landscapes, historical places, monuments, etc., the intangible qualities of the stories told makes the destination brand different, valuable, and positively perceived (Korez-Vide 2017).

The use of storytelling does not impact single destinations or local economies alone but also nations and their economies and this was confirmed by Hartman et al. (2019) who argued that strategic storytelling was an effective catalyst for stimulating transitions towards tourism destinations in the Netherlands.

In this vein, Olsson et al. (2016) found that storytelling by volunteers disseminated local knowledge and substantially contributed to the success of festivals in Nordic countries.

Figure 1. Stories in tourism system
(Source: Moscardo, 2020)

Also, countries have used stories in films to create exciting images as attractive destinations for tourists. Strielkowski (2017) found causality between the popularity of films and the number of tourists in destinations that are most impacted by those films. The study of the image created by stories in films has impacted the decisions of tourists all over the world positively in favour of the destinations within which these stories were told. The researcher went on to say that the two best examples of this fact are the popular film series "Harry Potter" and "Lord of the Rings" set in the United Kingdom and New Zealand. So far storytelling has been mostly used in tourism as a destination marketing tool, i.e. stories in destination marketing are primarily used to develop a destination brand but also to develop new products and experiences (Mei et al., 2018). However, its potential is huge in the process of conveying different messages to the target audience and creating unforgettable tourist experiences that impact the performance of the destination favourably (Korez-Vide, 2017; Yavuz et al., 2016).

Storytelling activities when placed beyond the main tourism season of Christmas, Summer, and other extensions of the season may impact the destination favourably. This may result in an extra-economic sustainable tourism trade-off season. Aside from this, it will increase employment opportunities in the area of guided tours. It will also create or preserve the culture, traditions, and value systems of the local area.

DESTINATION SUSTAINABILITY THROUGH STORYTELLING.

Many players in the tourism industry have become environmentally conscious due to the declining economic situation fuelled by globalization. These tourism stakeholders are encouraged to share in the global campaign of tourism destinations in maintaining the "naturalness" of their destination. There is therefore the need to create pockets of stories that encourage both locals and tourists to join in this campaign (Hassan, 2000). Discussions bordering on sustainability are essential for the wellbeing and protection of the monuments, heritage sites, and the environment for destination promotion and sustainability (Bramwell & Lane, 1999; WCED, 1987), especially among emerging tourism markets.

For most transitioning markets, the promotion of sustainable tourism has become a panacea for providing economic benefits to residents and visitors with superior destination experience and better environment engagement (Lee, 2013; Lepp, 2007; Shrestha, Stein, & Clark, 2007). It is important to note that tourism destinations that seek to have tourism as their core product in growing their economic basket must develop stories that strategically position their destination to address some of their sustainability issues (Puczko & Ratz, 2000).

From the WTO Regional Tourism Results (2019), Europe accounts for half of the world's international arrivals, followed by Asia and the Pacific with one in four arrivals. Again, Europe represents almost 40% of international tourism receipts followed by Asia and the Pacific with almost one-third. From the statistics above it is therefore estimated that about 70% of tourists visit the three major tourism destinations while 30% share the other destinations. This means that globally a few established destinations enjoy the bulk of the revenue generated in the global travel business, as such there is a need for the owners and management of undeveloped destinations to formulate some strategic plans aimed at attracting tourists to these destinations. In this vein, Oye et. al. (2013, p. 9) argued that "to compete in the tourism field, a destination needs not only comparative advantages but also competitive advantages". In other words, to compete, you must not only have a variety of products and tourism resources but also these products and resources must be managed efficiently to serve the medium- and long-term plans of the destination. This is mainly because destination attractions or location alone is not enough to attract tourists in today's global village as the internet makes available countless destinations to tourists. There must be something special or unique to draw tourists to a particular destination. Currently, tourists are drawn to destinations due to attractiveness that has a perceived ability to meet their needs. Destinations that portray this attractiveness go on to become popular and draw more visitors thereby sustaining themselves over time.

Sustaining tourism destinations is not only important to the destination managers alone but also to the government of the nations. Tourism has continually been seen as an avenue for creating employment, increasing foreign exchange earnings, poverty alleviation, and sustainable human development amongst others. Most countries today see tourism as a very important segment of the economy.

According to Statista Research Department (Dec 9, 2021), the number of international tourist arrivals worldwide declined sharply in 2020 over the previous year, as the Coronavirus pandemic (COVID-19) hit international tourism hard due to fear and also travel restrictions. Europe recorded the highest figure in 2020 with inbound arrivals in that region dropping from over 746 million in 2019 to roughly 236 million in 2020. In terms of region of origin, Europe provided the most international tourists as of 2019, proving that Europe as a region is a force to reckon with when it comes to international tourism. Other statistics show that the travel and tourism industry made a total contribution of 9.3 trillion U.S. dollars to the global economy, directly contributing 2.9 trillion in 2019. On the expenditure side, China's international tourist expenditure amounted to nearly 255 billion U.S. dollars in 2019. Even though the statistics show increasing trends in the growth of tourism globally there is also the issue of decreasing numbers of visitors, overnight stays, and sales in some tourism destinations across the globe. Just like in the marketing of any product, it is now a requirement in tourism to offer unique products to be a successful destination. A destination must be able to differentiate itself from the competition to be economically sustainable.

Accordingly, Choi (2016) concluded that stories about the culture and history of tourist destinations make them attractive to foreign tourists. There are various ways of influencing the consumption behaviour of tourists favourably towards the sustainability of a particular destination and one of such is the use of stories or storytelling. This concept has sparked renewed interest and debate in different fields of research and practice. Often stories convey information, explain problems, and evoke emotions (Fischer et al., 2020). Also, Fischer et al. (2020) posit that there is the need to probe into the probable benefit of storytelling in a sustainability context, however, there seems to be a struggle with conceptual ambiguity about what distinguishes storytelling from other pedagogical approaches, and how it best relates to sustainability issues.

It must be noted that "sustainable development is a concept that highlights the dynamic interaction processes of natural and social systems while focusing on intergenerational justice" (Clark & Dickson, 2003, p. 8059). On the other hand, several scholars have argued for narrative forms of communication as a promising approach to achieving these multifaceted demands (Reinermann, Lubjuhn, Bouman, & Singhal, 2014) of destination sustainability through storytelling. Storytelling has emerged as a narrative strategy for addressing issues related to sustainability where storytelling can provide concrete examples of otherwise abstract and inaccessible facts and evoke emotions by tapping into archetypical patterns for the transmission of knowledge and orientation (Wilson & Zipes, 2005). Storytelling is expected not only to convey content in an interesting, attractive, comprehensible, and motivating way (Dahlstrom, 2014), but also to provide interpretative patterns via the narrative framework that can facilitate reframing and building shared understanding (Black, 2013).

Stories are the best medium for influencing the mind and most of the time the actions of others. Stories have been used by our ancestors to teach morals and to shape behaviours over centuries. Denning (2005) stated that the 'tell' in storytelling does not only mean to 'speak', but also includes all other senses, such as the sense of touch and smell and the 'ing' has a sharing of the situation following the meaning of interactions. Accordingly, the concept of storytelling in tourism destination marketing should be understood from a broader perspective than just as a piece of information from a tour guide or cultural tourism interpretation. Stories are created and communicated in various ways to attract people to destinations and to promote the destination sustainably. Again, Kim and Young (2019) posit that developing rich stories aimed at different audiences stimulates positive emotions and enhance perceived genuineness and memorability.

There is therefore the need to propose a systematic understanding of how storytelling can be employed for the sustainability of tourism destinations (Fischer et al., 2020). Some certain historical artefacts and monuments can be preserved through the stories that are told about them. According to Lekić, (2017, p. 61) as quoted by Radić and Munir (2021), "destination management companies play an important role in this regard by creating unique and memorable experiences for tourists". In creating unique and complete tourist experiences, since there are no two destinations with equal resources and stakeholders involved in the experience creation process, differentiation based on the creation of a unique experience becomes an acceptable strategy to gain a competitive advantage (Mak, 2017). Moscardo (2020, p. 2) argues that "stories have always been part of tourism, most clearly in guided tours and attractions built around either a story in popular culture (such as Disneyland and tours to places associated with novels, films, and television) or presentations of destination history and culture". Disneyland is a great example of exclusively using storytelling to sell a destination to tourists. Moscardo (2020, p. 4) further argues that "the audit of stories in current tourism practice reveals, however, the rising explicit use of tourists' stories in tourism destination promotion is clear": from Abu Dhabi's "Your Extraordinary Story" campaign, and Korea's invitation to "Let your story begin" through to Poland's "Move your imagination – come and find your story."

Bassano et. al. (2019) concluded that storytelling in the tourism sector can be employed in defining the identity and improving the reputation of tourism service systems. In recent times, the dynamism of customer behaviour and the continuous research of new experiences have been modifying rules for packaging a story of a destination and its identity. In line with this conclusion, tourism destinations are gradually strategizing and securing the destination advantage and sustainability through storytelling to satisfy the motives and desire of visitors and to achieve the objective of management in the long run. According to Fischer et al. (2020), storytelling for sustainability must contribute to change towards sustainability, and build capacity in locals and tourists to act in a self-determined, competent way. It should foster awareness, challenge assumptions, clarify values and ideas of what kind of sustainability we want, and empower individuals and groups to act accordingly.

Most destinations come complete with amazing attractions that make them easier to sell or promote and the quality of these attractions is decisive for success, but in recent times the question of sustainability has been raised. Monuments and buildings like the Eiffel tower in Paris, the Statue of Liberty in America, the Colosseum in Rome and natural phenomena such as the Blue Lagoon in Iceland, the Yellow Stone National Park in the USA, the Victoria Falls Zimbabwe, Shai Hills in Ghana and amusement parks such as Legoland across the globe, and Astrid Lindgrens Värld in Sweden are all examples of well-known attractions that make destination development and marketing relatively easier. Across the globe there are many tourism destinations that lack these kinds of major attractions, leading to huge challenges due to limited resources and lack of investors. For such destinations, it is prudent to identify innovative ways of developing and marketing them. One of the most unique methods of achieving such an aim and sustaining them is using storytelling.

Storytelling is increasingly gaining global attention and setting marketing trends in the tourism industry with studies showing that real or fictional stories about the destinations have the potential of not only giving the destination a unique competitive advantage but also a unique and meaningful experience for tourists (Chronis, 2005; Mossberg, 2008). Creative storytelling is defined as "the development of unique narratives able to: 1) create a sense of community and belonging, 2) engage and interact with the audience, 3) transform followers into promoters, 4) inspire, show the creative process, ideas, invite the audience to be creative, etc." (OECD, 2014, p. 35).

Tourism offerings across the world are often portrayed as being unique but are merely copies of each other making stories and storytelling one of the very few unique approaches many destinations have at their disposal. Increasingly storytelling is gaining attention as a means of competitive branding that is unique to a destination (Kotler, Haider, & Rein, 1993). Storytelling as an approach to destination branding communicates the culturally embedded aspects of a destination that makes it unique and can hardly be replicated by other destinations (Olsson et al., 2016). This means that it's becoming inevitable for destination marketing organizations (DMOs) to aim at creating, developing, and sustaining destination brands strategically (King, 2002). However, studies have shown that this activity is not a concern for only branding practitioners or DMOs who create or shape marketing stories aimed at attracting tourists (Mossberg et al., 2010). Instead, the activity of storytelling in tourism also takes place on the side of the visitors, where consumers of tourism products share brand experiences with other consumers via social media, aside from the stories being told by tour guides and the DMOs. This is corroborated by Lund et al. (2018) who posited that, when people connect to social media and share content with their virtual friends and acquaintances, they are in essence all storytellers and the sharing of stories is how they perform socialites and if a story has sufficient appeal and interest, it can spread beyond people's networks in social media and potentially be shared around the globe.

Ababneh (2018) asserts that tour guides also contribute to destination sustainability through the role they play by providing visitors with information as they guide them through the destination's attractions. The performance of tour guides impacts the success of each destination as tourists tend to interpret everything they experience in their own way when not guided (Ooi, 2002). Tour guides help in shaping how tourists interpret the information they are exposed to at the destination. This is very important because they usually spend more time with tourists than any stakeholder.

Studies have shown that tour guiding is essential for the success of tourism promotion as tour guides are involved and present in the production and consumption of tourism products through storytelling. Their interpretation is a significant aspect of the tour that helps to determine visitor satisfaction. Their activities through storytelling provide education not only to the host community but to tourists as well (Best, 2012). They also serve a special purpose of promoting the cultural heritage, history, and tradition of the destination through storytelling in the way that tourists want, by getting tourists to actively participate, and connect with the local community as they get to know the story behind the destination.

Today's tourists are not only interested in the attraction of the destination but also the stories behind them (Sil, 2006). Aided by technology, creative storytelling is becoming an effective destination marketing tool. With the co-creation of experiences, long-term relationships are created leading to positive behaviour towards the destination which will eventually sustain the destination.

In the interest of sustainability, according to UNEP and WTO (2005), the tourist destination brand and associated images (including pictures and text) should:

- Be sufficiently strong to successfully command attention and generate new interest,
- Be sufficiently distinctive, to differentiate the destination from others,
- Match the values of the destination's target markets,
- Be authentic, i.e. it should relate to what visitors will find,
- Avoid stereotypes and images that may demean local communities and their values.

Again, the OECD (2014) indicates that the primary functions of storytelling as a destination marketing tool for sustainable tourism management are the following:

1. facilitating market access of all stakeholders, particularly smaller enterprises or community-based tourism initiatives with limited marketing resources,
2. promoting particular forms of tourism or specific tourist products that are more sustainable than others,
3. promoting history, cultural heritage, and traditions of the place, including implications both in terms of visitor interest and behaviour,
4. promoting natural environment, including special qualities and sensitivities to certain activities,
5. reducing seasonality, by promoting off-season images and opportunities,
6. raising the use of more sustainable transport,

7. maximizing value retained locally and
8. increasing spending per head and length of stay. Once the storyline has been defined, storytellers and various stakeholders must work on the dissemination process of the narratives.

The OECD continues to state that the multiple sources available for dissemination of stories in tourism due to technology offer the potential of targeting a wider audience and enable DMOs to explore narratives in different languages and forms (film, text, photography, art). Radić and Munir (2021) however posit that it is important to emphasize that although the role of storytelling in developing the destination brand is extremely significant, its potential in creating new products in the destination is even more significant. Tourism is about experiences and tourists often associate good and memorable experiences with products that are focused on discovering some local legends or on the revival of the spirit of past times that were driven by excellent storytelling by the tour guide and reflected throughout the destination. Again, Radić, Ravasi and Munir (2021), make it clear that destination management companies will use all the advantages of storytelling in the process of designing unique tourism products to promote the destination sustainably and create a competitive advantage for the destination without high fixed investments.

From the literature above, it can be deduced that using storytelling as a means of differentiating the attractions of a destination gives the destination a uniqueness that is not easily copied by other destinations. This approach helps in forming a positive perception among tourists towards the destination, leads to co-creation, and positive word of mouth that will generate revisits and create new interest. This will eventually translate into more resources for sustaining the destination and its local economy.

THE WAY FORWARD

Stories told about tourism destinations are aimed at enhancing destinations' competitiveness and making visitor travel experience worth cherishing (Mossberg, 2008). As a result, storytelling has been used as a strategic tool to generate destination value and increase tourists' attention to tourism destinations. (Akgün, Keskin, Ayar, & Erdoğan, 2015). It has been demonstrated in previous research that storytelling has a favorable effect on the behavioral intentions of visitors to tourist destinations (Akgün et al., 2015; Woodside & Megehee, 2009, Hsiao, Lu, & Lan, 2013).

According to Simmons (2006), stories affect consumers in their souls. In storytelling, the objective is to convey a message about a destination brand's attributes and identity to sustain the interest of tourists to encourage repeat visits (Ben Youssef et al., 2019). He further argues that storytelling can positively influence tourist decision-making in favor of a destination. This confirms the empirical findings from a study by Akgün et al. (2015), which concludes that storytelling can be used as a promotional tool in influencing consumers' behavioural intentions such as word-of-mouth and travel decisions. Findings of these studies suggest that it is imperative to develop a symbiotic relationship between the destination, the stories told about it, and the visitor, as this can favour co-creation. Hence storytelling should involve all stakeholders in the promotion of the destination. As Chronis (2012) and Mossberg et al. (2010) put it, stories told by a destination must be embraced and supported by all stakeholders involved with a narrative environment that enhances co-production among all stakeholders.

The link therefore between storytelling and place promotion, services, and experiences is by communicating stories that can attract tourists to visit the destination (Kim & Young, 2017). Destinations are encouraged to develop rich stories for a variety of audiences that elicit a positive response and enhance perceived authenticity and memorability (Harrison, 2002; Kim & Young, 2019) for destination promotion and sustainability. In achieving this aim, tourism destinations must strive to reduce the risk of complexity in storytelling most especially in the use of archetypal narrative forms. Stories told about the destination must be fluid and be able to carry tourists on the journey. Storytellers must also be abreast with information about the attributes of the destination to develop a value for the tourists that visit it.

This chapter illustrates that storytelling practices at tourist destinations have the potential to engage tourists and promote the unique attributes of the destination in a more responsible, thoughtful, and sustainable manner. However, for DMOs they must develop promotional campaigns that incorporate storytelling to remain relevant and competitive. Coupled with the above, storytelling could be used to create direct touristic experiences that explore the cultural artifacts of the destination while preserving them. Destination storytelling across different cultures enables tourists to explore diverse and common experiences and to develop a nuanced understanding of one another (Stevenson, 2019). Stevenson (2019, p. 10) further argued, "that stories enabled us to discuss experiences and feelings, opening up possibilities for shared cultural experiences through dance and conversation". Sharing multi-sensory stories about aspects of our lives and cultures developed interaction and discussion which supported a more in-depth understanding of one another derived from a co-created experience.

FUTURE RESEARCH DIRECTIONS

The travel industry is said to be blossoming due to an increase in wealth, transportation, and communication (Puhakka 2011), thereby giving rise to tourists finding creative ways of expressing their experience at destinations through storytelling. Storytelling has indeed been a common way of communicating throughout the history of mankind (Rauhala & Vikström 2014). Although there is a general notion out there that tourists travel to listen to interesting stories about a destination and create and tell their own stories, there seem to be few studies that have systematically investigated the character of storytelling as a tool in promoting destinations and how it enhances the touristic experience (Moscardo, 2018).

DMOs can delve into this topic using stories as a marketing tool in promoting tourism destinations. Again, there is the need for a shift of focus on how photographers use the images captured by them in telling stories that promote a destination. Storytelling can be employed as the "Pull Factor" in attracting people to certain tourism destinations as some destinations do not have physical attractions such as forts, castles, or waterfalls. Storytelling can be a strategic avenue in promoting such destinations.

CONCLUSION

This chapter highlighted the role played by storytelling in making destinations appealing and competitive. It is argued that stories have a "pull" factor that attracts tourists to visit a destination (See Dann 1981). This chapter shows the strategic impact of the storytelling process in defining the reputation of a destination, enhancing its competitiveness, and creating meaningful impressions. When tourists arrive at a destination, to promote the attributes of the destination, DMOs depend largely on tourism stakeholders such as tour operators, tour guides, and others that provide ancillary services at the destination in promoting its attractiveness. In promoting the destination, fascinating stories are told about it to make it appealing.

The chapter also explored how stories could be used in sustaining tourism destinations. Moscardo (2010) holds that in creating experiences for tourists framed with the context of storytelling, there is the need to develop engaging tales that have clear themes, and accessible, interactive, immersive storylines that provide an opportunity for destination preservation. Using a coherent narrative with a story arch and embedded humour, and integrating narrative with visuals, sound effects, and music, has the element of eliciting commitments in pro-environmental ways. This will then make the destinations appealing for tourists to visit and for destination marketers, it will provide them with the ease of having a strategic marketing campaign to boost tourists' arrival.

Tourism destination stories in most instances are employed as a destination feature to attract tourists and as a core element of tourist experience opportunities. The implementation of such strategies and its actual execution primarily lies not only on the storyteller, but also on the storytelling location (e.g., museums, farms, dairies, or where the story is located) as well as the various stakeholders at the destination that are not central to the storytelling activities (in some cases hotels, shops, cafés) (Moscardo, 2020). These may contribute to the storytelling offer in a coordinated way. For stories to be told and results achieved, the type of story being told and how it fits into the overall storytelling concept of the destination, the level of audience anticipation, and the physical environment where the story is being told constitute an integral part of the execution of storytelling (Mossberg et al., 2010). It is therefore imperative for destination promoters to employ storytellers that understand the destination's attributes and key features that will interest tourists and keep them engaged. When stories are well told, people immerse themselves into the plot and characters of the story (McCabe & Foster, 2006) and this is of importance since contemporary visitors do not merely encounter the physical space of a destination but create their own experiential space using their motivations and interpretations (Suvantola, 2018).

REFERENCES

Ababneh A. (2018) Tour guides and heritage interpretation: guides' interpretation of the past at the archaeological site of Jarash, Jordan. *Journal of Heritage Tourism, 13*(3), 257-272. doi:10.1080/1743873X.2017.1321003

Akgün, A. E., Keskin, H., Ayar, H., & Erdoğan, E. (2015). The influence of storytelling approach in travel writings on readers' empathy and travel intentions. *Procedia: Social and Behavioral Sciences, 207*, 577–586. doi:10.1016/j.sbspro.2015.10.129

Bassano, C., Barile, S., Piciocchi, P., Spohrer, J. C., Iandolo, F., & Fisk, R. (2019). Storytelling about places: Tourism marketing in the digital age. *Cities, 87*(November), 10–20. doi:10.1016/j.cities.2018.12.025

Ben Youssef, K., Leicht, T., & Marongiu, L. (2019). Storytelling in the context of destination marketing: An analysis of conceptualizations and impact measurement. *Journal of Strategic Marketing, 27*(8), 696–713. doi:10.1080/0965254X.2018.1464498

Best, J. (2012). *Damned lies and statistics*. University of California Press. doi:10.1525/9780520953512

Black, N. (2013). Patient reported outcome measures could help transform healthcare. *BMJ (Clinical Research Ed.), 346*(jan28 1), 346. doi:10.1136/bmj.f167 PMID:23358487

Bramwell, B., & Lane, B. (1999). Sustainable tourism: Contributing to the debates. *Journal of Sustainable Tourism*, *7*(1), 1–5. doi:10.1080/09669589908667323

Bruner, E. M. (2005). *Culture on Tour: Ethnographies of Travel*. University of Chicago.

Bryon, J., & Bryon, J. (2012). *Tour Guides as Storytellers – From Selling to Sharing Tour Guides as Storytellers – From Selling to Sharing*. doi:10.1080/15022250.2012.656922

Buhalis, D. (2000). Marketing the competitive destination of the future. *Tourism Management*, *21*(1), 97–116. doi:10.1016/S0261-5177(99)00095-3

Choi, S. S. (2016). A study on effect of tourism storytelling of tourism destination brand value and tourist behavioral intentions. *Indian Journal of Science and Technology*, *9*(46), 1–6. doi:10.17485/ijst/2016/v9i46/107387

Chronis, A. (2005). Coconstructing heritage at the Gettysburg storyscape. *Annals of Tourism Research*, *32*(2), 386–406. doi:10.1016/j.annals.2004.07.009

Chronis, A. (2008). Co-constructing the narrative experience: Staging and consuming the American Civil War at Gettysburg. *Journal of Marketing Management*, *24*(1-2), 5–27. doi:10.1362/026725708X273894

Chronis, A. (2012). Tourists as story-builders: Narrative construction at a heritage museum. *Journal of Travel & Tourism Marketing*, *29*(5), 444–459. doi:10.1080/10548408.2012.691395

Clark, W. C., & Dickson, N. M. (2003). Sustainability science: The emerging research program. *Proceedings of the National Academy of Sciences of the United States of America*, *100*(14), 8059–8061. doi:10.1073/pnas.1231333100 PMID:12794187

Dahlstrom, M. F. (2014). Using narratives and storytelling to communicate science with nonexpert audiences. *Proceedings of the National Academy of Sciences of the United States of America*, *111*(Supplement 4), 13614–13620. doi:10.1073/pnas.1320645111 PMID:25225368

Dann, G. M. (1981). Tourism Motivations: An Appraisal. *Annals of Tourism Research*, *8*(2), 189–219. doi:10.1016/0160-7383(81)90082-7

Davidson, R., & Maitland, R. (1997). *Tourism destinations*. Hodder & Stoughton.

Dawson, P., & Sykes, C. (2018). Concepts of time and temporality in the storytelling and sensemaking literatures: A review and critique. *International Journal of Management Reviews*, *00*, 1–18.

Denning, S. (2005). *The leader's guide to storytelling: Mastering the art and discipline of business narrative*. John Wiley & Sons.

Escalas, J. E. (2007). Imagine yourself in the product. Mental simulation, narrative transportation, and persuasion. *Journal of Advertising*, *33*(2), 37–48. doi:10.1080/00913367.2004.10639163

Fischer, D., Selm, H., Sundermann, A., & Storksdieck, M. (2020). Storytelling for sustainability. *Storytelling for Sustainability in Higher Education*, (April), 38–51. doi:10.4324/9780429291111-5

Gill, P. (2015). *Toward a scientific approach to identifying and understanding indicators of radicalization and terrorist intent: Eight key problems*. Academic Press.

Hall, C. M. (2000). *Tourism Planning: Policies, processes, relationships.* Prentice Hall.

Hanna, S., Del Casino, V. Jr, Selden, C., & Hite, B. (2004). Representation as Work in America's Most Historic City. *Social & Cultural Geography*, *5*(3), 459–481. doi:10.1080/1464936042000252822

Hanna, S., & Rowley, J. (2013). Place brand practitioners' perspectives on the management and valuation of the brand experience. *The Town Planning Review*, *84*(4), 473–493. doi:10.3828/tpr.2013.25

Hartman, S., Parra, C., & de Roo, G. (2019). Framing strategic storytelling in the context of transition management to stimulate tourism destination development. *Tourism Management, 75,* 90–98. doi:.2019.04.014 doi:10.1016/j.tourman

Harrison, J. (2002). *Being a tourist – Finding meaning in pleasure travel.* UBC Press.

Hassan, S. S. (2000). Determinants of market competitiveness in an environmentally sustainable tourism industry. *Journal of Travel Research*, *38*(3), 239–245.

Hee, L. T. (2001). Evaluation of brand equity of tourism destinations in Korea. *Tourism Research.*, *25*(3), 171–192.

Ho, C. I., & Mi, L. E. (2008). The place marketing using storytelling: Exploratory research. *Tourism Research.*, *32*(4), 411–429.

Hsiao, K.-L., Lu, H.-P., & Lan, W.-C. (2013). The influence of the components of storytelling blogs on readers' travel intentions. *Internet Research*, *23*(2), 160–182. doi:10.1108/10662241311313303

Iles, J. (2008). Encounters in the Fields: Tourism to the Battlefields of the Western Front. *Journal of Tourism and Cultural Change*, *6*(2), 138–154. doi:10.1080/14766820802376299

Järv, R. (2010). Fairy tales and tourist trips. *Fabula*, *51*(3–4), 281–294. doi:10.1515/FABL.2010.026

Kim, K., Hwang, J., Zo, H., & Lee, H. (2016). Understanding user's continuance intention toward smartphone augmented reality applications. *Information Development*, *32*(2), 161–174. doi:10.1177/0266666914535119

Kim, J., & Young, K. (2019). How to design and deliver stories about tourism destinations. *Journal of Travel Research*, *56*(6), 808–820. doi:10.1177/0047287516666720

King, J. (2002). Destination marketing organizations—Connecting the experience rather than promoting the place. *Journal of Vacation Marketing*, *8*(2), 105–108. doi:10.1177/135676670200800201

Korez-Vide, R. (2017). Storytelling in sustainable tourism management: Challenges and opportunities for Slovenia. *Journal of Advanced Management Science, 5*(5).

Lee, T. H. (2013). Influence analysis of community resident support for sustainable tourism development. *Tourism Management*, *34*, 37–46. doi:10.1016/j.tourman.2012.03.007

Lepp, A. (2007). Residents' attitudes towards tourism in Bigodi village, Uganda. *Tourism Management*, *28*(3), 876–885. doi:10.1016/j.tourman.2006.03.004

Lekić, R., Blajić, B., & Franjić, T. (2017, May). Interpretation of Mythical Landscape and Holy Geography in Creative Cultural Tourism. In *Proceedings of the 4th International Scientific Conference on ToSEE—Tourism in Southern and Eastern Europe, Opatija, Croatia* (pp. 4-6). Academic Press.

Lichrou, M., O'malley, L., & Patterson, M. (2010). Narratives of a tourism destination: Local particularities and their implications for place marketing and branding. *Place Branding and Public Diplomacy*, *6*(2), 134–144. doi:10.1057/pb.2010.10

Lund, N. F., Cohen, S. A., & Scarles, C. (2017). The power of social media storytelling in destination branding. *Journal of Destination Marketing & Management*, *5*, 1–10.

Lunda, N. F., Scott, A., Cohenb, S. A., & Scarlesb, C. (2018). The power of social media storytelling in destination branding. *Journal of Destination Marketing & Management*, *8*, 271–280. doi:10.1016/j.jdmm.2017.05.003

Mak, A. H. (2017). Online Destination Image: Comparing National Tourism Organisation's and Tourists' Perspectives. *Tourism Management*, *60*, 280–297. doi:10.1016/j.tourman.2016.12.012

McCabe, S., & Foster, C. (2006). The role and function of narrative in tourist interaction. *Journal of Tourism and Cultural Change*, *4*(3), 194–215. doi:10.2167/jtcc071.0

Mei, X. Y., Hågensen, A. M. S., & Kristiansen, H. S. (2018). Storytelling through experiencescape: Creating unique stories and extraordinary experiences in farm tourism. *Tourism and Hospitality Research*, *34*, 1–22. doi:10.1177/1467358418813410

Moscardo, G. (2010). The shaping of tourist experience: The importance of stories and themes. *The tourism and leisure experience: Consumer and managerial perspectives, 44*, 43-58.

Moscardo, G. (2018). Tourist experience design. In *Quality services and experiences in hospitality and tourism*. Emerald Publishing Limited. doi:10.1108/S2042-144320180000009007

Moscardo, G. (2020). Stories and design in tourism. *Annals of Tourism Research*, *83*(January), 102950. doi:10.1016/j.annals.2020.102950

Mossberg, L. (2008). Extraordinary experiences through storytelling. *Scandinavian Journal of Hospitality and Tourism*, *8*(3), 195–210. doi:10.1080/15022250802532443

Mossberg, L., Therkelsen, A., Huijbens, H. E., Björk, A-K., & Olsson, P. (2010). Storytelling and Destination Development. Nordic Innovation Centre (NICe) Project Number: 08041.

OECD. (2014). *Tourism and the creative economy, OECD Studies on Tourism*. OECD Publishing.

Ooi, C. S. (2002). *Cultural tourism and tourism cultures: The business of mediating experiences in Copenhagen and Singapore*. Copenhagen Business School Press DK.

Olsson, A. K., Therkelsen, A., & Mossberg, L. (2016). Making an effort for free – volunteers' roles in destination-based storytelling. *Current Issues in Tourism*, *19*(7), 659–679. doi:10.1080/13683500.2013.784242

Oye, N. D. (2013). Reducing corruption in African developing countries: The relevance of E-Governance. *Greener Journal of Social Sciences*, *3*(1), 6–13. doi:10.15580/GJSS.2013.1.103112183

Perovic, Đ., Moric, I., Pekovic, S., Stanovcic, T., Roblek, V., & Pejic Bach, M. (2018). The antecedents of tourist repeat visit intention: Systemic approach. *Kybernetes*, *47*(9), 1857–1871. doi:10.1108/K-12-2017-0480

Pike, S. (2007). Consumer-based brand equity for destinations. *Journal of Travel & Tourism Marketing*, *22*(1), 51–61. doi:10.1300/J073v22n01_04

Polletta, F., Chen, P., Gardner, B., & Motes, A. (2011). The sociology of storytelling. *Annual Review of Sociology*, *37*(1), 109–130. doi:10.1146/annurev-soc-081309-150106

Puczko, L., & Ratz, T. (2000). Tourist and resident perceptions of the physical impacts of tourism at Lake Balaton, Hungary: issues for sustainable tourism management. *Journal of Sustainable Tourism*, *8*(6), 458-478.

Puhakka, R. (2011). *Matkailukysynnän trendit vuoteen 2030 mennessä.* http://www.kulmat.fi/images/tiedostot/Artikkelit/tuleva_trendit_2030.pdf

Radić, M., Ravasi, D., & Munir, K. (2021). Privatization: Implications of a shift from state to private ownership. *Journal of Management.*

Rauhala, M & Vikström, T. (2014). *Storytelling työkaluna.* Talentum Media.

Reed, A. (2002). City of Details: Interpreting the Personality of London. *Journal of the Royal Anthropological Institute*, *8*(1), 127–141. doi:10.1111/1467-9655.00102

Reinermann, J. L., Lubjuhn, S., Bouman, M., & Singhal, A. (2014). Entertainment-education: Storytelling for the greater, greener good. *International Journal of Sustainable Development*, *17*(2), 176–191. doi:10.1504/IJSD.2014.061781

Rhodes, N., Toole, J., & Arpan, L. M. (2016). Persuasion as reinforcement: Strengthening the pro-environmental attitude-behavior relationship through ecotainment programming. *Media Psychology*, *19*(3), 455–478. doi:10.1080/15213269.2015.1106322

Rojek, C., & Urry, J. (1997). *Touring cultures: Transformations of travel and theory.* Routledge.

Salazar, N. (2005). Tourism and Glocalization: "Local" Tour Guiding. *Annals of Tourism Research*, *32*(3), 628–646. doi:10.1016/j.annals.2004.10.012 PMID:32572282

Shrestha, R. K., Stein, T. V., & Clark, J. (2007). Valuing nature-based recreation in public natural areas of the Apalachicola River region, Florida. *Journal of Environmental Management*, *85*(4), 977–985. doi:10.1016/j.jenvman.2006.11.014 PMID:17275161

Sil, C. H. (2006). *Creating a culture of collaboration: The International Association of Facilitators handbook* (Vol. 4). John Wiley & Sons.

Simmons, A. (2006). *The Story Factor: Secrets of Influence through the Art of Storytelling.* Basic Books.

Sole, D., & Wilson, D. (2002). *Storytelling in Organizations.* LILA: Harvard, Graduate School of Education.

Statista. (2021). *Number of Monthly Active Facebook Users Worldwide as of 2nd Quarter 2021(in Millions).* Available at: https://www. statista.com/statistics/264810/number-of-monthly-active-facebook-users-worldwide/

Steinemann, S., Iten, G., Opwis, K., Forder, S., Frasseck, L., & Mekler, E. (2017). Interactive narratives affecting social change. *Journal of Media Psychology, 29*(1), 54–66. doi:10.1027/1864-1105/a000211

Stevenson, N. (2019). Developing cultural understanding through story-telling. *Journal of Teaching in Travel & Tourism, 19*(1), 8–21. doi:10.1080/15313220.2018.1560528

Strielkowski, W. (2017). Promoting tourism destination through film-induced tourism: the case of Japan. *Market-Tržište, 29*(2), 193-203. doi:10.22598/mt/2017.29.2.193

Suvantola, J. (2018). *Tourist's experience of place*. Routledge. doi:10.4324/9781315184821

Therkelsen, A., Halkier, H., & Jensen, O. B. (2010). Branding Aalborg: Building community or selling place? In *Towards effective place brand management*. Edward Elgar Publishing. doi:10.4337/9781849806398.00014

Tobin, P. (2006). Teaching IT through storytelling. *Proceedings of The Conference on Information Technology in Tertiary Education.*

Tussyadiah, I. P., & Fesenmaier, D. R. (2008). Marketing places through first-person stories—An analysis of Pennsylvania roadtripper blog. *Journal of Travel & Tourism Marketing, 25*(3-4), 299–311. doi:10.1080/10548400802508358

UNEP & WTO. (2005). *Making Tourism More Sustainable. A Guide for Policy Makers*. Available at: http://www.unep.fr/shared/publications/pdf/DTIx0592xPATourismPolicyEN.pdf

WECD. (1987). *Our Common Future: The World Commission on Environment and Development*. Oxford University Press.

Wilson, M., & Zipes, J. (2005). *Storytelling and theatre: Contemporary professional storytellers and their art*. Palgrave Macmillan.

Woodside, A. G., & Megehee, C. M. (2009). Travel storytelling theory and practice. *Anatolia, 20*(1), 86–99.

Woodside, A., & Martin, D. (2015). Introduction: The tourist gaze 4.0: Uncovering non-conscious meanings and motivations in the stories tourists tell of trip and destination experiences. *International Journal of Tourism Anthropology, 4*(1), 1–12.

Yavuz, M., Sumbul, M., Ergec, N., Derdiyok, C. (2016). Storytelling in Destination Brand Communication: A Qualitative Analysis. *International Interdisciplinary Business-Economics Advancement Journal, 1*, 63–72.

Yu, H., & Chang, Y. (2013). How to influence the brand attitude of the audience by micro-films. *Journal of Promotion Management, 19*(5), 674–686.

Zavattaro, S. M., Daspit, J. J., & Adams, F. G. (2015). Assessing managerial methods for evaluating place brand equity: A qualitative investigation. *Tourism Management, 47*, 11–21.

KEY TERMS AND DEFINITIONS

Destination: Denoting a place that people will make a special trip to visit. It can be a city, town, or an area, that is significantly dependent on revenue from tourism. Destinations can be natural areas, town or cities, sport areas, cultural and heritage monuments, and resorts.

Destination Marketing Organizations: An organization that promotes a location as an attractive travel destination. It is the backbone of a tourism destination as it promotes destinations, attracts visitors, and develops the local economy. These organizations have local knowledge, expertise, and resources needed in designing and implementing events, tours, transportation, and logistics.

Destination Promotion: A marketing strategy aimed at increasing customer awareness of a destination with the objective of influencing them to start thinking of visiting the destination and also favorably positioning the destination in the mind of the consumer for future decision making. Some promotional activities can be using influencers, creating and sharing destination videos, interacting and sharing new happenings with customers on social media as well as sales promotions.

Destination Sustainability: A concept that looks at measures that ensure all activities undertaken at a destination lead to progress in line with the economic development, social development, and environmental protection of the destination for the future benefit of stakeholders. Here negative impacts are minimized as positive impacts are maximized.

Stories: Narratives that give an account of events or incidents that may be factual or fictitious and also past or present. They educate, arouse interest, socialize, entertain, amuse, or instruct the audience. They are told through words (written or spoken), imagery (still or moving), body language performance, music, or any form of communication.

Storytelling: The social activity of sharing stories with an audience through various mediums. This activity is deeply rooted in human culture in many forms. Storytelling will aid in building and maintaining a good image for the destination.

Chapter 8
Destination Storytelling as a Destination Branding Tool

Abdullah Ülkü
Harran University, Turkey

Günay Erol
https://orcid.org/0000-0001-8288-998X
Nevşehir Hacı Bektaş Veli University, Turkey

ABSTRACT

This study discusses how storytelling can be used as a destination branding tool by examining some of the most visited sites in Turkey as an example of destination storytelling. Stories have been used to transfer knowledge and culture from generation to generation. Stories have become an innovative communication channel for destinations. Tourists no longer want to just buy an experience; they also want to have the stories behind the destination experience. From this point of view, storytelling has become indispensable for destinations. Destinations have symbolic meanings for tourists, and thanks to branding, destinations communicate to tourists some messages about themselves and are accordingly positioned in their minds. In this study, the formation process of destination storytelling, the effects of stories on the formation of the destination brand, and the destination brand are explained.

INTRODUCTION

Stories have a very important place in human life (Gottschall, 2012). People often communicate by telling stories to each other. Stories are spread over a deep and wide area in the human mind. The story is at the centre of human understanding and without the story, there is no identity, self, or anything else (Lewis, 2011). Stories have always fascinated people and are easier to remember than facts. Well-told stories about a brand can affect feelings as well as induce cognitive and behavioural responses about the brand (Lundqvist et al., 2012). Considering that consumers are intuitively looking for brands that offer more than products or services, the importance of storytelling for businesses is well understood (Smith and Wintrob, 2013). The way tourism businesses present their stories can have an impact on the behaviour

DOI: 10.4018/978-1-6684-3436-9.ch008

of tourists. For this reason, storytelling is needed for tourism brands to build a strong identity. Since brand management includes a communication process, brand stories are one of the most powerful ways to express the essence of the brand, its history, and what it represents (Ryu et al., 2018).

Stories have been used to transfer the knowledge and cultures of human beings from generation to generation (Reamy, 2002). This process has become faster and easier thanks to the many possibilities developed in today's world. As competition between destinations increases, the transfer of culture, knowledge, and experiences about destinations has gained importance for destination marketing and branding activities. This is where storytelling comes into play. Stories have become an innovative communication channel for destinations (Hagen, 2008). Accommodation, transportation, food and beverage activities are the base of tourism classically. But nowadays it has become necessary to promise a memorable experience to potential tourists. Tourists no longer want to just buy a service, they also want to have the stories behind the service or the destination. From this point of view, storytelling has become indispensable for destinations (Mossberg et al., 2010).

Storytelling is one of the most impressive ways of telling the story of the destination (Soon Choi, 2016). Destinations have started to adopt storytelling to tell tourists what they have to offer (Youssef et al., 2018). Storytelling helps connect destination brands and tourists. Also, it has an impact on tourist experiences. Stories that place destinations in the minds of potential tourists can be highly influential and shape the actual journey (Moin et al., 2020). Stories make destinations memorable and enable tourists to dream about the destination (Salazar, 2012).

Destinations convey to people a cultural, natural, or historical feature that makes it unique and not available in other destinations through stories (Olsson et al., 2013). Stories include location, action, attributes, issues, and characters that give meaning, create empathy and provide a convenient means for communicating brand messages to consumers (Moin et al., 2020). When this situation is considered in terms of destination branding, potential tourists can be informed about the character, place, etc., about the destination. Storytelling significantly affects the brand value of the tourism destination. This brand equity is also effective on the behavioural intentions of tourists (Soon Choi, 2016).

Stories establish a relationship between the destination and potential tourists, as in brand communication (Moscardo, 2010). Stories help creating a perception of the destination through image, identity, and personality (Vermeulen, 2002); they provide tangible benefits and added value for the destination (Kavaratzis and Ashworth, 2006), increase the probability of choosing the destination thanks to the themes in the stories (Wood & Lego Muñoz, 2007), increase the competitiveness of the destination brand (Lund et al., 2018), make the destination brand distinguishable among its competitors (Pawaskar & Goel, 2014), convey the symbolic meanings of the destination to potential tourists (Garcia et al., 2012), and affect the visit intention of destination (SKIFT, 2020).

In this chapter, for some of the most visited popular touristic destinations of Turkey, storytelling is explored as a destination branding tool. With this aim, based on the stories of these destinations reflecting their own characteristics, suggestions are made for the use of these stories. Suggestions to tourism stakeholders are developed by considering the use of stories as a destination branding tool, and also which variables can be addressed in the storytelling process and story themes. In general, the story elements of top touristic destinations in Turkey are examined with the data obtained from the literature review, and a situation assessment is made for the use of these stories as a destination branding tool.

Brand Storytelling

According to the American Marketing Association (AMA), a brand is "*a name, term, sign, symbol, or design, or a combination thereof, intending to identify and distinguish the goods and services of a seller or group of sellers*" (Maurya and Mishra, 2012: 123). Branding is part of the alignment of image, vision, and culture. Interdepartmental coordination with stakeholders, brand communication, consumer intentions, and evaluations are important factors for branding (Chan, Peters, and Marafa, 2015). Businesses use some branding strategies to build strong relationships with their customers and increase their recognition. For example, Unilever is the main company that includes other brands, such as Lipton and Domestos. Although these two brands seem to be different brands with their own identities, they are actually subsidiaries of Unilever. The focus point of these companies is to create strong and central stories so that products and brands of the main company do not conflict with each other. Thanks to these stories, these companies establish a bond with their customers, and Unilever, the main company, is introduced (Keskin et al., 2016).

Many businesses and brands are using storytelling as a new marketing method. According to Da Costa (2019):

"Businesses need to connect with audiences and engage with them on a much deeper level than ever before to survive. This is where telling stories about brands come into play… There are three important reasons for brands to tell stories about themselves; first, it's not enough to have a quality product or service, you need to know how to talk about it in a way that will set it apart from the others. That is why brand storytelling is so important. Second, when crafting their stories, brands should consider what the consumer really needs in addition to a product or service… Third, a brand story can do a lot more than connect you to your ideal audience, get noticed in the noise of your competitors' messages, and turn a profit – it also has the potential to make a huge impact."

In communication practice, storytelling is a form of narrative in which a company establishes its philosophy and identity as it creates events and products. A story can create a strategic position for the brand by defining the brand and enabling to communicate. In addition, storytelling can connect brands with values and emotions, add a sense of authenticity and quality to products (Febby and Hifni, 2018).

Today, brands are telling stories about themselves. Thus, relations with consumers are strengthened and the brand is positioned in the mind of the consumer. Especially the opportunities offered by the internet are important for brands. Millions of people can be reached through storytelling on social media or digital applications. So, if brands have the right storytelling capabilities, the potential dissemination and reach of stories can impact millions of people. Thus, storytelling leads to impact and power; it can affect discourses and even behaviours (Lund et al., 2018). Not only touristic establishments but also consumers via social media (e.g. discussion forums, travel and consumer blogs, social platforms) have a greater role in brand storytelling than ever before, and these stories can spread quite quickly. This shows us that brand owners no longer have full control over their brands and their meanings. That is why brands are starting to pay attention to their storytelling roles more than ever before. Brands are trying to tell stories about themselves created with the right methods (Delgado-Ballester and Fernández-Sabiote, 2016).

Destination Branding and Destination Storytelling

There are six key characteristics of brands (Arnold, 1992; Gardner and Levy, 1995; Aaker, 1997; Kotler et al., 1996; Stern, 2006): visual, perceptual, positioning, added value, image, and personality. By using these six characteristics, branding can be applied to destinations, as well as tourism businesses (Kerr, 2005). The idea that product branding should be adapted to destination branding has attracted attention in recent years. It is thought that a destination's cognitive (natural and cultural resources, atmosphere, social environment, etc.) and affective components (pleasant, exciting, etc.) should be seen as an "entity" (Souiden, Ladhari, and Chiadmi, 2017). Researchers acting with this idea have conducted studies on destination personality and image (Ekinci and Hosany, 2006), visitors' brand attitudes, and destination branding (Campelo et al., 2013).

According to Ritchie and Ritchie (1998: 17), destination branding is "*a name, symbol, logo, word or other graphic that both defines and differentiates the destination; it also conveys the promise of an unforgettable travel experience that is uniquely associated with the destination; it also serves to reinforce the recollection of pleasant memories of the destination experience*". Anholt (2004 as cited in Kerr, 2005) says it is "*the practice of applying brand strategy, marketing techniques and disciplines to the economic, political and cultural development of destinations*". A destination brand is defined as a powerful tool capable of creating emotional appeal, and destination image is considered important for a tourism destination's marketing success, stimulating visitors' attitudes towards the destination, and enhancing their bond with the destination (Leisen, 2001). People make decisions when buying an ordinary product or relocating their businesses, and some of these decisions are rational and others are emotional. The brand or image of the destination affects the emotional and rational part of the making of decision process concerning the destination (Anholt, 2006).

The meaning of a brand to consumers is based on a story that comes from the fiction consumers create about a brand. Destinations are known to have similar resources (natural, historical or cultural attraction), so there is a possibility of substitutability of destinations. This situation encouraged destinations to implement new methods and activities to become prominent among their competitors. Destinations are trying to create a positive perception and image for themselves in highly competitive environment in the world (Pawaskar and Goel, 2014). Especially the Covid-19 pandemic has increased competition (Rasoolimanesh et al., 2021; Zarezadeh et al., 2021).

While managing the destination branding process, there are some ways destination marketers can convey destination messages. One of these ways is destination storytelling. People create stories from their personal experiences, and these stories are the basis for: "(a) grasping new experiences; (b) making decisions about the people, objects and events to which the stories refer; and (c) developing general attitudes and beliefs about these references" (Adaval and Wyer, 1998: 208). Taking this into consideration, storytelling can be seen as a promising tool in the branding process. People's use of storytelling in their communications, along with experiences and ideas about destinations, can positively affect potential tourists on destination preference. Storytelling and stories in the tourism sector are associated with many elements (Moscardo, 2021). The factors related to the storytelling method, which is used to persuade tourists to buy the product or service and to give remarkable messages about the touristic establishments, are presented in Figure 1.

Figure 1. Storytelling in tourism

Ekinci (2003: 22) discussed the importance of destination image and destination branding in the following way:

"The process of destination branding begins, in reality, when the evaluation of destination image includes a strong emotional attachment. These two concepts share some common ground but destination branding represents the emotional component of the destination image... In the ever more competitive tourism marketplace, destinations are increasingly adopting branding techniques to craft an identity that emphasizes the uniqueness of their product... Destination image is more cognitive and involves subjective knowledge of a destination (e.g. expensive, exotic, urban, developed) and therefore its evaluation may vary from one person to another. For example, some tourists may consider Turkey to be a European holiday destination while others disagree with this image."

According to Ekinci (2003), there are some methods to form destination identity and image that are important for destination branding. Destinations have symbolic meanings for tourists. Thanks to branding, destinations communicate messages to tourists, and simultaneously position in the minds of tourists (Garcia et al., 2012). One of the tools used for destination branding is storytelling. Destination storytelling is one of the most powerful marketing tools in branding and promotion of a destination (Tussyadiah and Fesenmaier, 2008).When stories are well structured, they can create an emotional connection between the destination and the target tourist group. Thus, a destination image can be formed (Moscardo, 2010).

Destination branding and destination image have important communication potential. When people encounter a destination, they have an idea about the destination through perceptions and images (Vermeulen, 2002). Destination storytelling is one of the most important factors in the perception of destination image and identity. Destination branding, which will be provided by destination storytelling, meets the need for historical and cultural resources of the city where the destination is located. It can recreate the gap in the social structure of the city with positive meanings.

According to Kavaratzis (2004), some questions need to be answered in order to ensure that destination branding is built on a solid foundation. The following five important questions arise for successful storytelling in destination branding: (1) what the destination is; (2) what the destination perceives; (3) what the destination says; (4) whom the destination aims to serve; and (5) what is offered and expected?. The answers to these questions refer to the observable, tangible, and visible aspects of the destination. It also includes promotional tools that the destination promotes and adopts to market itself. Finally, there is a communication via people's voices (word, mouth, advice) and social media (blogs, social networks). In the implementation of destination branding, successful storytelling can provide tangible benefits and added value for residents and foreigners alike (Kavaratzis and Ashworth, 2006).

Seth Godin said, "*people don't buy products or services; they buy relationships, stories, and magic*" (The Economic Times, 2017). People who visit or are likely to visit the destination will also establish an emotional connection with it thanks to the stories. The possibility of telling these stories to others will arise and this may affect their visiting to the destination. According to Skift (2020), 74% of the people participating in the research, the intention to visit a destination is affected by stories.

A real or imagined story about the destination gives it a competitive advantage and can provide a more meaningful experience to the tourist. Destinations can be viewed as story landscapes, i.e. commercial environments where narratives are negotiated, shaped and transformed through the interaction of producers and consumers. Stories can act as a mediator between tourist, potential tourist, and destination, because they can tell the core values and attractions in a destination, known or unknown, in an understandable and memorable way. Through the story, the destination can create the desired meaning about the city or region, museum, restaurant, hotel, or a particular event. The story becomes a verbal and visual metaphor describing the tourism offering, but for this, a good story must have a structure that focuses on catchy characters and a continuous storyline (Mossberg et al., 2010).

Bassano et al. (2019) mentioned the importance of storytelling in destination management. In their study, it has been stated that the contributions of local governments and tourism stakeholders are important in the storytelling process. Stories go through three stages when being created: building a place story (story building); telling stakeholders to deploy (storytelling), and listening to story feedback (story listening). The preliminary stage of story creation is divided into three sub-stages. The first is to link the unique aspects (culture and values) and important events that make up the destination's image and identity with the region in a memorable way. The second, to base destination stories on different aspects and critical success factors, and to construct the story with the definition of the preferred plot–narrative structure (melodrama, drama, comedy, etc.) to enable the destination to differentiate itself from its competitors. The third, to convey the essence of destination stories to stakeholders and strengthen the ties that bind people to that place. The selection of communication channels is also available in Figure 2. The stories created can be conveyed to people, especially through visual and digital media. As a result of this, perception and image of the destination are created.

Figure 2. Storytelling management

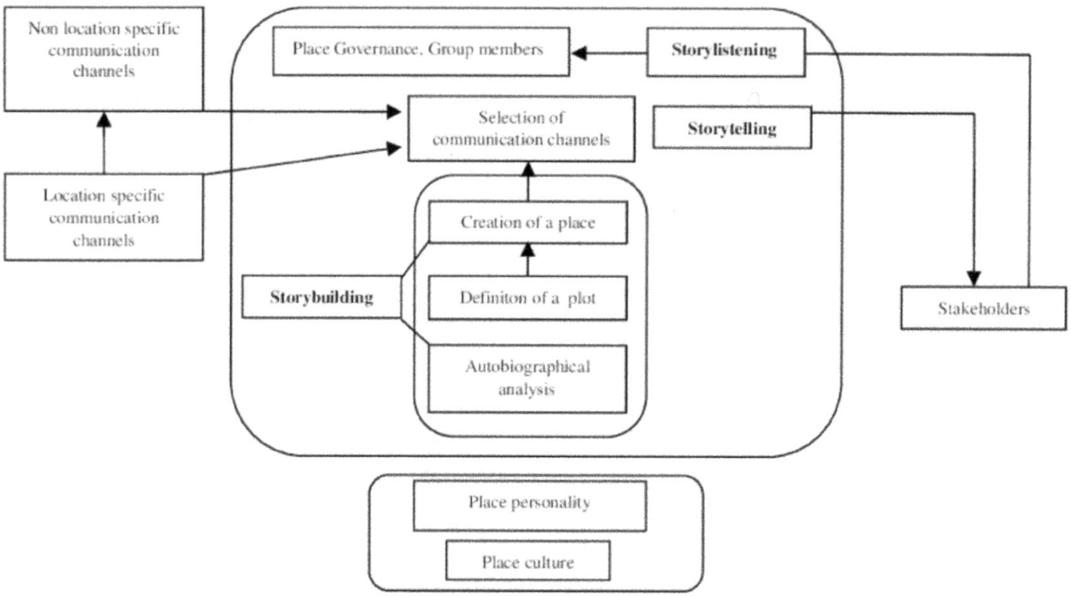

According to Soon Choi (2016) it can be said that storytelling has a significant impact on tourism destination brand value, with its detailed description, vivid narration, lots of attractions, and varied opportunities for experience, it tends to take the story much more easily. Tourism destinations should have an interesting history and unique features that should offer a variety of interesting and entertaining stories such as folk tales, fairy tales, and oral traditions. If destinations have an educational, stimulating, sensitive, and attractive story in terms of society, culture and history, tourist revisit intentions and recommendations will increase.

Figure 3. Relationship between storytelling, destination brand and tourist behaviour

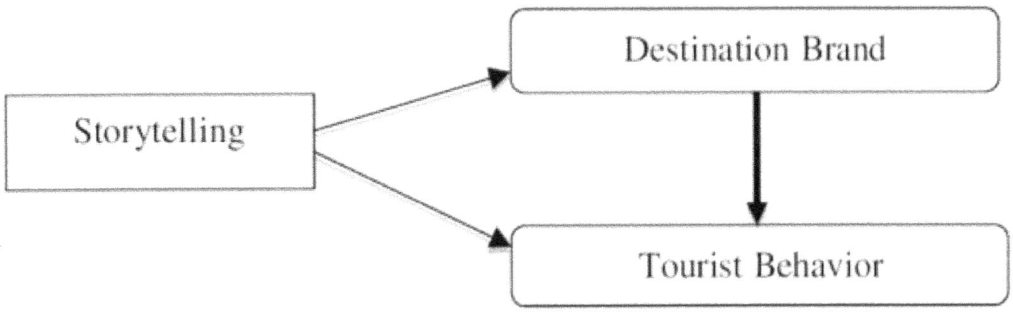

Themes are used to make the stories interesting and remarkable. The concept of the theme used by tourism businesses and destinations is an effective tool for businesses to position themselves in the minds of consumers. The theme is a phenomenon that we can see in some shopping malls, destinations, hotels, museums, retailers, amusement parks, events, and restaurants. Themed environments are built with symbolic meanings and they give some messages about the business to the tourist. Tourists understand the stories told about the relevant touristic establishment and destination and can choose that touristic establishment and destination thanks to the messages (Wood and Muñoz, 2007). Examples of accommodation establishments and destination stories are shown in Table 1.

Table 1. Storytelling about hotels and destinations

Hotels	Location	Theme	Story
Keunkan	Japan	History	Historical stories
Pera Palace Hotel	Istanbul	Mystery	It is said that he wrote Murder on the Orient Express, one of his most famous books, in room 411, which is the favorite of the famous crime novel writer Agatha Christie.
Hotel California	There are multiple hotels with the same name.	Love	Two young people in love, whose families have forbidden them to meet, and the difficulties they experience in meeting.
Chelsea Hotel	New York	Art, Music, Magazine	Private lives of these famous names and some scandals and memories. These memories have been the subject of many movies, songs, and books.
Ritz Paris	Paris	Exclusiveness	The French elite of the last century stayed here and their memories are told.
Tile Prison	Latvia	Fear	Nazi stories and accounts of a female ghost who learns that her fiancé has been hanged in this hotel.
Destinations	**Location**	**Theme**	**Story**
Balıklıgöl	Şanlıurfa	Religion	Some events were experienced by the Prophet Abraham.
Hagia Sophia	İstanbul	History/Mystery	Stories of Islamic and Christian beliefs. Stories about being seen as a symbol of conquest. Secrets and legends that are said to be unsolved.
Ali-Nino Statue	Batum	Love	The events of two lovers struggling to reunite with each other.
Maiden's Tower	İstanbul	Love/Mystery	The king, whose girl is said to die, took his daughter under protection here, but could not prevent his daughter's death. Narratives about the love of Hero and Leandros.
The Taj Mahal	Agra	Love	Stories about the love of Shah Cihan for his wife.
Pyramids	Eygpt	History/Mystery	Secrets and legends about their construction.
Juliet's house	Verona	Love	Stories about the romance of Romeo and Juliet.
Nemrut Mountain	Adıyaman	History/Mystery	The stories about the King of Commagene. Legends of the time it was constructed.
Göbeklitepe	Şanlıurfa	Mystery	Legends and stories about its construction and the period in which it was constructed.

In destinations, developing a storytelling concept requires close contact between various stakeholders. Stakeholders include destination management organizations (DMO), public administration at the municipal or regional level, public and private sector attractions, various tourism-related service providers, and real storytellers (Mossberg et al., 2010; Olsson et al., 2013). Public tourism organizations, local municipalities, and district councils are likely to be central stakeholders in the agenda-setting, as the storytelling is explored in the context of local and regional tourism destinations. In addition, central private and public tourism stakeholders (e.g. restaurants, hotels, food producers, tourist attractions, museums, nature consultants) can play a role in storytelling. Stories can have economic, environmental, and socio-cultural implications for destination branding and destination development. With storytelling, the destination brand is strengthened and tourism activity can be achieved in the region outside the main season. This can benefit not only the stakeholders involved in storytelling activities but also the tourism sector in general. In the case of successful storytelling concepts, the destination image will be positively affected (Mossberg et al., 2010).

The stories of Turkey's popular touristic destinations are presented below. Based on the stories of these top destinations in Turkey, suggestions for storytelling processes and outputs are presented according to Mossberg et al. (2010). These suggestions relate to stakeholders, storytelling process, themes and destination branding.

Table 2. Storytelling process and outputs

Stakeholders	Storytelling Process	Theme	Destination Branding
Republic of Turkey Ministry of Culture and Tourism Municipalities Provincial Culture and Tourism Directorates Tourism Associations Tourism Managements	The theme of the story Platforms to tell stories Tangible story elements	Selection of themes	Image Identity Personality Destination Preference Recommendation

Cappadocia

Cappadocia is a destination called "a fairy tale country" as a unique region of the world. It is considered as a reflection of the harmony between nature and human beings. It is the town of light and dreams, where the historical and the mythical, the real and the imagined come together (Pekin, 2018).

It is often stated that the word 'Cappadocia' comes from the Persian "Katpatuka", which means "land of beautiful horses". Contrary to this popular belief, it does not only mean the land of beautiful horses. According to historians, Cappadocia is also referred to as a region where mules and sheep are famous, as well as horses. Although the origin of the name 'Cappadocia' is debated, it is also said that it may come from the word from "Kappadox" (Delice brook), a branch of the Red River used to surround the region and determine its borders, and from "Cappadox", the son of the Assyrian King Ninias. It is also stated that it is included in the new testament (works of the ambassadors, 2/9-11) (Korat, 2018; Pekin, 2018). Today, Cappadocia is known as the land of fairy chimneys where horses and valleys meet and horseback riding tours are among the important tourist attractions of the region.

Fairy chimneys, which consist of conical, pyramidal, interesting columns with a hat-like guard on top with various geographical influences, sometimes have double, triple caps or bodies at the top. This appearance makes them seem to belong to other worlds, especially the lunar surface. The people living in this area gave the name "fairy chimney" to these mysterious formations with rock caps. This mysterious nature, with its dreamy and fascinating scenes, has also been a shooting area for many movies and TV series (Pekin, 2018). French travel writer Paul Lucas described Cappadocia as a land of fairy tales. It was even claimed that Paul Lucas' mental health deteriorated in his country upon this statement, and those who came to the region to check this situation later confirmed the existence of different pyramids in the region (Uysun, 2011).

Numerous myths are common among the people in Cappadocia. A myth about rock formations is as follows (Pekin, 2018: 32):

"Once upon a time, people lived a life away from fighting and noise in Göreme. Since they did not know what a fight is, they did not have an army. However, when they woke up, they found an armed army of soldiers standing in rows outside of the village to attack. In great excitement, they began to wait to know the fact of death. But meanwhile, there was no movement in the army surrounding Goreme. After a few days of overwhelming waiting, an old shepherd could not stand it and said, "Whatever happens, I will go and look after this army." What should he see when he gets to the army: rows of soldiers turned into stones and turned into fairy chimneys. These fairy chimneys still stand at the foot of Aktepe and overlook Goreme from far away."

Another place in Cappadocia that Paul Lucas called "phantasmagoric" is Devrent valley. There are the most interesting fairy chimneys and petrified animal figures here. The most famous of these is the camel-shaped fairy chimney, and again in the valley, in the form of dinosaurs, turkeys, and seals (Pekin, 2018). The other name of this valley is *Pink* valley, but it is commonly called *Imagination* valley. By using your imagination, you can make the fairy chimneys in the valley look like many shapes (Turkey Tourkeys, 2020).

Paşabağları, which has three-headed fairy chimneys in Cappadocia, was known as "the valley of the priests" in the past. Fairy chimneys, both multi-body and multi-headed, have hosted clergy who led a reclusive life. For instance Gülyaz and Ölmez (1998: 52) wrote that *"St. Simeon, who lived a solitary life near Aleppo in the 5th century, escaped the public's attention after rumours that he had created miracles and lived in these columns, first two meters high and then 15 meters high. He would only go downstairs to get the food brought by his disciples"*. Later, Cappadocian hermits carved a fairy chimney and began to live in rock beds at a height of 10-15 meters (Gülyaz and Ölmez, 1998: 52). There are many stories and legends about fairies helping people, presenting sections from the lives of humans and fairies, and talking about fairies living in this region.

Table 3. Suggestions of storytelling process and outputs of cappadocia

Stakeholders	Storytelling Process	Theme	Destination Branding
UNESCO Republic of Türkiye Ministry of Culture and Tourism Nevşehir Municipality Nevşehir Provincial Directorate of Culture and Tourism Cappadocia Touristic Hoteliers and Operators Association Hotels and Travel Agencies Nevşehir Tourist Guides Association	The theme of the story: Fairy chimneys, landforms of the region, Christian saints Platforms for Stories: websites of public institutions, social media, printed and visual media, movies, and series Concrete Story Elements: fairy chimneys, churches, chapels, valleys	Mystery History Christianity Geography Culture	Image Identity Personality Destination Preference Recommendation

Themes such as Christianity, nature, culture, and mystery draw attention as important elements in the stories about Cappadocia. In Table 3, storytelling processes about Cappadocia are listed as fairy chimneys, volcanic formations, and Saints. Cappadocia has been used as an important belief centre for Muslims as well as Christians, and includes Islamic elements such as Hacı Bektaş Veli Dervish Lodge, Ottoman and Seljuk religious works, rock-hewn mosques. In addition, Cappadocia-themed stories, fairy chimneys, hot air balloons, and other natural formations are used as important visuals in many TV series and movies shot in the region. This situation increases the interest in Cappadocia, and the use of these concrete elements, especially for domestic tourists, can affect tour preferences and tour contents.

Ephesus and the Virgin Mary

Ephesus is located in the West of Turkey, called Minor Asia, next to Selçuk, a district of İzmir. Ephesus is also known as the land of Artemis, which remained under the rule of many empires such as the Ionians and Romans. Also, it is known that St. John and St. Paul also lived in Ephesus. Even when Jesus was crucified, Jesus said St. John "this is your mother" and said Meryem "this is your son" and this brings up many events related to Ephesus (Keskin, 2000).

In one of the Ephesus's foundation stories, the hunt of Androklos, the brave and nature-loving son of King Kodros of Athens, on the Aegean coast is described as follows (Türkoğlu, 2015: 11-12):

"Androcles took his friends and sailed to the shores of the Aegean... One day, the gods brought their ships to such a cove that Androklos was surprised when he saw that his wishes came true... They forgot the places they saw, even Athens and Greece (Hellas). Androcles, who decided to establish a city here, consulted with the oracles...The oracles at the Apollo temple said, "A fish will point you to this place, a pig will guide you"...I wonder what the oracle meant? One day, a boar began to flee when a fish fell into the bushes while they were roasting live fish. At that moment, lightning flashed in Androklos' brain and he followed the pig with his horse and dog. Finally, he drew his bow and knocked the pig to the ground. So he was going to establish the city here... Ephesus..."

The existence of the Virgin Mary's house in Ephesus is based on a different and equally interesting narrative. Although these narratives are supported by some concrete and abstract findings, it is also confirmed by the Papacy that the Virgin Mary came to Ephesus from Jerusalem and died here or ascended to the sky. Finding the house of the Virgin Mary is based on the perceptions of a priestess named Anne Catherine Emmerich. According to Catherine, who is in a trance state, the house of the Virgin Mary is:

"When entering Ephesus, it was on the left of the road coming from Jerusalem and on a mountain close to Ephesus... This house was square with high windows ... When you entered through the door, a fireplace with fire could be seen. That's why the walls were a little dark... There was a young woman with Virgin Mary who helped and brought her food... I saw a man who I thought was St. Jean entering in and going out of the house. Mary became very weak in the last days of her life. That's why everyone in Jerusalem thought that she would die and the grave was prepared but Virgin Mary got better... After coming to Ephesus, the Virgin Mary became weak and died at the age of 64. The saints and holy people around her organized a funeral for her and placed her specially prepared coffin in a cave two kilometers (?) from the house." Catherine also states that the Apostle Thomas was taken to the tomb after the funeral, when he was upset and wanted to see Mary, and said, "... but when the lid of the coffin was opened, they froze in amazement; Mary's body was not in the shroud. But the shroud was still intact ". It is expressed as "God had lifted Mary to heaven". Then the mouth of the cave where the grave was found was closed and the house was converted into a church (Türkoğlu, 2015: 158-160).

Later, Catherine's perceptions were published in 1878, some research was carried out to find the home of the Virgin Mary. With the encouragement of Marie de Mandat Grancey, Head Nurse of the French Hospital in Izmir, new research was started. Thus, a delegation headed by Henry Jung began to investigate it. They came to a small plain on Nightingale Mountain, tired and thirsty, with the book containing Catherine's perceptions in their hands. They wanted water from the women working in the tobacco field. The women said that there was a ruined monastery and that they could find water there. They forgot their thirst and ran in that direction and began to inspect the ruined monastery. The structure of the ruin conformed to what Catherine had described. One of them went up a hill and said that he saw the sea and Ephesus. The Christian Greeks in the region said that they held even a religious ceremony in memory of the Virgin Mary on 15 August. They called it Panaya Kapulu. It is thought that Catherine described a place so accurately she had never seen was not an accident (Türkoğlu, 2015).

Table 4. Suggestions of storytelling process and outputs of Ephesus

Stakeholders	Storytelling Process	Theme	Destination Branding
UNESCO Republic of Türkiye Ministry of Culture and Tourism İzmir Metropolitan Municipality Selçuk Municipality İzmir Provincial Directorate of Culture and Tourism Aegean Touristic Enterprises and Accommodation Association Hotels and Travel Agencies İzmir Tourist Guides Association	The theme of the story: Ancient city, greek mythology, Virgin Mary Platforms for Stories: websites of public institutions, social media, printed and visual media, movies Concrete Story Elements: Ephesus Ancient City, Virgin Mary's House/Church	Ancient Period Mythology History Christianity	Image Identity Personality Destination Preference Recommendation

Along with the importance of the ancient period in the stories about Ephesus, also the Virgin Mary. Jesus and his Apostles are mentioned. These stories contain important elements for destination branding, especially for the Christian world. In addition, the ancient city of Ephesus contains many treasures from the Ancient Period with its ruins. This destination includes story processes identified with the House of the Virgin Mary (as seen in Table 4). Visiting Virgin Mary's house and the ancient city of Ephesus are considered as the most important concrete elements of the destination. Thus, it can be said that these concrete elements will come to the fore in the Ephesus destination and it is remarkable for visitors, especially Christians.

Troy

Troy was excavated by Schliemann, Dörpfeld and Blegen, and, in particular, Schliemann's desire to find Priam's Troy and the city described in Homer's epic is mentioned. Thus, Schliemann thought that he had found traces of the Trojan War in his excavations and that he had the treasure of Priam. Later excavations were expanded and it was determined that the 1st Troy dates back to 3000 BC, the 9th Troy dates back to the Roman period (Akurgal, 2020), and the 10th Troy to the Byzantine period (Turkey Culture Portal, 2018). Like the foundation of Troy, the foundations of the Trojan War are based on a mythological story. At the same time, the Trojan War takes place in Homer's *Iliad* and includes the 51-day part of this war. The stories and myths about Troy are so common and extensive that they have an important place in Roman and Greek mythology. It is possible to see the subject of this war in many epics called kyklos (circle) in Greek (Erhat and Kadir, 2019). According to the legend, the reason for the Trojan war is explained as follows (Erdemoğlu, 2015: 18-21):

"At the wedding of King Peleus and the goddess Thetis (from this marriage, Achilles is born), the Goddess of Fighting Eris throws an apple with the inscription "To the Most Beautiful" among the guests at the feast, at the request of Zeus. Thereupon, a beauty debate begins between the goddesses, Hera, Athena, and Aphrodite. In such a case, Zeus, avoiding judgement, chooses -a mortal for this job- Paris, the son of the Trojan King Pyriamos. Paris, gives the apple to Aphrodite, who says that he will present the beautiful Helen, wife of Menelaos, as his wife... Paris goes to the royal palace of Menelaos, son of Atreus, in Sparta. Paris, who was hosted here, kidnaps Helena and returns to Troia. Upon this was realized, Menelaos seeks help from his brother, King Agamemnon of Mycenae. Agamemnon creates a huge army and navy with the support of Hera... They reach Troy and engage in a long-term siege. The gods were also divided into two regarding this war. Some support the Trojans, some support the invaders. In the tenth year of the war, with the advice of Athena, the city is captured by the wooden horse trick performed by Odysseus."

The Trojan War, dated to 1180 BC, was discussed by different authors after the *Iliad*, especially after Hector's death episode. Accordingly (Turkey Culture Portal, 2018):

"Achille dies when Paris, hiding behind a rock, shoots him from the heel with an arrow. In the ongoing war, when the two sides could not provide exact dominance to each other, the Achaeans played a very cunning game against the Trojans, and Odysseus, one of the wisest kings of the Achaeans, came up with the idea of making a wooden horse. According to the plan, the Achaeans seem to be withdrawing from the war and leave behind a very large wooden horse. While Odysseus and other distinguished commanders

hide inside the horse, the others go to sea and hide their ships behind Tenedos (Bozcaada), where the Trojans cannot see them. The Trojans, burning with longing for peace, take the wooden horse, which is a symbol of peace, inside. When the night is enthralled with the celebrations of peace and suffocating under the influence of alcohol, they are caught by surprise by the Achaean warriors coming out of the horse. Meanwhile, with the reinforcement of the Achaean army, which approached the walls of Troy, the city of Troy was completely destroyed."

Table 5. Suggestions of storytelling process and outputs of Troia

Stakeholders	Storytelling Process	Theme	Destination Branding
UNESCO Republic of Türkiye Ministry of Culture and Tourism Çanakkale Municipality Çanakkale Provincial Directorate of Culture and Tourism Çanakkale touristic hoteliers and investors Association Hotels and Travel Agencies Çanakkale Tourist Guides Association	The theme of the story: Ancient city, Greek mythology, War Platforms for Stories: websites of public institutions, social media, printed and visual media, movies, and series Concrete Story Elements: Troy Ancient City, The Trojan horse	Ancient Period Mythology History War Epic Heroism	Image Identity Personality Destination Preference Recommendation

Troy owes its importance today to the fact that it was the subject of many events in ancient times and epics such as the *Iliad* and *Odyssey*. Today, many stories and myths about Troy are told and the war of Troy is the subject of movies. This situation enabled Troy, which is also on the UNESCO world heritage list, to take place in stories with themes such as war, antiquity, mythology, and heroism. The Trojan Horse (Table 5), which is considered as a concrete element, is used as a symbol in many different areas today. In addition, the presence of Troy in world-famous and most-read epics is also important in terms of destination branding.

Göbekli Tepe

Göbekli Tepe, located in Şanlıurfa. Its reality was revealed by the excavations that started after 1994. Göbekli Tepe, with its history dating back to 12000 years ago, has been determined to be a cult centre consisting of structures with T-shaped columns built approximately 7500 years before the Egyptian Pyramids and Stonehenge in England. Göbekli Tepe provides data that can reshape history and it is understood that it is a unique sanctuary with its location, dimensions, and monumentality of buildings. This place, where a great organization and imagination came together at a time when people lived as hunter-gatherers, and where buildings that require advanced architecture are located, is called the "zero point of history". Göbekli Tepe, which has many unknowns, is also home to still questioning many figures and structures (Turkey Culture Portal, n. d.).

Klaus Schmidt, whose name is remembered with Göbekli Tepe, describes it as a rediscovery in the following way (Schmidt, 2018: 19-20):

"It was October 1994, and it was covered with the intense colours of the autumn sun. Our path was passing through a land that was slightly elevated, indistinguishable at a glance, with lots of basalt blocks piled on top of each other. There was no trace of prehistoric humans; neither a wall, no potsherds nor stone tools...an old man in the village, who was interested in our questions, unexpectedly said "yes" to our question whether there was a mountain called Göbekli Tepe nearby and whether there were flintstones there...Consisting of many rounded hills and pits in between, there was a hill rising from the plateau. It was almost like a postcard picture...". Again, Schmidt talks about a wishing tree, which he defines as a "holy place to be visited", visited by the public and tied to its branches. "In the largely barren and treeless South East Anatolia Region, it is mostly just that tree on the top of the mountain... We are looking at the tree at the top of the hill, which indicates the existence of a visiting place. We were sure we had found Göbekli Tepe..."

According to Schmidt's (2010) excavation report, there is a prevailing view that Göbekli Tepe was a great meeting place 12000 years ago, covered with monumental structures built for ceremonial purposes. In addition, it is stated that it will be possible to reach new information and data about the symbolic world of the people of this period with the excavations that will continue (Schmidt, 2010). On the obelisks, which reach up to 5 meters in length, there are various animal depictions of snakes, foxes, wild boars, and birds made in relief. It is thought that some of the obelisks may have human figures due to their arm and hand. The presence of wild, predatory animal statues in Göbekli Tepe brings to mind Kerberos, the protector of the underworld known from antiquity, and it is thought that these statues may be protective guards (Governorship of Şanlıurfa, n. d.).

Table 6. Suggestions of storytelling process and ouputs of Göbekli Tepe

Stakeholders	Storytelling Process	Theme	Destination Branding
UNESCO Republic of Türkiye Ministry of Culture and Tourism Şanlıurfa Metropolitan Municipality Şanlıurfa Provincial Directorate of Culture and Tourism Şanlıurfa Tourism Corporation Hotels and Travel Agencies Şanlıurfa Tourist Guides Association	The theme of the story: Neolithic age, figures, and steles (t-shaped stone) Platforms for Stories: websites of public institutions, social media, printed and visual media, movies, and series Concrete Story Elements: Excavation site, temples, mound, Göbeklitepe archaeological site	Mystery Neolithic Age Mythology History Human History Unknowns	Image Identity Personality Destination Preference Recommendation

Göbekli Tepe is still a destination open to new discoveries and new stories. Today, this archaeological site, which has values that can change what is known with the history of humanity, is the subject of TV series and movies that contain mystery and obscurity. As seen as concrete elements in Table 6, Göbekli Tepe ruins, mounds and other undiscovered areas can be visited today. Especially after the TV series "Atiye" shot in Göbekli Tepe, the interest in the destination increased and tour companies used the stories in the series as a marketing element and sloganized "Are you ready to discover Atiye's mysterious journey" (Kaya and Gümüş, 2020). Göbekli Tepe's known and unknown are also seen as an important theme in terms of storytelling. Because the unknown in the stories can also attract people and increase the interest in the destination.

Pamukkale/Hierapolis

Hierapolis was founded by Eumenes the Second, King of Pergamum in the 2nd century AD and takes its name from Hiera, the wife of Telephos, the legendary founder of Pergamum. Hierapolis, which survived for a long time with its Hellenistic urbanization structure, later lived through the Roman and Byzantine periods. It has been the centre of Christianity after the 4th century AD. The reason why it is important for Christians is due to the murder of St. Philip, one of Jesus' apostles, here in the 80s AD (Central Directorate Of Rotary Capital Management, 2014; Republic of Türkiye Ministry of Culture and Tourism, n. d.).

Hell's Gate/Plutonium in the ancient city is considered to be the "gateway to the country of the dead". This name comes from the cave where the thermal waters and the gas (carbon monoxide) that causes the death of those approaching it. This cave is considered to be the entrance to the underworld where God Pluto and his wife Persophone ruled. During the excavations, the source of the thermal waters, which created the famous Pamukkale travertine pools, was brought to light. They believed the cave was the entrance to Hell, and they sacrificed bulls to these gods who were suffocated by the carbon anhydride gas that leaked from the cave. There are two statues on the entrance of the cave. One of them is a snake and the other is a statue of Kerberos. *"The snake, which symbolizes everything that lives underground, is coiled and its head is turned towards the person approaching. The other Kerberos is 1.30 m., the three-headed "Hellhound of Hades", which is a unique work in the world in terms of its height and size, stands at the Gate of Hell. No living creature can approach here without being killed. Only the hero Heracles succeeded in catching him."* (D'Andria, 2021).

The ruins of the ancient city of Hierapolis are located in Pamukkale, which means "cotton castle" in Turkish and takes its name from its perfect whiteness and the flower-like appearance of the cotton plant (D'Andria, 2010). Since the waters of Pamukkale region contain high amounts of calcium carbonate, they form white pools (travertines). Travertine is a versatile rock formed by precipitation as a result of a chemical reaction and it is also called by this name because it has a cotton colour (white) (Governorship of Denizli, n. d.). The name of the Hierapolis thermal settlement as the Holy City in the archaeological literature is due to the presence of temples, structures, and hot water sources known in the city. Therefore, together with Hierapolis, medicine was considered an important science in the cities of Laodikeia, Attouda, Karura, Herakleia Salbace, and Eumeneia in this area, and thermal waters were used for treatment and rehabilitation in this environment (Denizli Provincial Directorate of Culture and Tourism, n. d. a). The Pamukkale legend about these thermal waters is as follows (Denizli Provincial Directorate of Culture and Tourism, n. d. b):

"Long ago, there was a poor woodcutter family living on the skirts of Çökelez Mountain. The daughter of this family was so ugly that the mothers of the boys changed their ways when they saw her. She didn't care about her poverty, but she was obsessed with her ugliness. She jumped to the void from the skirts of Çökelez Mountain. She fell quickly into the pool full of water and sediment. Here she was unconscious in the water for a long time. At that time, this water drowned that ugly girl in beauty. The son of Denizli Bey (Governor of Denizli), who was passing by, saw the beautiful girl covered in blood. He took the woodcutter girl to his horse and took her home. The girl recovered and they got married. After that day,

women started to visit these hot springs to be beautiful. All women who want to be beautiful have thrown themselves into this water so far."

Table 7. Suggestions of storytelling process and outputs of Pamukkale/Hierapolis

Stakeholders	Storytelling Process	Theme	Destination Branding
UNESCO Republic of Türkiye Ministry of Culture and Tourism Denizli Metropolitan Municipality Pamukkale Municipality Denizli Provincial Directorate of Culture and Tourism Denizli touristic hoteliers and investors Association Hotels and Travel Agencies	The theme of the story: Ancient city, Greek mythology, regional landforms (cotton castle) Platforms for Stories: websites of public institutions, social media, printed and visual media, movies Concrete Story Elements: Hierapolis ancient city, Hell's Gate, travertines, thermal springs	Mystery Ancient Period Mythology History Geography Christianity	Image Identity Personality Destination Preference Recommendation

The ancient city of Hierapolis and Pamukkale are close to each other and these destinations stand out with their thermal springs. This thermal spring gives Pamukkale the appearance of a castle made of cotton and this water is approached with a mystical feature. This situation is associated with the metaphor of turning the stone into cotton and the ugly into beautiful in the stories and has a mysterious and geographical theme (Table 7). The importance of Hierapolis increases even more when it is combined with mythology. In addition, the murder of St. Philip here also makes the destination religiously important. When all these are combined, the destination has stories that contain many themes, and this richness can be explained by many variables that can affect travel motivation.

Hagia Sophia

Hagia Sophia has survived from the 4th century and when it was first built, it was called Megale Ekklesia (Great Church), and since the 5th century, its name has been changed to Sophia. This name comes from Theia Sophia, the Holy Wisdom, which is the second of the holy trinity in Christianity. It was known as the Great Church by the Byzantine people, and after 1453, its name was partially changed and called Hagia Sophia (Erdoğan, 2012). First construction of Hagia Sophia dates back to Emperor Constantine and the name of the present building is often mentioned with Emperor Justinian. It is said that when Justinian entered Constantinople to inaugurate Hagia Sophia, *"he escaped from the patriarch's hands, ran from the imperial gates to Ambon alone and raised his arms: "Thank God, he allowed me to perform such a work! I beat you, Solomon," he* said (Chuvin, 2018: 12). When Hagia Sophia was built, there was no building bigger than it. That is why it is compared to the largest church in the world, even the largest structure: the Temple of Jerusalem and Solomon who founded it (Chuvin, 2018). Hagia Sophia is very important for both Christians and Muslims. The myths, histories, and sources about this structure also reflect the values and beliefs of each empire (Yerasimos, 1993).

There are many sources about the legends, stories, and myths of Hagia Sophia, and it is known that these sources cover Byzantine, Roman, Ottoman, and a period close to the present day and starting before its construction and covering its entire history. About the Hollow Column (Wishing Column) in Hagia Sophia, two legends of the Virgin Mary are common. The first of these is about the relief of

Emperor Justinian's headache, and the second is about the crucifixion of Jesus. According to these legends (Arslan, 2009):

"During the construction of Hagia Sophia, Emperor Justinian often came to Hagia Sophia to control the construction. One day, while the Emperor was walking around Hagia Sophia, he fell ill and rested his head on this column and prayed. After a while, he lifted his head, and then the disturbance and the head pain completely passes. The emperor looked at the column carefully then a small hole occurred in the column and saw water filtering like tears from the hole. He thinks that this tear is the tear of the Virgin Mary and was sent by God to heal him. The public would be informed of these signs and accepts the holy column. After that, those who want their illness to be healed begin to come to Hagia Sophia. They believe in that putting their fingers into this hole on this column and by spreading water to the ill parts of their body will heal them."

According to other legends (Arslan, 2009), "the crying column in Hagia Sofia is a column of the Virgin Mary's house. One day, somebody says to the Virgin Mary that Jesus was caught and he was tortured. Virgin Mary cries so much that with that sadness her tears melt the column as soon as they drop. While the Hagia Sophia was being built, the emperor brought this column from the house of the Virgin Mary and had it put in Hagia Sophia to consecrate the church."

Anyone who has a wish put his/her fingers inside the hole which was drilled with Mary's tears and turns his/her fingers and wishes into something. Mostly, girls who want to make a fortune and women who want to have a son come here to make wishes.

Hagia Sophia also has special importance for Muslims, and there are many miracles and legend, both about the conquest of Istanbul and about the sanctity of Hagia Sophia. A conversation about Hagia Sophia between Hz. Muhammad and Gabriel in Miraj are as follows (Arslan, 2009: 186-187):

"One night Gabriel comes, He invites Hz. Muhammad to Miraj. Gabriel and Hz. Muhammad begins to wander through the sky layers and the heavenly layers. Suddenly, they enter the heaven of Firdevs. There they see a lofty maqam resembling a mosque. Inside this building, there are forty ruby pillars, the inside of which is covered with emerald and turquoise stones. Muhammad said, "O my brother Gabriel! Where is this beautiful place?" he asks. Gabriel said, "O Muhammad! Allah has created that station for your ummah who has believed and united with Allah. They call it Camiü'l-Kübra (Great Mosque). Similar to this office is in the city of Constantinople, which is surrounded by sea on three sides and land on one side. In this city, there is a beautiful place of worship called "Sofiya" and a lofty office. On behalf of this, it is called Camiü's-Suğra (Small Mosque), and it is the symbol of the great maqam you saw in the world. Your nation will be blessed to worship in it he replies."

In addition, it is told that Muslims worshiping in this building will be given the good news of Paradise and it is very important for Muslims to see Hagia Sophia as sacred and to be able to worship there. According to another legend (Arslan, 2009: 175), *"there is a picture of an angel in each of the four corners of the dome. One of these pictures represents Gabriel, one Michael, one Israfil, and one Azrael. When Hz. Muhammad was born, these angels spoke with their mouths on their navels: The image of Gabriel is in the East and reports the signs and events that will happen. Michael's image informs that the enemy will appear in the West and there will be famine and poverty. The image of Israfil informs about the events that will take place in the North. The Azrael image gives the news of the death of various sultans all over the world. After the Hz. Muhammad the talismans of these pictures are broken."* Despite this, when looking at the historical facts, four angel figures are seen to be perfect to each other on the pendants that provide the transition to the main dome. According to the researchers, these angels represent the Seraphim, one of the highest order six-winged angels who guard the throne of God as stated in the Bible. According to another belief, these angels will protect Hagia Sophia from all kinds of disasters until the Day of Judgement (Arslan, 2009).

Table 8. Suggestions of storytelling process and outputs of Hagia Sophia

Stakeholders	Storytelling Process	Theme	Destination Branding
UNESCO Republic of Türkiye Ministry of Culture and Tourism İstanbul Metropolitan Municipality İstanbul Provincial Directorate of Culture and Tourism Hotel Association of Turkey Tourist Guides Association of Turkey Association of Turkey Travel Agencies	The theme of the story: Hagia Sophia Mosque, Cathedral, Byzantium, Christianity, Islam Platforms for Stories: websites of public institutions, social media, printed and visual media, movies Concrete Story Elements: Hagia Sophia Mosque, Column, Figures, Frescoes	Mystery Belief Mythology History Christianity Islam	Image Identity Personality Destination Preference Recommendation

With its stories dating back to before its construction, Hagia Sophia also has been the subject of many stories after its construction. Hagia Sophia, which is very valuable in terms of beliefs, has a different position and importance for Christians and Muslims. In the story process, it is accepted as a great cathedral for Christians and as a great mosque for Muslims. In addition, as seen in Table 8, Hagia Sophia has a theme that includes different beliefs as well as mystery in the stories.

Conclusion and Discussion

Storytelling has cognitive, emotional and behavioural implications. Storytelling attracts the attention of audiences, promises more than technical or commercial content, increases the interest in the destination at low cost, positively affects the visibility and image perception of the destination, and makes the destinations more memorable. In addition, storytelling provides distinctiveness in destination brand communication and increases the brand competitiveness of destinations (Youssef, 2018). Promotion of destination brands can be faster thanks to word-of-mouth communication. Stories also facilitate word of mouth promotion (Mossberg et al., 2010; Akgün et al., 2015).

Stories and myths have attracted people's attention throughout history and are seen as an important attraction for tourists to visit destinations. In fact, stories about destinations have an impact on the revisit of tourists (Youssef et al., 2018). Therefore, it is seen as an important factor in the development of destinations (Mossberg et al., 2010), destination marketing (Pinzaru, 2012), and increasing visits as a tourist attraction factor (Soon Choi, 2016). Especially, when destination storytelling is done through social media, stories can reach more people (Lund, 2017; Pearce et al., 2019). Also authenticity and originality of the brand is considered as important brand reputation elements (Bosch et al., 2005). Considering that destination stories reflect the authenticity and originality of the destination (Järv, 2010). These stories can be used as a means of destination marketing, awareness and policy in destination branding.

Storytelling is considered one of the pillars of strategic destination branding tools by giving the destination a personality and identity (Yavuz et al., 2016). It is known that there are stories and myths about many destinations and touristic attractions such as these destinations in Anatolia. For this reason, it is considered important to examine the stories about other touristic destinations in addition to the top touristic destinations in Anatolia and to evaluate touristic destinations as tools that can help them gain identity. In addition, these stories can be used both electronically and verbally on social media platforms for destination marketing. Destinations are more easily remembered with a story and this story can help developing that destination's identity. An object, a rock, a volcano, a ruin, or a person as an element of a story can make the destination memorable, and thus destination identity can be created more easily because the destination storytelling process not only provides strategic communication (Bassano et al., 2019), it also helps tourists understand destinations better and the destination's personality (Moin et al., 2020).

As a result of this study, the use of destination storytelling as a destination branding tool is presented in Figure 4. The process, which starts with listening and reviewing the existing stories, continues with the selection of the mediators that can be used to spread the reputation of the destination brand and the story themes to be created. In the next stage, the story creation process begins. At this stage, tangible and intangible values that can be used in stories are examined. As a result of the storytelling of these values, the perception of the destination identity, image, and personality can be formed. As a result of presenting this perception and features through the right marketing and communication channels, destination branding can be achieved.

Figure 4. Destination storytelling as destination branding tool

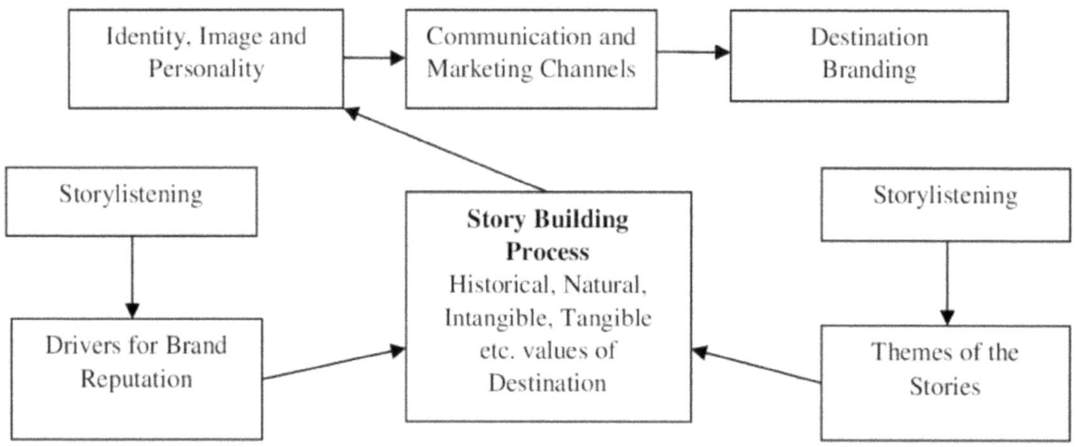

Anatolian lands have been the subject of not only physical events but also many mythological events throughout history. From this point of view, this land contains tangible (ancient cities, museums, ruins, etc.) and intangible (myths, stories, epics, etc.) cultural richness that makes people to experience it. In this study, Cappadocia, Pamukkale, Ephesus, Troy, Hagia Sophia, Göbekli Tepe, which are the most visited places in Turkey, important destinations for visitors from around the world, and also the subject of many stories, are discussed within the scope of storytelling and destination branding. In the tables above, the storytelling process and output information for these destinations are presented as tools to use in destination branding. Authors suggest about the themes of the destination stories, tangible and intangible outputs to be used in the stories and stakeholders playing a role in the destination branding. In addition, some stories about these destinations were presented and the impact and potential benefit of these stories on destination branding were evaluated together with the tangible elements in the destination. Therefore, these places, which already contain historical ruins that can be visited and physically witnessed, can be made more attractive with stories and can get more attention from visitors. Thus, instead of marketing these places physically, Turkey can follow a more effective marketing strategy enriched with intangible elements.

FUTURE RESEARCH DIRECTIONS

Storytelling is seen and implemented as a new and effective method for destination branding. Touristic destination managers try new ways to differentiate from other destinations and create attractiveness, and they use storytelling in this way. This draws attention to storytelling in destination branding in order to highlight the destinations' own characteristics and their own symbolic variables. It is seen that there are very few studies in Turkey related to the use of storytelling in destination branding. In fact, it can be said that these studies have started to be studied all over the world recently. Therefore, in future studies, the role of storytelling in the branding process of many destinations both in Turkey and in the world can be mentioned.

In this study, the storytelling processes of a limited number of destinations in Turkey were discussed. In future studies, it may be useful to consider a specific destination in its entirety in terms of addressing the destination with all its variables. In addition, in future studies, interviews with local people in tourist destinations can contribute to the emergence of more specific stories and authentic values in branding can be evaluated in a more original way. Stories about a specific destination can be handled in many ways and used in the destination marketing process. And also few research about digital storytelling has been done about the destinations to be visited within the scope of culture and religious tourism. Each of the digital platforms may have different features. Considering these features, story creation and marketing processes can be evaluated in order to communicate with tourists and to brand the destination.

REFERENCES

Aaker, J. L. (1997). Dimensions of brand personality. *JMR, Journal of Marketing Research*, *34*(3), 347–356. doi:10.1177/002224379703400304

Adaval, R., & Wyer, R. S. Jr. (1998). The Role of narratives in consumer information processing. *Journal of Consumer Psychology*, *7*(3), 207–245. doi:10.120715327663jcp0703_01

Akgün, A. E., Keskin, H., Ayar, H., & Erdoğan, E. (2015). The influence of storytelling approach in travel writings on readers' empathy and travel intentions. *Procedia: Social and Behavioral Sciences*, *207*, 577–586. doi:10.1016/j.sbspro.2015.10.129

Akurgal, E. (2020). *Anadolu Uygarlıkları*. Phoenix Yayınevi.

American Marketing Association. (n.d.). *Definition of Brand*. Available at: https://www.ama.org/the-definition-of-marketing-what-is-marketing/

Anholt, S. (2004). Editor's foreword to the first issue. *Place Branding and Public Diplomacy*, *1*(1), 4–11. doi:10.1057/palgrave.pb.5990001

Anholt, S. (2006). The Anholt-GMI City Brands Index. How the world sees the world's cities. *Place Branding*, *2*(1), 18–31. doi:10.1057/palgrave.pb.5990042

Arnold, D. (1992). *The handbook of brand management*. Century Business: The Economist Books.

Arslan, F. (2009). *Legends of Hagia Sophia (determine-examine)* [Unpublished doctoral dissertation]. İstanbul University, İstanbul, Turkey.

Åstrøm, J. K. (2019). Why theming? Identifying the purposes of theming in tourism. *Journal of Quality Assurance in Hospitality & Tourism*, 1–22. doi:10.1080/1528008X.2019.1658147

Bassano, C., Barile, S., Piciocchi, P., Spohrer, J. C., Iandolo, F., & Fisk, R. (2019). Storytelling about places: Tourism marketing in the digital age. *Cities (London, England)*, *87*, 10–20. doi:10.1016/j.cities.2018.12.025

Bosch, A. L. M., John, M. D. T., & Elving, W. J. L. (2005). How corporate visual identity supports reputation. *Corporate Communications*, *10*(1), 108–116. doi:10.1108/13563280510596925

Campelo, A., Aitken, R., Thyne, M., & Gnoth, J. (2013). Sense of Place. *Journal of Travel Research*, *53*(2), 154–166. doi:10.1177/0047287513496474

Central Directorate Of Rotary Capital Management. (2014). *Hierapolis (Pamukkale) Örenyeri*. Available at: http://www.dosim.gov.tr/muze/200

Chan, C. S., Peters, M., & Marafa, L. M. (2015). Public parks in city branding: Perceptions of visitors vis-à-vis residents in Hong Kong. *Urban Forestry & Urban Greening*, *14*(4), 1157–1165. doi:10.1016/j.ufug.2015.10.016

Chuvin, P. (2018). Introduction. In Mabeyinci Pavlos (Paulos Silentarios), Ayasofya'nın Betimi (pp. 11-26). Kırmızı Kedi Yayınevi.

D'Andria, F. (2010). *Hierapolis (Pamukkale)*. Ege Yayınları.

D'Andria, F. (2021). *Pamukkale Hierapolis*. Available at: https://denizli.ktb.gov.tr/TR-211745/pamukkale-hierapolis.html

Da Costa, C. (2019). *3 reasons why brand storytelling is the future of marketing*. Available at: https://www.forbes.com/sites/celinnedacosta/2019/01/31/3-reasons-why-brand-storytelling-is-the-future-of-marketing/?sh=7fd1667455ff

Delgado-Ballester, E., & Fernández-Sabiote, E. (2016). "Once upon a brand": Storytelling practices by Spanish brands. *Spanish Journal of Marketing*, *20*(2), 115–131. doi:10.1016/j.sjme.2016.06.001

Denizli Provincial Directorate of Culture and Tourism (n.d.a). *Tarih boyunca su ve sağlık turizmi şehri; Denizli.* Available at: https://denizli.ktb.gov.tr/TR-211794/termal-saglik-turizmi.html

Denizli Provincial Directorate of Culture and Tourism (n.d.b). *Efsaneler.* Available at: https://denizli.ktb.gov.tr/TR-211827/efsaneler.html

Ekinci, Y. (2003). From destination image to destination branding: An emerging area of research. *e-Review of Tourism Research (eRTR), 1*(2), 21-24.

Ekinci, Y., & Hosany, S. (2006). Destination Personality: An Application of Brand Personality to Tourism Destinations. *Journal of Travel Research, 45*(2), 127–139. doi:10.1177/0047287506291603

Erdemoğlu, B. (2015). *Troia ve Çevresi* (E. Yazıcı, Ed.). Uranüs Fotoğraf Ajansı ve Yayıncılık.

Erdoğan, E. G. (2012). Bizans dönemi'nde Ayasofya, tarihçesi ve mimari özellikleri hakkında genel bilgiler. *İstanbul. Journal of Social Sciences, 1*, 1–7.

Erhat, A., & Kadir, A. (2019). Introduction. In İlyada (pp. 1-84). Türkiye İş Bankası Kültür Yayınları.

Febby, A., & Hifni, A. (2018). Destination brand storytelling: Analisis naratif video the journey to a wonderful world kementerian pariwisata. *Jurnal Komunikasi Indonesia, 7*(1), 73–86.

Garcia, J. A., Gomez, M., & Molina, A. (2012). A Destination-Branding model: An empirical analysis based on stakeholders. *Tourism Management, 33*(3), 646–661. doi:10.1016/j.tourman.2011.07.006

Gardner, B. B., & Levy, S. J. (1955). The product and the brand. *Harvard Business Review, 33*, 33–39.

Gottschall, J. (2012). *The storytelling animal: How stories make us human.* Houghton Mifflin Harcourt.

Governorship of Denizli. (n.d.). *Pamukkale travertenleri.* Available at http://www.denizli.gov.tr/pamukkale

Governorship of Şanlıurfa (n.d.). *Göbeklitepe.* Available at http://www.sanliurfa.gov.tr/gobeklitepe

Gülyaz, M. E., & Ölmez, İ. (1998). *Kapadokya.* Dünya Kitap.

Hagen, O. (2008). Driving environmental innovation with corporate storytelling: Is radical innovation possible without incoherence? *International Journal of Innovation and Sustainable Development, 3*(4), 217–233. doi:10.1504/IJISD.2008.022227

Järv, R. (2010). Fairy tales and tourist trips. *Fabula, 51*(3–4), 281–294. doi:10.1515/FABL.2010.026

Kavaratzis, M. (2004). From city marketing to city branding: Towards a theoretical framework for developing city brands. *Place Branding, 1*(1), 58–73. doi:10.1057/palgrave.pb.5990005

Kavaratzis, M., & Ashworth, G. J. (2006). City branding: An effective assertion of identity or a transitory marketing trick? *Place Branding, 2*(3), 183–194. doi:10.1057/palgrave.pb.5990056

Kaya, B., & Gümüş, M. (2020). The importance of storytelling as a postmodern tool in the context of tourism. *Journal Of Applied Tourism Research, 1*(2), 115–132.

Kerr, G. (2005). From destination brand to location brand. *Journal of Brand Management, 13*(4-5), 276–283. doi:10.1057/palgrave.bm.2540271

Keskin, H., Akgün, A., Zehir, C., & Ayar, H. (2016). Tales of cities: City branding through storytelling. *Journal of Global Strategic Management, 10*(1), 31–41. doi:10.20460/JGSM.20161022384

Keskin, N. (2000). *Efes*. Keskin Color.

Korat, G. (2018). *Taş kapıdan taç kapıya Kapadokya*. Doğan Kitap.

Kotler, P., Armstrong, G., Saunders, J., & Wong, V. (1996). *Principles of marketing*. Prentice Hall Europe.

Leisen, B. (2001). Image Segmentation: The Case of a tourism destination. *Journal of Services Marketing, 15*(1), 49–66. doi:10.1108/08876040110381517

Lewis, P. J. (2011). Storytelling as Research/Research as Storytelling. *Qualitative Inquiry, 17*(6), 505–510. doi:10.1177/1077800411409883

Lund, N. F., Cohen, S. A., & Scarles, C. (2017). The power of social media storytelling in destination branding. *Journal of Destination Marketing & Management, 8*, 271–280. doi:10.1016/j.jdmm.2017.05.003

Lundqvist, A., Liljander, V., Gummerus, J., & van Riel, A. (2012). The impact of storytelling on the consumer brand experience: The case of a firm-originated story. *Journal of Brand Management, 20*(4), 283–297. doi:10.1057/bm.2012.15

Maurya, U. K., & Mishra, P. (2012). What is a brand? A perspective on brand meaning. *European Journal of Business and Management, 4*(3), 122–133.

Moin, S. M. A., Hosany, S., & O'Brien, J. (2020). Storytelling in destination brands' promotional videos. *Tourism Management Perspectives, 34*, 100639. doi:10.1016/j.tmp.2020.100639

Moscardo, G. (2010) The shaping of tourist experience: the importance of stories and themes. In M. Morgan., P. Lugosi & B. Ritchie (Eds.), The Tourism and Leisure Experience: Consumer and Managerial Perspectives. New York: Channel View Publications. doi:10.21832/9781845411503-006

Moscardo, G. (2021). The story turn in tourism: Forces and futures. *Journal of Tourism Futures, 7*(2), 168–173. doi:10.1108/JTF-11-2019-0131

Mossberg, L., Therkelsen, A., Huijbens, E., Björk, P., & Olsson, A. (2010). *Storytelling and destination development.* Nordic Innovation Centre (NICe) project number: 08041. Available at: https://www.nordicinnovation.org/2010/storytelling-and-destination-development

Olsson, A. K., Therkelsen, A., & Mossberg, L. (2013). Making an effort for free – volunteers' roles in destination-based storytelling. *Current Issues in Tourism, 19*(7), 659–679. doi:10.1080/13683500.2013.784242

Pawaskar, P., & Goel, M. (2014). A Conceptual Model: Multisensory Marketing and Destination Branding. *Procedia Economics and Finance, 11*, 255–267. doi:10.1016/S2212-5671(14)00194-4

Pearce, J., & Moscardo, G. (2019). The role of stories in travel posts to social media. In M. Mansson, A. Buchmann, C. Cassinger, & L. Eskilsson (Eds.), *Routledge Companion to Media and Tourism*. Routledge.

Pekin, F. (2018). *Kapadokya: kayalardaki şiirsellik*. İletişim Yayınları.

Pinzaru, F. (2012). From conjunctural urban storytelling to city branding: An empirical model for bucharest. *Management Research And Practice*, *4*(2), 33–44.

Rasoolimanesh, S. M., Seyfi, S., Rastegar, R., & Hall, C. M. (2021). Destination image during the COVID-19 pandemic and future travel behaviour: The moderating role of past experience. *Journal of Destination Marketing & Management*, *21*, 100620. doi:10.1016/j.jdmm.2021.100620

Reamy, T. (2002). *Imparting knowledge through storytelling*. Available at: https://www.kmworld.com/Articles/Editorial/Features/Imparting-knowledge-throughstorytelling-Part-1-of-a-two-part-article-9358.aspx

Republic of Türkiye Ministry of Culture and Tourism. (n.d.). *Hierapolis archaeological site*. Available at: https://muze.gov.tr/muze-detay?SectionId=PMK01&DistId=PMK

Ritchie, J. R., & Ritchie, J. B. (1998). The branding of tourism destinations. *Annual Congress of the International Association of Scientific Experts in Tourism*.

Ryu, K., Lehto, X. Y., Gordon, S. E., & Fu, X. (2018). Compelling brand storytelling for luxury hotels. *International Journal of Hospitality Management*, *74*, 22–29. doi:10.1016/j.ijhm.2018.02.002

Salazar, N. B. (2012). Tourism imaginaries: A conceptual approach. *Annals of Tourism Research*, *39*(2), 863–882. doi:10.1016/j.annals.2011.10.004

Schmidt, K. (2010). *Göbekli Tepe Kazısı 2010 Yılı Raporu*. Directorate General of Cultural Assets and Museums, Publishing Number: 155-3. Available at: http://www.kulturvarliklari.gov.tr/Eklenti/4653,33kazi3.pdf?0

Schmidt, K. (2018). *Göbekli Tepe: en eski tapınağı yapanlar* (R. Arslan, Trans.). Arkeoloji ve Sanat Yayınları. (Original book published 2006)

Skift. (2020). *Destination storytelling 2020: A guide to telling great stories*. Available at: https://skift.com/destination2020/

Smith, K., & Wintrob, M. (2013). *Brand storytelling: a framework for activation.* Available at: https://onlinelibrary.wiley.com/doi/full/10.1111/drev.10227

Soon Choi, S. (2016). A study on the effect of tourism storytelling of tourism destination brand value and tourist behavioural intentions. *Indian Journal of Science and Technology*, *9*(46). Advance online publication. doi:10.17485/ijst/2016/v9i46/107387

Souiden, N., Ladhari, R., & Chiadmi, N. E. (2017). Destination personality and destination image. *Journal of Hospitality and Tourism Management*, *32*, 54–70. doi:10.1016/j.jhtm.2017.04.003

Stern, B. B. (2006). What does brand mean? Historical-analysis method and construct definition. *Journal of the Academy of Marketing Science*, *34*(2), 216–223. doi:10.1177/0092070305284991

The Economic Times. (2017). *Do people buy goods and services, or stories and magic?* Available at: https://economictimes.indiatimes.com/do-people-buy-goods-and-services-or-stories-and-magic/articleshow/57016683.cms

Turkey Culture Portal. (2018). *Troya antik kenti*. Available at: https://www.kulturportali.gov.tr/portal/troyaantikkenti

Turkey Culture Portal. (n.d.). *Göbeklitepe*. Available at: https://www.kulturportali.gov.tr/portal/gobeklitepe

Turkey Tourkeys. (2020). About Cappadocia. Available at: https://www.turkeytourkeys.com/about-cappadocia/

Türkoğlu, S. (2015). *Efes'in öyküsü*. Arkeoloji ve Sanat Yayınları.

Tussyadiah, I., & Fesenmaier, D. (2008). Marketing places through first-person stories—An analysis of Pennsylvania roadtripper blogs. *Journal of Travel & Tourism Marketing*, *25*(3–4), 299–311. doi:10.1080/10548400802508358

Ülkü, A. (2021). *Dijital dönemde hikaye/hikaye anlatıcılığı ve destinasyonlar* [Conference Presentation]. 2nd International Congress of New Generations and New Trends in Tourism, Sakarya, Türkiye.

Uysun, M. (2011). *Kapadokya kaya kiliseleri ikonografisi*. Doğu Ofset.

Vermeulen, M. (2002). The Netherlands, Holiday Country. In V. Patteeuw (Ed.), *City Branding: Image Building and Building Images*. NAI Uitgevers.

Wood, N. T., & Muñoz, C. L. (2007). 'No rules, just right' or is it? The role of themed restaurants as cultural ambassadors. *Tourism and Hospitality Research*, *7*(3–4), 242–255. doi:10.1057/palgrave.thr.6050047

Yavuz, M. C., Sümbül, M., Ergeç, N. E., & Derdiyok, Ç. (2016). Storytelling in destination brand communication: A qualitative analysis. *International Interdisciplinary Business-Economics Advancement Journal*, *1*(2), 63–72. doi:10.5038/2640-6489.1.2.1008

Yerasimos S. (1993). *Türk metinlerinde Konstantiniye ve Ayasofya Efsaneleri* (Ş. Tekeli, Trans). İletişim Yayınları. (Original book published 1990)

Youssef, K. B., Leichtb, T., & Marongiuc, L. (2018). Storytelling in the context of destination marketing: An analysis of conceptualizations and impact measurement. *Journal of Strategic Marketing*, *26*(2), 1–18. doi:10.1080/0965254X.2018.1464498

Zarezadeh, Z. Z., Rastegar, H. R., & Gretzel, U. (2018). Reviewing the Past to Inform the Future: A Literature Review of Social Media in Tourism. *Czech Journal of Tourism*, *7*(2), 115–131. doi:10.1515/cjot-2018-0006

ADDITIONAL READING

Bassano, C., Barile, S., Piciocchi, P., Spohrer, J. C., Iandolo, F., & Fisk, R. (2019). Storytelling about places: Tourism marketing in the digital age. *Cities (London, England)*, *87*, 10–20. doi:10.1016/j.cities.2018.12.025

Da Costa, C. (2019). *3 reasons why brand storytelling is the future of marketing*. Available at: https://www.forbes.com/sites/celinnedacosta/2019/01/31/3-reasons-why-brand-storytelling-is-the-future-of-marketing/?sh=7fd1667455ff

Lund, N. F., Cohen, S. A., & Scarles, C. (2017). The power of social media storytelling in destination branding. *Journal of Destination Marketing & Management*, *8*, 271–280. doi:10.1016/j.jdmm.2017.05.003

Moin, S. M. A., Hosany, S., & O'Brien, J. (2020). Storytelling in destination brands' promotional videos. *Tourism Management Perspectives*, *34*, 100639. doi:10.1016/j.tmp.2020.100639

Moscardo, G. (2010) The shaping of tourist experience: the importance of stories and themes. In M. Morgan., P. Lugosi & B. Ritchie (Eds.), The Tourism and Leisure Experience: Consumer and Managerial Perspectives. Channel View Publications. doi:10.21832/9781845411503-006

Mossberg, L., Therkelsen, A., Huijbens, E., Björk, P., & Olsson, A. (2010). *Storytelling and destination development*. Nordic Innovation Centre (NICe) project number: 08041. Available at: https://www.nordicinnovation.org/2010/storytelling-and-destination-development

Olsson, A. K., Therkelsen, A., & Mossberg, L. (2013). Making an effort for free – volunteers' roles in destination-based storytelling. *Current Issues in Tourism*, *19*(7), 659–679. doi:10.1080/13683500.2013.784242

Ritchie, J. R., & Ritchie, J. B. (1998). The branding of tourism destinations. *Annual Congress of the International Association of Scientific Experts in Tourism*.

KEY TERMS AND DEFINITIONS

Destination Branding: A marking that differentiates a touristic destination from its competitors and enables the marketing of destination-specific features to ensure that the destination is positioned and remembered in people's minds.

Destination Branding Process: Operations related to the branding of the destination. These are operations to create destination-specific branding, such as improving the variables that establish a relationship between the destination and the tourist, creating a positive image specific to the destination, increasing the preferability of the destination, and bringing all stakeholders together.

Destination Identity: Symbolic, experiential, and emotional values of the destination. At the same time, it is a factor that creates the image of the destination and ensures that the destination is perceived by the tourists.

Story Theme: The main factor that creates an existing story with symbolic meanings and enables it to be positioned in people's minds.

Storytelling Management: The process of creating, maintaining, differentiating, and transferring stories about a place, person, event, or an object.

Chapter 9
Storytelling by the Sea:
The Legend of Ílhavo as a Reinvented Tourism Destination

Paulo Teixeira Costa
https://orcid.org/0000-0003-2449-8555
Instituto Superior de Ciências da Informação e da Administração, Aveiro, Portugal

Ana Sofia Pereira
https://orcid.org/0000-0003-1829-0205
Instituto Superior de Ciências da Informação e da Administração, Aveiro, Portugal

ABSTRACT

Global competition has challenged destinations in devising new ways to enhance image and awareness, so they can be perceived as attractive to visitors. Storytelling is gaining increasing importance in destination marketing by adding emotional value to products. Stories impact positively on tourist involvement in experiences, as well as on destination differentiation. Furthermore, visitors' identification with destination stories seems to increase whenever stories are perceived as authentic. Ílhavo's history as a fishing community deeply connected with the sea developed from traditional cod fishing techniques. The history of this Portuguese community has helped shaping a new and distinctive narrative that sustained its positioning as a cultural destination around the identity "the Portuguese cod fishing capital", the gem of Portuguese cuisine, and a major symbol of the country itself.

INTRODUCTION

Marketing has always used some sort of storytelling technique to engage audiences. Stories are a useful tool for brands and agencies as they help engage with people, influence their decision and work as triggers for purchase. Destination marketing is no exception. Nonetheless, this field of work has its own specificities when it comes to the use of storytelling.

According to studies on destination marketing and the role of storytelling, marketing strategies aiming at promoting a destination, more than focusing on beautiful landscape, imagery and locations,

DOI: 10.4018/978-1-6684-3436-9.ch009

should focus on personal and local stories derived from myths and tradition to emotionally engage with travellers (see Kim et al., 2020; Moscardo, 2020a; Mossberg et al., 2010; Wang, 1999; Youssef et al., 2019; among others). Tapping into tradition and memories of the community, telling a nuanced story on heroic and memorable events engages with potential travellers in a way that facts and scenery cannot.

Stories and the art of storytelling by using local tradition and myths is precisely what helped put Ílhavo, a coastal municipality in the centre of Portugal, on the map as a showcase of successful destination marketing. The question thus arises: "Why Ílhavo? What makes it special?". In addition to being known as a Sun & Sea destination, Ílhavo is, first and foremost, the Portuguese town of cod fishing - the gem of Portuguese cuisine. Cod fishing entails the use of specific tools in a hard fishing process that has greatly influenced Ílhavo's culture and tradition. One might even relate cod fishing to the perils of the times of Portuguese Discoveries. In Ílhavo, just as the sailors that set forth into the sea to find new and exciting worlds, fishermen threw themselves completely alone into the distant, unknown, treacherous and icy sea with nothing more than a small wooden boat and lines and hooks to fish cod ("bacalhau"). Ílhavo used its history as a fishing community, as well as its contemporary immaterial and material sea heritage, to craft a new and enticing narrative that has boosted the municipality's visibility and exposure as a travelling and cultural destination.

In this chapter, the case of Ílhavo is analysed as a way to capture the destination's specific approach to destination storytelling. It will be discussed the process of how Ílhavo took from its memory and tradition to build a unique story that created an emotional impact, both within the community and beyond it. Furthermore, the study will examine the process that led Ílhavo to being accepted as a cultural destination. From the analysis, contributions to managerial practices and knowledge in the field are expected.

LITERATURE REVIEW

"Tell me a fact and I will learn, tell me a truth and I'll believe. But tell me a story and it will live in my heart forever" (Native American proverb) (Stevenson, 2019).

Story Marketing

Given the globalized world in which we now live in, that entails the growing mobility of people and businesses and the concurrent challenges, territories have increasingly been made aware of the need to create and use marketing strategies to become more attractive to tourists. This has led territories to try to: (i) create awareness and favourable images that better sell each region (Keller, 2008); (ii) use new information and communication technologies to reach a wider, and sometimes more demanding, audience (Salvado Alves, 2008); (iii) envision strategies that could satisfy the needs of different groups of users in each territory (Almeida, 2004). The purpose is to make territories increasingly welcoming and receptive to new challenges, new companies and investments (Aragonez & Caetano Alves, 2012), the arrival of new residents and tourists (Keller, 2008), new qualified migrants (Hay et al., 2021), while capturing the organization of media impact events in these destinations (Fonseca & Ramos, 2006).

To make a territory attractive to the consumer, it is fundamental to associate it with a set of features and benefits, that grant it a competitive advantage over other territories (Pasquinelli, 2013). However, the successful implementation of this strategy largely depends on the territory's ability to define its competitive advantages, making use of its endogenous and unique characteristics (Teixeira Costa & Veríssimo, 2020).

It is in this context that storytelling has gained increasing importance, both in strategic planning and in territorial management in general (Hartman et al., 2019). Storytelling allows for the communication of a territories' unique cultural features, which are difficult to replicate by others, persuasively and effectively (Olsson et al., 2016), granting it favourable and exclusive associations (Keller, 1993) while conveying an image of authenticity (Youssef et al., 2019). Additionally, recent findings related to brain activity have validated the enormous potential of storytelling in marketing, also referred to as story marketing, proving that consumers are more likely to remember stories which may linger factually, visually and/or emotionally in their memory (Youssef et al., 2019). Stories talk about human needs, transfer values and principles, communicate knowledge and give credibility to a message. A good story that entertains remains in memory (Mossberg & Eide, 2017). As a result, storytelling is more convincing to consumers than statistics or facts because it reframes the message into a format that is easier to understand and that is relatable (Kim et al., 2020).

Story marketing is, therefore, a valuable factor of success in sales, adding emotions to products. This most effective communication tool, with proven positive effects on consumer involvement in tourism (Li, 2014), has become a popular resource in promoting business and tourist destinations (Moscardo, 2020a; Mossberg et al., 2010; Olsson et al., 2016), particularly in designing strong and competitive territorial brands (Kotler et al., 1993). Stories give brands a personality (Lund & Kimbu, 2021).

At a time when the market for local brands is highly competitive (Kotsi et al., 2018), destination marketing has become an essential asset in regional management, helping build up brands by identifying and developing competitive advantages that stem from the distinctiveness of each destination (Franco et al., 2019). Knowing which values should be connected to the destination's identity is key to effectively position it in the minds of tourists. Such is the role of destination marketing (Correia & Brito, 2011). Still, capturing the multidimensionality of tourism products is a complex process (Lund & Kimbu, 2021).

So, stories come into play. They attract consumer interest (Escalas, 2004), reveal positive effects on their behaviour (Li, 2014), and even lead them to be less critical as they believe the "heroes" (brand, employees or customers) within them (Youssef et al., 2019). Learning how to use stories, and traditional stories, as a theme for storytelling, can thus reinforce brand's values and identity (Camprubí & Planas, 2020) and help with destination marketing.

Destination Storytelling

Hartman et al. (2019) emphasise a dual role of stories in tourism, when used as a strategic tool, arguing that stories are used not only to inform, share or teach, that is, as a means of transferring knowledge from the destination to visitors; but also to persuade people to visit the destination. Through storytelling, the destination can increase the appeal of its city, region, museum, restaurant or specific event (Mossberg et al., 2010), which can be considered the best approach to allow potential consumers to identify themselves with a certain destination (Li & Liu, 2020).

Stories have always been part of tourism, used in guided tours and attractions built around stories of popular culture, or in presentations about the history and culture of a destination. However, this practice has grown considerably in recent years (Kim & Youn, 2017; Moscardo, 2020b), as stories have increasingly been perceived as a decisive factor in influencing tourists' cognition and behaviour. Moscardo

(2020b) identifies three reasons for such a rise: development of the experience economy, rise of digital mobile technologies and changes in the profiles of international tourists.

Mossberg et al. (2010) identify the following advantages in the use of storytelling for destinations and companies: (i) increased visibility in the general marketing of a region; (ii) ease of communication between companies sharing the same network built around a storyline; (iii) the collaboration around a storyline can create the ideal scenario for companies to develop and test new products; (iv) all companies involved in storytelling and storyline development can be visible, regardless of size, economy and reputation; and (v) a story about the destination can give a unique competitive advantage. Also, according to the same authors, placing storytelling activities outside the main season will provide other benefits for tourism, such as combating seasonality, thus contributing to the economic sustainability of the destination.

Storytelling's success as a marketing tool to attract tourists to destinations lies mainly in the fact that people like to listen to stories, which can touch their emotions (Li & Liu, 2020), their heart (Li, 2014) and even change them (Escalas, 2004), functioning as an enhancer of the travel experience (Youssef et al., 2019), and hence allowing for more meaningful and memorable experiences (Kim & Youn, 2017; Mossberg et al., 2010).

Building an entire business, or parts of it, around a story, be it a hotel, a restaurant, a tourist attraction, an event or a destination, has become a global trend in this new industry (Mossberg, 2008). These stories can be linked to a wide range of different consumption experiences (Moscardo, 2020a), and different media can be used for that, which Ryan (2016) refers to as *transmedia storytelling*.

The experience economy (Pine & Gilmore, 1998) is characterized by consumers looking for unique experiences and products or services that appeal to their emotions and dreams. Stories help create such opportunities (Youssef et al., 2019) and can even be considered as central concepts in tourist experiences (Moscardo, 2020b).

Moscardo (2020a) argues that some themes or story topics, which should reflect the ways in which destination communities wish to be presented to tourists and highlight distinctive features of that destination, will be more effective in attracting and holding the tourist's attention than others, e.g., those connected with human survival, overcoming challenges, the creation and maintenance of personal, family, and collective relationships, altruism, struggles against injustice and heroism. The effectiveness of experience scenarios that offer tourists the opportunity to act out their own stories, thus personalizing their experience and giving them the opportunity for active interaction and participation (Moscardo, 2017), is also noteworthy.

Tourist Involvement and Participation

This leads to the concept of co-creation, as tourists actively co-create their experiences while interacting with the attributes of the destination (Kim & Youn, 2017). To this end, developing a relationship between destination, the consumer and the story itself, involving all types of stakeholders (Youssef et al., 2019), such as local communities, who contribute in particular to the coherence of the stories (Moscardo, 2020b) and to their credibility (Kotsi et al., 2018), is highly desirable. Nonetheless, these local communities may feel alienated from the destination strategy when they are not conveniently and effectively involved in decision making. Managers must ensure that the strategy is attractive and aligned with the needs and desires of external and internal stakeholders to encourage and maintain long-term commitment and support (Hay et al., 2021).

Tourist participation in storytelling can then range from a passive role, as someone who simply listens to or reads a story told or written by someone else to a markedly active role, someone who creates or co-creates his own story, and can him/herself become the storyteller for others, based on his/her own experiences (Moscardo, 2020a), as a result of their interaction with the places and people who live there (Tussyadiah & Fesenmaier, 2008). The creation of immersive emotional experiences, in which tourists feel that they are protagonists, is vital for the success of a destination (Lund & Kimbu, 2021). Moin et al. (2020) advocate not only for the placing of tourists in the role of protagonists, but as heroes, which will facilitate the understanding of the messages and objectives. Thereby, both the creation and the narration of stories about destinations is no longer of the responsibility of planners and managers only, but also of the consumers who share their experiences as tourists (Youssef et al., 2019). In fact, consumers can now, through the use of digital technologies, create and publish stories whenever and wherever they want (Tussyadiah & Fesenmaier, 2009).

Destination marketing organizations (DMO) have also begun to incorporate storytelling and the interactive nature of first-person stories as a resource on their websites, using these stories or blogs as a digital version of word-of-mouth communication (Tussyadiah & Fesenmaier, 2008). Travel blogs have revolutionized the ways tourists communicate with each other, serving as a space to exhibit and exchange information, opinions and experiences about destinations (Franco et al., 2019).

Co-creation opportunities can include: (i) stories about the destination told to tourists face to face by guides, through audio-visual presentations, through text on boards or travel guides and digital platforms, such as mobile applications; (ii) activities designed and structured to encourage tourists to represent their own personal history; (iii) transmedia story worlds explored across different media and locations (Moscardo, 2020a).

According to Mossberg et al. (2010), the story on which storytelling is based must be: (i) accepted by those involved in the process; (ii) based on locations; and (iii) be susceptible to conceptualization and commercialization. Moscardo (2020a) argues the success of storytelling in tourism depends on: (i) the existence of credible and authentic characters, described in detail, with similarities and/or likely to be admired by tourists; (ii) consistent and plausible plots; and (iii) strong expressions of emotion. Kim et al. (2020), on the other hand, identify five essential conditions to this end: (i) authenticity; (ii) educability (the degree to which the public can obtain knowledge, information and facts through stories); (iii) enjoyability (the degree to which the audience enjoys themselves through the stories); (iv) descriptiveness (the degree to which the intended ideas are structurally organized and well presented); and (v) emotionality (the degree to which the audience is emotionally stimulated through the stories).

For a storytelling strategy to be and remain effective, storytellers, staff and volunteers need to be up to date with the story's facts, continuously develop their ability to act out a story, and be able to maintain their interest and commitment, otherwise, the risk of entering an "autopilot" situation arises (Olsson et al., 2016).

Authentic Stories

As proposed by Moscardo (2020a) and Kim et al. (2020), authenticity is key in storytelling marketing: the more authentic the stories are perceived, the more the visitors identify with them and, consequently, with the destinations (Li, 2014), contributing to their differentiation from competitors (Youssef et al., 2019), and visit intention (Overend, 2012).

Despite its importance, defining "authenticity" is a challenge, especially in a globalized, diverse, ever-changing world, such as the one we live in, which explains the varied approaches to the subject. Wang (1999) points out that tourists recognize authenticity, or inauthenticity, in art, festivals or gastronomy, based on whether they are created by natives following local customs or traditions. Authors, such as Kolar and Zabkar (2010), consider that authenticity is an identifiable property, as well as a social and individually constructed perception (of objects or cultures). Chiu et al. (2012) argue that the reference to numerous details in a story, such as characters' names and dates, gives it authenticity, leading people to interpret the information contained in it as being credible. Youssef et al. (2019) associate the uniqueness of a given destination, and the novelty and mystery that entails, to the concept of authenticity. Mossberg and Eide (2017) mention that it is difficult to copy a unique story linked to a place and that, if the concept is based on local food products and local culinary heritage, it can be even harder.

To be perceived as real, a story needs to be based on real events. Still, tourists also appreciate fictional stories, just as long as the characters are relatable and the narrative not misleading as this will lead to a breach of trust (Youssef et al., 2019). Hereof, Cho et al. (2014) refer to the concept of perceived realism, which is based not on whether the story is fiction or non-fiction, but on the perception that the characters and their reactions are similar to those of real people and that the sequence of causal events is consistent and sensible.

In this complex and increasingly significant universe of stories in tourist destinations, Moscardo (2020a) introduces the concept of story world, defining it as a set of stories and activities linked to a central story or a particular character. According to the author, the story world of a destination is more than just stories told by tourism managers, it also includes stories told in books, films, television, games and news media, as well as tourists' own stories related to previous visits and stories told to them by other tourists through social media, the internet or directly. The story world then stems from a central focus, offering the audience the opportunity to immerse themselves in an entire world built around the central story or character.

Given the positive effects of using storytelling in tourism, and its evolution in recent years, Moscardo (2020b) predicts that the future will be marked by the growing and increasingly sophisticated use of stories in promoting destinations and in designing experience opportunities, with a rise of tourists' involvement and interaction with the destinations' stories, namely through the creation of their own stories. Transmedia storytelling that is attractive to tourists and accepted by destination communities, combined with the development of a destination story world linked to its distinctive features, is envisioned to be the future of destination marketing.

However, the use of storytelling to stimulate the development of destinations is not easy. The development of regions as tourist destinations depends on the actions of several stakeholders, dispersed at various levels, and often with different ambitions, interests, and worldviews. Therefore, the greatest challenge is to align these actors in destinations (Hartman et al., 2019) while assuring a strong leadership (Mossberg et al., 2010).

RESEARCH METHODOLOGY

This chapter adopted a qualitative approach, making use of an extensive literature review, as well as a case study to better reach the objective of analysing a successful storytelling destination marketing strategy.

The study considers Yin's (2003) assumptions that case studies: (i) investigate contemporary phenomenon within its real-life context, especially when the boundaries between phenomenon and context are not clearly evident; (ii) can be applied when the researcher has little control over events; and (iii) can be examined through detailed contextual analysis of a limited number of events or conditions, and their relationships. Considering these postulations, the municipality of Ílhavo presents itself as a noteworthy case study to better explore destination storytelling as a contemporary phenomenon, and to allow for further reflection on good examples.

As a single, longitudinal and exploratory case (Zainal, 2007), the study provides a systematic method of observing the events, collecting data, analysing information, and reporting the results over a long period of time, focusing on a small geographical area, yet underpinning an in-depth analysis on why and how Ílhavo is highlighted as a successful example of destination marketing. To this end, data collection was conducted both through direct observation and analysis of secondary data, such as reports, press releases, records, and other documents of interest, spanning from 1998 to 2021.

SAILING THROUGH THE STORY OF ÍLHAVO – A CASE STUDY

1. Narrating the Story Background – Cod Fishing and "Bacalhau" in Portugal

The Scandinavians were the first European people to maintain a large-scale commercial trade in cod (*Gadus Morhua*) in the North Atlantic, probably tracing back to the end of the first millennium AD (Silva, 2015). In Portugal, the trade and consumption of cod began to have some predominance in the 14th century (Salvado & Joukes, 2019). At the end of the 15th century, Portuguese boats were already involved in cod fishing in the North Atlantic, Newfoundland, and along the east coast of Canada (Sobral & Rodrigues, 2013). Some Portuguese ports, which nowadays play a fundamental role in fishing and trade, such as Aveiro or Viana do Castelo, already had boats in those waters (Sobral, 2016).

Basque fishermen, who fished in the North Seas, were, in the Middle Ages, the first to spread a new curing technique in these regions, consisting in salting the fish before drying it, which was used to cure the fish in the Mediterranean for thousands of years. Portuguese, Norman, Breton and English fishermen, who, around 1500, joined the Basque fishermen in Newfoundland, also adopted this curing technique (Silva, 2015). In the Portuguese case, this process was made easier by the control of salt-rich areas, namely Aveiro, considered to be of the best quality (Sobral & Rodrigues, 2013). This technique is still used today, and currently Portugal is the main international market for dry-salted cod, followed by Brazil (Pinto & Neto, 2017).

The Portuguese cod fishing process was carried out by line in small, one-man wooden boats, the "dóris" (Silva, 2015), in the dangerous environment of the icy seas of the north. This technique would be used by the Portuguese until the 1970s, making them the last of the fishermen to use it (Silva, 2015). This, along with the dire conditions of this activity, and the mythical image of the White Fleet, immortalized by the colour with which the Portuguese ships distinguished themselves from the others during World War II, bestows an epic tone to cod fishing, called "Faina Maior". This allowed this activity to be placed among those with the greatest symbolic value (Peralta, 2010), evoking images only comparable to those of the "Great Adventure" of the Discoveries (Peralta, 2003). The dramatic tone of "Faina Maior" is also mentioned by Silva (2015):

"At the end of a 1967s National Geographic Society documentary, The Lonely Dorymen, Portugal's Men of the Sea, the narrator, Alexander Scourby, concluded: 'Soon again, husbands and wives must part to await the end of the long summer alone. To them, it is accepted as a way of life; for life is synonymous with the sea. It is their home, their heritage. A host that has sustained them during all the time they have lived on its shore. It is the very breath of Portugal and their indomitable men of the sea'" (p. 130).

The consumption of fish, in Portugal as in other European societies, is closely connected to religious motivations, to the practices of fasting and abstinence as a means of purification and as a sign of obedience to the divinity. Fish thus become a substitute for meat (associated to sin and lust), even on the plate of the most fortunate (Silva, 2015; Sobral, 2016; Sobral & Rodrigues, 2013). In Portugal, specifically, cod has long stood out among the fish. In the 18th century, cod was already part of the daily diet of religious communities, and later its consumption became popular among the general population, notably because it was a cheap source of protein (Silva, 2015). It was at the end of the 19th century that cod achieved a privileged status in Portuguese cuisine, being consumed by all social classes, often forming part of culinary recipes, and becoming a symbol of national identity (Salvado & Joukes, 2019).

Among its consumption patterns, past and present, the celebration of Nativity on Christmas Eve stands out: cod is the favourite meal of Portuguese families (Sobral, 2016) and its consumption during this festive season represents about a third of the total annual consumption (Silva, 2015). Given that, in Portugal, cod is primarily consumed salty, dry and flattened, it can be preserved for long periods of time and easily transported over long distances (Sobral, 2016).

Cod has a unique double status in the Portuguese cuisine: common ingredient and symbol of national identity (Sobral & Rodrigues, 2013). As such, simultaneously, it plays an important economic, religious, social, political and cultural role (Salvado & Joukes, 2019). It is, amid all the typical dishes, considered to be the most typical of all, referred to by the Portuguese as something that links them to their origins (Peralta, 2003). "Over the centuries, cod has been transformed from simple foodstuff to a symbol of Portuguese identity, from food socially connoted with situations of abstinence and even typical of the poor, to an expensive and prestigious food in the gastronomic field" (Sobral & Rodrigues, 2013, p. 622). The "Bacalhau" identity heritage is thus a symbolic marker of gastronomic cultural identity in the Portuguese territory (Salvado & Joukes, 2019).

Furthermore, Silva (2015) also considers cod to be more than a traditional ingredient, but rather a national symbol just like Fado, the Sanctuary of Fátima or Cristiano Ronaldo, contributing to the construction of a collective identity, unequivocally referring to the idea of "Portugueseness". This idea is supported by Sobral and Rodrigues (2013), who argue that the identification between salted and dried cod and Portugal and the Portuguese is clear and evident, no other food having been invested in such a way by popular imagination.

Perhaps that is why we often find codfish "off the plate", namely in literature, theatre, music, caricature, photography and cinema, perpetuating the memories of fishing in the icy and distant seas, but also of the drying process on the Portuguese coast, and of its consumption, both in Portugal and in Portuguese emigrant communities spread around the world (Silva, 2015). Codfish plays an iconic role for the Portuguese diaspora, having motivated, for example, the creation, in 1968, in South Africa, of the so-called Codfish Academies ("Academias do Bacalhau"), an important network of societies. Today it has about 50 branches spread across Europe, Africa, North and South America, Australia and Portugal, and, under the emblematic invocation of the "faithful friend" (the name given to cod in Portugal), promotes socializing between Portuguese people, mutual aid and philanthropic activity (Silva, 2015; Sobral & Rodrigues, 2013).

Portuguese culture has in its gastronomic traditions a powerful asset, which has contributed in a very relevant way to the enhancement of the national tourist offer. At a time when tourists are increasingly looking for the values of authenticity, which can be imparted in particular by tradition and memories, gastronomy is becoming ever more important (Araújo, 2021). Franco et al. (2019), in a study that aimed at understanding the sensations experienced and remembered by foreign tourists who visit Portugal, considered the importance of the five senses for tourist marketing and concluded that, in terms of taste, tourists essentially valued fish (36%), mentioning cod specifically (30%). After centuries of history, the codfish tradition is presented, inside and outside the country, as a common heritage shared by all Portuguese and as a distinctive element of its national character (Silva, 2015).

2. Connecting Story to Place - Ílhavo, Land of "Bacalhau Men"

Ílhavo is a municipality located in the centre of Portugal, in the Aveiro region. It has approximately 75 km2 and 38,598 inhabitants (Teles & Anselmo de Castro, 2015). Geographically, Ílhavo is bathed by the Atlantic Ocean and by the Aveiro Lagoon, an extensive saltwater mass that crosses the territory from north to south through two channels.

Figure 1. Ílhavo's location in the map of Portugal
Source: Ílhavo City Hall (2019)

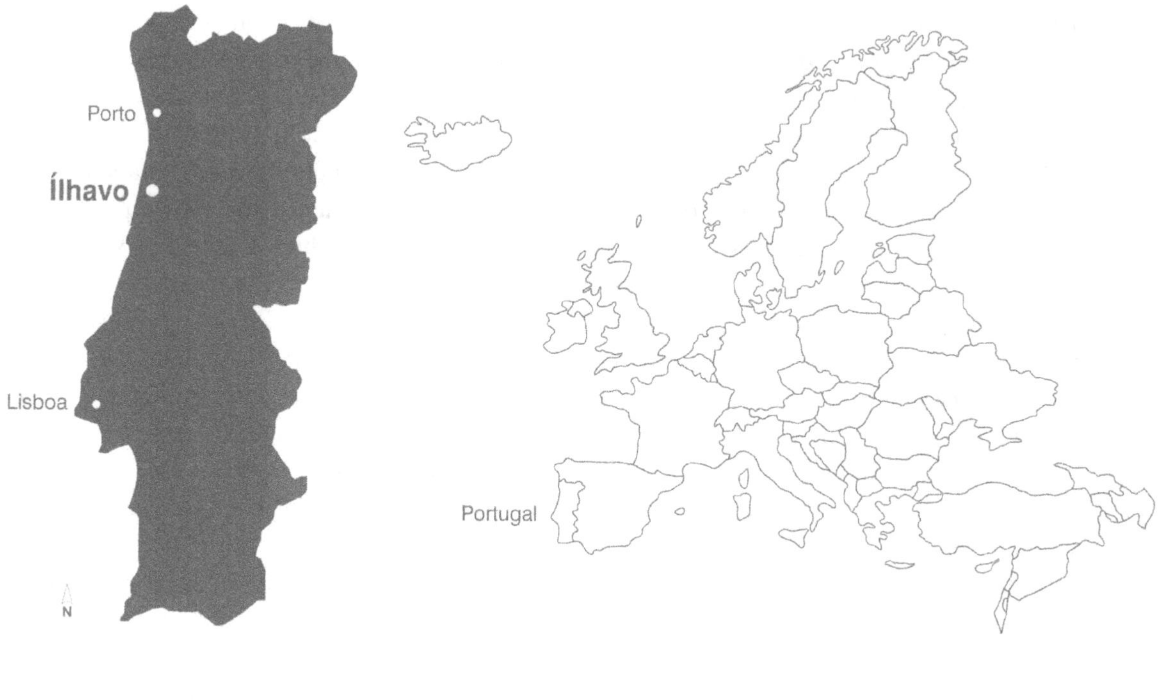

Historically, and especially in recent decades, Ílhavo's economy has been based on activities related to the blue economy, such as fishing. This has greatly influenced the municipality's traditions and culture. Such is the case of cod fishing, particularly with lines and hooks in the dories (as introduced before), that marks Ílhavo's past as a fishing village, as well as its current immaterial and material heritage (which spans from traditional gastronomy - based on cod fish and sea food -, to facilities – like the Maritime Museum, that documents the municipality's fishing history).

Nowadays, Ílhavo remains anchored to the sea due to other activities, such as oyster farms, water sports and the Aveiro port, one of the most important in the country, located in the municipality. Furthermore, tourism, a fast-growing activity, has been basing its strategy on Ílhavo's coastal and maritime heritage, its beaches, seafood gastronomy, among other tourist attractions.

Cod Fishing in Ílhavo

Ílhavo has had a long and intimate connection to the sea, and specifically to cod fishing, having had its people board cod fishing vessels for centuries. As stated before, by the end of the 15th century there were Portuguese boats involved in cod fishing in the North Atlantic, Newfoundland and off the east coast of Canada, with the port of Aveiro, now mostly located in Ílhavo, having boats in those waters. Some of these ports, which once participated in the "great fisheries", are also now competing locally for the right to transform this memory into heritage (Silva, 2015).

It is estimated that between 1935 and 1974, more than 17,000 men (members of the Guild of Ship-owners - "Grémio dos Armadores"), from just 16 municipalities in Portugal, including Ílhavo, were on board cod fishing vessels ("lugres") (Garrido, 2016). In fact, between 1935 and 1979, Ílhavo was the most relevant municipality in recruiting crew members, with approximately 10% of the total (Garrido, 2004). In 1936, for instance, in the 50 cod fishing vessels with line and hooks, there were 1709 fishermen, 419 crew members and 42 captains from Ílhavo (Aveiro, 2009). In the 1950s, all the ships in the fleet were commanded by people from Ílhavo, who also occupied other prominent roles on board (Peralta, 2010).

In the 19th century, as Fonseca (2007) states, people from Ílhavo occupied, almost exclusively, the command posts of the "lugres" that fished in Newfoundland, monopolizing the hierarchical positions from captain, to pilots, boatswains and cooks. In the 62 sailing ships that left in 1922 (with Aveiro being the most represented port), there were 54 captains and 45 pilots from Ílhavo (Marques, 2001). Silva (2001, p. 264) also refers to this fact in the first person:

"At any time in our history, if we do not find half of the Portuguese ships governed by officers from Ílhavo, we will certainly find two thirds, say, this considering all our ships in the Merchant Navy, because in my time at sea, that is, from 1952 until 1988, in the distant water fishing fleet, only myself and five or six others were not from Ílhavo".

Valdemar Aveiro, another renown cod fishing captain, born in Ílhavo, "son of a nursery land for sailors" (Aveiro, 2014, p. 15) and author of several books on this tradition, says:

"I was born in a land par excellence of sailors. It was not the only one, but it was undoubtedly the most important Portuguese town, percentage wise, in terms of the number of male children dedicated to this fishery" (Aveiro, 2016, p. 19).

The perceived importance people from Ílhavo have had in cod fishing over centuries is understood and shared by the population, who claim for the locality the status of "land of cod", or "capital of cod", in Portugal (Peralta, 2010).

After the decline of the White Fleet, the codfish industry concentrated in the municipality of Ílhavo, more precisely in Gafanha da Nazaré, one of its parishes, an important shipyard during the Portuguese dictatorial regime (1933-1974). The Portuguese Association of Cod Industries (AIB) was founded here in 1993, with its members representing about 75% of the companies. Despite its main goal being the promotion and development of the industrial activity of cod, and the protection and promotion of the interests of companies in the sector (Martin, 2011), AIB has also become an influential lobby that actively contributes to the preservation of this tradition, collaborating with local non-governmental organizations such as the Codfish Brotherhood and the Friends of the Maritime Museum of Ílhavo (Silva, 2015), as well as with the City Council.

Still, Portugal has seen a sharp decline in its cod fishing quota since the 1980s, as well as a reduction in the fleet, as consequence of European Union policies. Nevertheless, the important local industry has not yet been affected by these problems, and the scrapping of vessels has been offset by investment in import routes and the modernization of processing plants, which has maintained the competitiveness of the local industry in this sector (Peralta, 2010).

3. From Story to Tourism Strategy – Sea, Tourism and Storytelling

As a tourism destination, Ílhavo has been traditionally known as a Summer spot in the central region, with a marketing strategy that has, up until the late 90s, been based on promoting its two beaches, Barra and Costa Nova, the Sun, Sand and Sea. Yet, this “old” strategy shed light on several issues that prevented promotion effectiveness: seasonality, lack of accommodation, local tour operators and other related services.

Hence in 1998, with the change in the municipal governance, the strategy was redesigned (CMI, 2008). One of the first and decisive steps towards the definition and effective materialization of this new strategy was the implementation of the Strategic Plan for the Municipality of Ílhavo (2000-2020). This plan, carried out by the University of Aveiro, identified the strong and distinctive historical cultural identity of the territory, which is directly linked to the sea, especially to cod fishing. This valuable asset was thus integrated in the new development strategy (Pires et al., 2000). Considering a sea-focused plan as Ílhavo’s main feature, tourism strategy followed the same path, looking at the sea as an anchor product, not only in terms of beaches, but mainly by considering it as the foundation of Ílhavo’s historic and contemporary heritage.

Consequently, a new tourism strategy was designed, one that reinforced the sea in its historical and cultural aspects, mainly through the use of cod fishing as a distinctive, differentiating tradition, and as a key element for the new “Ílhavo” brand (Teixeira Costa & Veríssimo, 2020).

Figure 2. Lagoon and Port of Aveiro

Figure 3. Lighthouse in Barra beach

4. Representations of the Story

Between 1998 and 2021, several actions were undertaken to implement the new strategy, anchored largely on cod fishing, thus supported by the rich narrative around it, and set on 7 main pillars: (i) Visual Identity; (ii) Cultural Facilities/Mediation; (iii) Public Space; (iv) Cultural Events; (v) Religion; (vi) Publications; and (vii) Internationalization.

Figure 4. Pillars of Ílhavo's strategy (own elaboration)

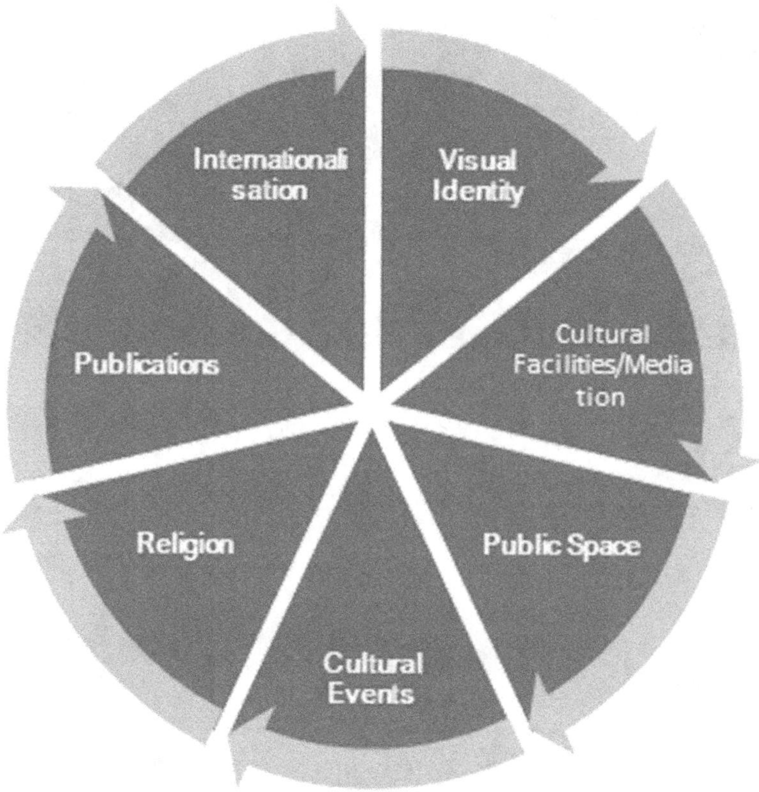

The aim was to disseminate the narrative around the sea, and in particular around cod fishing, among visitors and residents, in a coherent, sustained and engaging manner. The focus was placed on historical aspects, while still underpinning emotional qualities associated with the destination: adventure, bravery, courage.

Ílhavo thus began to tell a story, that of its brave sailors who, for decades on end, faced the distant, icy, foggy and dangerous northern seas, alone in small and fragile wooden boats, on voyages that could last between six to nine months, thousands of miles away from their homes, where their children and their wives, known as "living widows", remained anxiously waiting for their return… or not.

Fundamental to this new strategy was the strong and clear leadership from City Hall and the support and active participation of various stakeholders, among them the local community, who, given their emotional connection to the history of cod fishing in Ílhavo, accepted and nurtured it from the outset.

VISUAL IDENTITY

Public visibility was one of the first concerns in this strategy, which led to the creation of a logo in 1999. The aim was to differentiate Ílhavo from other municipalities, incorporating the main assets of the territory's tradition (CMI, 2008a).

Additionally, the slogan "The Sea by Tradition" was created in the same year with the purpose of "(...) having a meaning on its own; at the same time, to be a reference to the past, relating itself to Ílhavo's deep connection with the Sea. A very visible relationship that evolves each day into an intense economic activity, as well as into a perspective for the future, which intends to maintain the Sea as a constant social, cultural and economic reference in the territory's life" (CMI, 2008, p.42).

Figure 5. Logo (created in 1999)

CÂMARA MUNICIPAL DE ÍLHAVO

However, shortly after, in 2004, the logo was redesigned, as it proved not to add real value to the creation of the territorial brand "Ílhavo", nor to the differentiation and definition of its positioning. If, on the one hand, the slogan took us to the more cultural and memory-based aspect of the Sea, the logo did not support this effort and, aesthetic value aside, was somewhat innocuous.

Figure 6. New Logo and visual identity redesigned in 2004

www.cm-ilhavo.pt ílhavo Câmara Municipal o mar por tradição

The new logo design was a very important step that strengthened the image of Ílhavo as a municipality linked to the Sea, to its past, its present, but also its future. The sails, representing freedom, speed, boldness, courage and also a source of clean energy, extended their positive influence far beyond memories and Tourism.

"The Ílhavo brand introduces the Lugre profile as an iconographic evocation of the epic of the Faina Maior with all the symbolic and affective charge that the great sailing ships of the 19th century arouse all over the world. Ílhavo is thus closer to its origin of adventure, intelligence and enterprise, as means of preserving its freedom (economic, cultural and political)" (Providência, 2004).

Furthermore, the Strategic Plan for the Municipality of Ílhavo (2000-2020) was followed by a new one, developed in 2015, spanning to 2035. This new strategic plan, that, again, was carried out by the University of Aveiro, set new strategic objectives, such as the valorization of natural resources for tourism, the affirmation of a territorial identity, and the promotion of culture (Teles & Anselmo de Castro, 2015).

CULTURAL FACILITIES/MEDIATION

Among the several investments that were made in infrastructures, given its strategical and financial importance, special mention should be made to the expansion of the Maritime Museum of Ílhavo (MMI), and the creation of its extension, the Ship-Museum Santo André.

The construction of the MMI, concluded in 2001, mainly signified a profound change in its philosophy, placing the Sea, and particularly the history of line cod fishing, a singular expression of cod fishing that found its permanent representation here (Peralta, 2003), definitely and unequivocally in the foreground, as its grandest and greatest message. This transition from the old "Maritime and Regional Museum of Ílhavo", founded in 1937, with its ethnographical and regional vocation, to a "maritime museum" *par excellence* was successfully completed (Garrido, 2018), despite creating some controversy at the time. It meant removing from the exhibitory discourse of the museum several elements that, regardless of their importance, did not clearly align with this historical, social, cultural and economic distinctive narrative of the destination. This lead to a profound change in the museum that established itself as an institution that promoted the maritime culture, plural in memories and identities, where cod fishing became a patrimonial symbol (Costa, 2016).

The new message spread by the MMI and by the municipality was greatly reinforced by the extension to the Ship-Museum Santo André. Opened in 2001, this is a transformed old cod fishing vessel that is anchored in Oudinot Garden, Gafanha da Nazaré, a port city in Ílhavo's municipality in which, to this day, numerous cod fishing processing industries still work, in a location that is very close to the entrance of the Aveiro Port. The Ship-Museum Santo André enabled different types of remote cod fishing to be properly illustrated (Peralta, 2003).

Figure 7. Maritime Museum of Ílhavo

Figure 8. Ship-Museum Santo André

Roughly a decade after these investments, a new vital and differentiating value emerged within the MMI, i.e. the Centre for Research and Entrepreneurship of the Sea (CIEMar), comprised of an extensive archive, a research unit, a training centre and an incubator for businesses to connect to the blue economy.

Shortly after CIEMar's opening, in 2013, a new investment was made: the Codfish Aquarium, a unique structure in the country, and rare in Europe and around the world, where dozens of codfish live and that has helped complement and enrich the exhibitory discourse of the museum, and greatly increase its activity. In 2018, the MMI reached the impressive mark of one million visitors (since its expansion in 2001), reinforcing its stance as one of the main cultural references in the region (CM-Ílhavo, 2018).

Today, MMI's mission is to preserve the memory of working in the sea, promote the Portuguese maritime identity and culture, organizing, to that effect, its cultural project and program around the concept of "memory preservation" (Costa, 2016), and making use of "program imagination, constantly recreated trough research projects that are able to craft narratives out of immaterial heritages. Furthermore, the partnerships with several museums, universities and research centres have given MMI a wider network and scale that promotes an active role within society" (Garrido, 2018, p. 99).

The MMI is home to numerous initiatives aiming, mainly, at promoting discourses and knowledge on cod fishing, fishing, and fishermen in general, through cinema, theatre, photography, to name but a few. These activities then extent beyond the scope of the MMI and into the community, other institutions, museums and Spanish maritime communities (Garrido, 2018).

Three of these projects deserve special attention due to how they embody, preserve and disseminate the narrative of cod fishing. Firstly, the temporary exhibition "Memory Box – A tribute to the men who went fishing for cod". Spanning from 2005 to 2012, this project entailed building an enormous lightbox with the names and faces of fishermen, taken from the over 20.000 registration forms from crewmen in cod fishing ships between 1934 and 1974. The Memory Box travelled from north to south in Portugal, allowing MMI to create an unprecedented dialogue with its community, the maritime community of the Portuguese coast and the Azores islands (Garrido, 2018).

Secondly, the "Memory Archive of Cod Fishing", aimed at preserving, archiving and building the memory of cod fishing. This was accomplished by collecting testimonies from men that were connected to cod fishing (Costa, 2016), a never-ending practice constantly being updated. Beyond simply amassing information, the project wanted to help reimagine and reconstruct exhibitory discourses, working as a tool in this process (Carvalho, 2013).

Thirdly, the stories about cod fishing and their heroes were told in Educational Contexts on a day-to-day basis to children enrolled in schools that regularly visited the MMI. This sparked in them curiosity, amazement, and often pride, especially on those brought up in fishermen families.

PUBLIC SPACE

The history of cod fishing has also been recounted on streets, squares and gardens of Ílhavo. Several streets and boulevards have been given the names of people connected to cod, from captains to simple fishermen, who, due to their feats, deserved to have their memories, and with it, their activities, preserved.

This memory can also be seen in numerous monuments, such as the Statue of the Man from the Sea, among others, that embellish the destination. In 1999, the St. John's Square was inaugurated in memory of the legendary Canadian port that, for decades, harboured fishermen from Ílhavo. In 2004, on the other hand, Ílhavo's Square was inaugurated on St. John's, within the Friendship Agreement celebrated between the two municipalities.

Currently, Ílhavo's City Hall, through Ílhavo's Centre for Documentation, is developing a project named "If this street was mine…" conceived to communicate the toponymy of the municipality, and that should soon result in a published itinerary. This project invites people to share their memories through photographs, tales or curiosities about each street. A regular program in the local radio station was created to promote the initiative and to share told memories.

Figure 9. Statue of the Man from the Sea

Figure 10. Codfish Festival

CULTURAL EVENTS

Throughout the years, several events were promoted using the sea, and memories associated to line cod fishing, as its main motto, particularly by City Hall. Among these, we highlight three: the Codfish Festival, Ílhavo's Sea Festival, and the Onboard Gastronomy Festival.

Codfish Festival

The first edition of the Codfish Festival took place in 2008, organized by Ílhavo's City Hall and the Codfish Brotherhood. From the outset, it was a grand success, bringing in tens of thousands of people, turning it into one of the major gastronomical events in the country. As such, the Codfish Festival became a valuable asset in promoting the municipality, its cod related history, along with its cuisine, economy, industry, commerce, and other tourism resources.

In 2014 the festival went through a significant transformation, setting as its main goals to: (i) reinforce its cultural dimension; (ii) increase national and Iberian notoriety; (iii) conquer new audiences; and (iv) increase its impact in the local economy (CMI, 2014). This transformation gave cod a wider and more diverse emphasis, focusing on its remembrance. The event's program thus started to include guided tours to the ship museum, re-enactments involving the community, stories for children, exhibitions, the fish market, to name but a few.

The Codfish Festival hence emerges as a tool that consolidates the Municipality of Ílhavo's differentiation from its competitors, using, to that effect, its cuisine and predominantly the codfish culture, as well as the many stories around it, which the festival aimed at affirming as a major national reference.

Ílhavo's Sea Festival

Since the Municipality of Ílhavo's participation, in 2008, as a host port in Funchal 500 Tall Ships' Regatta (connecting Falmouth - UK, Ílhavo and Funchal, and organized by Sail Training International), a new era for events was born. It was remarkedly the biggest and most complex event held in the municipality to date. That first year, 1.500 crewmen signed in, there were twenty sailboats from 13 different countries (some almost 100 meters long), and more than 350.000 visitors (CMI, 2008b).

In 2012 and 2016 history would repeat itself at Ílhavo's Sea Festival, with the event bringing in dozens of ships, hundreds of crewmen, and hundreds of thousands of visitors (CMI, 2012; CMI, 2016).

Onboard Gastronomy Festival

This festival, that is held in November in various restaurants, had its first edition in 2018. With a program that spans for several days, this festival brings to the dinner table delicacies inspired by iconic codfish dishes and its by-products, such as, for instance, the famous "Chora", a soup made from the cod's head, codfish spine stew, or fried tongues.

The Onboard Gastronomy Festival aims at heralding the cuisine that, traditionally, was made on board the cod fishing ships. However, the selling point of this initiative is not only how the food tastes, but the stories that go with each meal, the guided visits and other experiences, that highlight an iconic chapter in Portuguese history that has had a huge impact on the families from Ílhavo throughout the last centuries: cod fishing (CMI, 2019).

RELIGION

Faith has always played a prominent role for fishermen and their families. Thousands of fishermen received an annual blessing in Lisbon, the country's capital, before setting forth to the sea for yet another long and dangerous journey. Ílhavo was no exception: concerns and fears that came with this perilous activity, along with anxiety towards the news that reached home, were frequently mitigated by participating in religious celebrations in honour of saints with a strong and specific allegiance to the sea and its community, saints that, to this day, are still celebrated for their protection.

This connection between religion, the sea and fishing, was magnified in Ílhavo in August 2021 with the inauguration of the Maritime Religious Centre. This new museum, an extension of the MMI, is the first maritime themed religious centre in Portugal, and it focuses on the preservation and exhibition of a rich collection of art pieces and cultural objects, of religious nature, that showcase the relationship between the people from Ílhavo and the sea (CMI, 2022).

PUBLICATIONS

Throughout the years, the City Hall has released numerous publications connected to cod fishing, namely research papers and recipe books, mainly through the MMI. Furthermore, it has given financial support towards the publishing of books written by researchers and various authors, particularly local. Many of them are old fishermen, or their offspring, that share and immortalize their memories, enriching the history of cod fishing.

INTERNATIONALIZATION

The fact that Ílhavo has a long emigration history, particularly due to its fishing related to past and the cod industry, was clearly reflected in the choices made for international partners. Backed by this reality, but also aiming at strengthening cultural and economic relationships with other destinations, from 1998 onwards, several Friendship Agreements were signed between the Municipality of Ílhavo and cities from various countries from around the world, such as Canada, Unites States, Iceland, Germany and Brazil. These Friendship Agreements were mostly driven by a relevant emigrant community or commercial relationships related, primarily, to cod fishing. These partnerships have allowed Ílhavo to regularly participate in several world-wide events, conduct international workshops and seminars in its own territory, welcome visits from members of the government and international students.

The Friendship Agreement with the city of St. John's, the oldest, celebrated in 1998, should be highlighted. Throughout the years, numerous initiatives have been implemented, such as the students' exchange that allowed dozens of young people from both communities to know more about the relationships that, for decades, marked the day-to-day of both people, sharing cod fishing as backdrop.

The Friendship Agreement with Grindavik, Iceland, an important cod fishing port, has allowed for the strengthening of relationships between the two communities, and even between the two countries. Subsequent to this agreement, in 2016, Iceland was chosen to be invited to the Codfish Festival, a new initiative that aimed at opening the doors of the festival to other countries and cultures, using codfish as a link for this connection. In the subsequent years, other countries followed, such as Spain, France and Italy.

FUTURE RESEARCH DIRECTION

The existence of a community of cod fishermen naturally presupposes the existence of their families, made up of children and women, who we refer to in this chapter as "living widows". They stayed on land during the fishing seasons, often in anxiety and suffering, ensuring the education of their children and the support of the family. It would therefore be important to further investigate the role of fishermen's families, especially women, in this story, namely the ways in which, for centuries, they have been the ones recounting the fishermen's stories.

Moreover, looking at the growing importance of digital technologies in our daily lives, it would be equally important to deepen the research about their role and their use in destination storytelling.

CONCLUSION

More than a Sun & Sea destination, linked to the Laguna and the Sea, storytelling has helped Ílhavo in being perceived as much more than it was in the past. Combining most of the features presented in the literature review, Ílhavo now offers an authentic experience (due to community engagement, the use of cod as historical patrimony and delicacy, cultural events, etc.), an unparalleled human and dramatic story (of hero fishermen that risked their lives at sea), and thus a greater emotional connection with the public and with its own community.

Using memory, tradition and the stories told, harshly lived by people, Ílhavo fostered its own identity, honed its own distinctive story and mythology, and accomplished something very interesting: the transformation of cod line and hooks fishing into a legend. And legends and myths, as Aristotle (2018) established, are at the heart of stories that connect emotionally with human beings.

This, however, would not have been possible if not for the participation of stakeholders and City Hall in this storytelling strategy. Community engagement and involvement as both agents and ambassadors of Ílhavo's intangible heritage, along with the strong leadership provided by City Hall, has been vital in making this narrative successful and authentic.

Nevertheless, there are still features that have not yet been used to their full potential in Ílhavo. The use of new technologies, for instance, is still lacking and in need of further investment. Also, there is evidence of low co-creation orientation and activities (with tourists building their own stories in the destination) that could help promote tourist engagement and maintain the storyline dynamics.

In a world where novelty, uniqueness, even stories, have a short life span, where people's attention deficits are increasing, we must question the limitations of Ílhavo's narrative. Simultaneously, new storytelling and digital strategies must be considered to avoid the failure of the model and to sustain and enhance Ílhavo's visibility.

Ílhavo's outstanding story, the one that is made of brave fishermen, and the one that is being made now as the municipality turned story into legend, is one that deserves to be known due to the richness of its traditions and how they reflect Portuguese History. As showcased in this chapter, Ílhavo is a story of success, one that might help researchers to analyse applications of storytelling, examine tourist and community engagement, explore research methodologies, and try to envision a model for using storytelling that might be applied elsewhere and benefit different tourism destinations in the future. If Ílhavo, a small municipality in Portugal, turned into a success story, what is preventing other locations from doing the same?

REFERENCES

Almeida, C. da C. (2004). O marketing das cidades. *Gestao e Desenvolvimento, 12*(12), 9–45. doi:10.7559/gestaoedesenvolvimento.2004.104

Aragonez, T., & Caetano Alves, G. (2012). Marketing territorial: o futuro das cidades sustentáveis e de sucesso sucesso. *Tourism and Management Studies International Conference, 1*, 316–319.

Araújo, M. J. (2021). "Portugal at Table" - Food Heritage, Identity and Tradition: The tourist gaze. *Journal of Tourism & Development, 35*, 243–257. https://dialnet.unirioja.es/servlet/articulo?codigo=7891681

Aristóteles. (2018). *Poética* (A. M. Valente, Trans.). Fundação Calouste Gulbenkian.

Aveiro, V. (2009). *Histórias Desconhecidas dos Grandes Trabalhadores do Mar*. Editorial Futura.

Aveiro, V. (2014). *Ecos do Grande Norte*. Âncora Editora.

Aveiro, V. (2016). *Nómadas do Oceano*. Âncora Editora.

Ben Youssef, K., Leicht, T., & Marongiu, L. (2019). Storytelling in the context of destination marketing: An analysis of conceptualisations and impact measurement. *Journal of Strategic Marketing*, *27*(8), 696–713. doi:10.1080/0965254X.2018.1464498

Camprubí, R., & Planas, C. (2020). Storytelling in tourism destination branding: The case of Girona. *Cuadernos de Turismo*, *46*, 617–619. doi:10.6018/turismo.451841

Carvalho, M. (2013). Construído a memória num espaço museológico: O Arquivo de Memórias da Pesca do Bacalhau do Museu Marítimo de Ílhavo. In *ARGOS – Revista do Museu Marítimo de Ílhavo* (pp. 94–101). Câmara Municipal de Ílhavo / Museu Marítimo de Ílhavo.

Chiu, H. C., Hsieh, Y. C., & Kuo, Y. C. (2012). How to Align your Brand Stories with Your Products. *Journal of Retailing*, *88*(2), 262–275. doi:10.1016/j.jretai.2012.02.001

Cho, H., Shen, L., & Wilson, K. (2014). Perceived Realism: Dimensions and Roles in Narrative Persuasion. *Communication Research*, *41*(6), 828–851. doi:10.1177/0093650212450585

CM-Ílhavo. (2018). *Museu Marítimo de Ílhavo ultrapassa 1 milhão de visitantes e conquista importante prémio nacional*. https://www.cm-ilhavo.pt/pages/388?news_id=3137

CMI. (2008a). *10 Anos de Mudança - Município de Ílhavo 1998-2008*. Câmara Municipal de Ílhavo.

CMI. (2008b). *Boletim Informativo "Município de Ílhavo" #23 (dezembro)*. CMI.

CMI. (2012). *Boletim Informativo "Município de Ílhavo" #38 (dezembro)*. CMI.

CMI. (2014). *Boletim Informativo "Município de Ílhavo" #45 (setembro)*. CMI.

CMI. (2016). *Boletim Informativo "Município de Ílhavo" #53 (setembro)*. CMI.

CMI. (2019). *Boletim Informativo "Município de Ílhavo" #63 (dezembro)*. CMI.

CMI. (2022). *Centro de Religiosidade Marítima*. https://www.cm-ilhavo.pt/viver/areas-de-intervencao/cultura/equipamentos-culturais/museus/centro-de-religiosidade-maritima

Correia, R., & Brito, C. (2011). A Importância do Marketing para o Desenvolvimento Turístico: O caso de Montalegre. *Revista Turismo & Desenvolvimento*, *16*, 127–143.

Costa, N. M. (2016). A memória documental dos pescadores bacalhoeiros de Ílhavo. *Vox Musei Arte e Patrimônio, 1*, 39–44.

da Fonseca, S. (2007). *Ílhavo Ensaio Monográfico: do século X ao século XX*. Papiro Editora.

da Silva, A. J. M. (2001). Gazela Primeiro - o navio e a sua gente. In Á. Garrido (Ed.), *A Pesca do Bacalhau: História e Memória* (pp. 255–272). Editorial Notícias.

da Silva, A. J. M. (2015). The fable of the cod and the promised sea - About Portuguese traditions of Bacalhau. *Heritages and Memories from the Sea, 48296*(January), 130–143.

Escalas, J. E. (2004). Imagine yourself in the product : Mental simulation, narrative transportation, and persuasion. *Journal of Advertising, 33*(2), 37–48. doi:10.1080/00913367.2004.10639163

Fonseca, F. P., & Ramos, R. A. R. (2006). O planeamento estratégico de marketing territorial em busca de potenciar o território: o caso de Almeida. *"PLURIS 2006 : Anais Do 2º Congresso Luso-Brasileiro de Planeamento Urbano Regional Integrado Sustentável".* http://repositorium.sdum.uminho.pt/handle/1822/7224

Franco, A., Correia, R., & Meneses, R. (2019). Os 5 sentidos associados a Portugal na perspectiva dos turistas. *Revista Ibérica de Sistemas e Tecnologias de Informação, 24*, 272–286.

Garrido, Á. (2004). *O Estado Novo e a Campanha do Bacalhau.* Círculo de Leitores.

Garrido, Á. (2016). *Portugal no Mar: Homens que foram ao bacalhau - 2.ª Edição* (CMI/MMI). Âncora Editores.

Garrido, Á. (2018). Espaço Público e memórias Inclusivas: A Experiência de Gestão do Património da Pesca do Bacalhau no Museu Marítimo de Ílhavo. In A. Polónia, C. M. Costa, & F. Mouta (Eds.), Boas prática para políticas públicas de memória, ciência e património (pp. 89–106). Media XXI – Publishing, Research and Consulting Reservados.

Hartman, S., Parra, C., & de Roo, G. (2019). Framing strategic storytelling in the context of transition management to stimulate tourism destination development. *Tourism Management, 75*(May), 90–98. doi:10.1016/j.tourman.2019.04.014

Hay, N. A., Chien, P. M., & Ruhanen, L. (2021). Tell me your story: Branding destinations through residents' (place) stories. *Journal of Vacation Marketing*, 1–16. doi:10.1177/13567667211060567

Keller, K. L. (1993). Conceptualizing, Measuring, and Managing Customer-Based Brand Equity. *Journal of Marketing, 57*(1), 1–22. doi:10.1177/002224299305700101

Keller, K. L. (2008). Strategic Brand Management - Building, Measuring, and Managing Brand Equity (D. Parker, Ed., 3rd ed.). Pearson Education.

Kim, J. H., & Youn, H. (2017). How to Design and Deliver Stories about Tourism Destinations. *Journal of Travel Research, 56*(6), 808–820. doi:10.1177/0047287516666720

Kim, S. H., Song, M. K., & Shim, C. (2020). Storytelling by medical tourism agents and its effect on trust and behavioural intention. *Journal of Travel & Tourism Marketing, 37*(6), 679–694. doi:10.1080/10548408.2020.1795046

Kolar, T., & Zabkar, V. (2010). A consumer-based model of authenticity: An oxymoron or the foundation of cultural heritage marketing? *Tourism Management, 31*(5), 652–664. doi:10.1016/j.tourman.2009.07.010

Kotler, P., Haider, D. H., & Rein, I. (1993). *Marketing Places: attracting investment, industry, ans tourism to cities, states, and nations.* Free Press.

Kotsi, F., Balakrishnan, M. S., Michael, I., & Ramsøy, T. Z. (2018). Place branding: Aligning multiple stakeholder perception of visual and auditory communication elements. *Journal of Destination Marketing & Management, 7*, 112–130. doi:10.1016/j.jdmm.2016.08.006

Li, C. H., & Liu, C. C. (2020). The effects of empathy and persuasion of storytelling via tourism micro-movies on travel willingness. *Asia Pacific Journal of Tourism Research, 25*(4), 382–392. doi:10.1080/10941665.2020.1712443

Li, Y. M. (2014). Effects of story marketing and travel involvement on tourist behavioural intention in the tourism industry. *Sustainability (Switzerland), 6*(12), 9387–9397. doi:10.3390u6129387

Lund, N. F., & Kimbu, A. N. (2021). Applying the Hollywood scriptwriting formula to destination branding. *Current Issues in Tourism, 24*(8), 1058–1078. doi:10.1080/13683500.2020.1739005

Marques, F. (2001). O Creoula: a pesca do bacalhau no crepúsculo da nevegação à vela. In Á. Garrido (Ed.), *A Pesca do Bacalhau: História e Memória* (pp. 225–253). Editorial Notícias.

Martin, J. I. (2011). *A pesca em Portugal*. Academic Press.

McAdams, D. P. (2013). The Psychological Self as Actor, Agent, and Author. *Perspectives on Psychological Science, 8*(3), 272–295. doi:10.1177/1745691612464657 PMID:26172971

Moin, S. M. A., Hosany, S., & O'Brien, J. (2020). Storytelling in destination brands' promotional videos. *Tourism Management Perspectives, 34*, 100639. Advance online publication. doi:10.1016/j.tmp.2020.100639

Moscardo, G. (2017). Exploring mindfulness and stories in tourist experiences. *International Journal of Culture, Tourism and Hospitality Research, 11*(2), 111–124. doi:10.1108/IJCTHR-11-2016-0108

Moscardo, G. (2020a). Stories and design in tourism. *Annals of Tourism Research, 83*(January), 102950. doi:10.1016/j.annals.2020.102950

Moscardo, G. (2020b). The story turn in tourism: Forces and futures. *Journal of Tourism Futures, 7*(2), 168–173. doi:10.1108/JTF-11-2019-0131

Mossberg, L. (2008). Extraordinary experiences through storytelling. *Scandinavian Journal of Hospitality and Tourism, 8*(3), 195–210. doi:10.1080/15022250802532443

Mossberg, L., & Eide, D. (2017). Storytelling and meal experience concepts. *European Planning Studies, 25*(7), 1184–1199. doi:10.1080/09654313.2016.1276156

Mossberg, L., Therkelsen, A., Huijbens, E., Björk, A-K., & Olsson, P. (2010). *Storytelling and Destination Development*. Nordic Innovation Centre (NICe).

Olsson, A. K., Therkelsen, A., & Mossberg, L. (2016). Making an effort for free – volunteers' roles in destination-based storytelling. *Current Issues in Tourism, 19*(7), 659–679. doi:10.1080/13683500.2013.784242

Overend, D. (2012). Performing sites: Illusion and authenticity in the spatial stories of the guided tour. *Scandinavian Journal of Hospitality and Tourism, 12*(1), 44–54. doi:10.1080/15022250.2012.678070

Pasquinelli, C. (2013). Competition, cooperation and co-opetition: Unfolding the process of inter-territorial branding. *Urban Research & Practice*, *6*(1), 1–18. doi:10.1080/17535069.2012.727579

Pellowski, A. (1977). *The World of Storytelling*. R. R. Bowker Company.

Peralta, E. (2003). O mar por tradição: O património e a construção das imagens do turismo. *Horizontes Antropológicos*, *9*(20), 83–96. doi:10.1590/S0104-71832003000200005

Peralta, E. (2010). "We're all seafarers": An ethnography of the (in)visibilities of power in the social representation of the local past in Ílhavo. *Etnográfica (Lisboa)*, *14*(3), 443–464. doi:10.4000/etnografica.249

Pine, J., & Gilmore, J. H. (1998). Welcome to the Experience Economy. *Harvard Business Review*, *76*(4), 97–105. PMID:10181589

Pinto, R., & Neto, D. C. (2017). The Brazilian demand for dried and salted cod - from 1989 to 2014. *Smart and Inclusive Development in Rural Areas - Iberian Conference on Rural Studies.*

Pires, A. da R., Castro, E. A. de, Esteves, C., & Santos, C. (2000). *Ilhavo 2020: construir uma Visão Partilhada das Opções e dos Percursos de Desenvolvimento*. Academic Press.

Providência, F. (2004). *Manual de Identidade Gráfica - Câmara Municipal de Ílhavo*. Academic Press.

Rutledge, P. (2019). Transmedia Psychology: Creating Compelling and Immersive Experiences. In M. Freeman & R. R. Gambarato (Eds.), *Routledge Companion to Transmedia Studies* (pp. 350–363). Toutledge.

Ryan, M. L. (2016). Transmedia narratology and transmedia storytelling. *Artnodes*, *2016*(18), 1–10. doi:10.7238/a.v0i18.3049

Salvado, J., & Joukes, V. N. P. M. (2019). Bacalhau, um marcador simbólico da identidade alimentar portuguesa: Um olhar sobre o território de Trás-os- Montes e Alto Douro. *Revista Portuguesa de Estudos Regionais*, *53*, 55–68.

Salvado Alves, A. J. F. (2008). A competitividade dos territórios num contexto de globalização : Uma utopia ou uma realidade? *VI Congresso Português de Sociologia - Mundos Sociais: Saberes e Práticas.*

Sobral, J. M. (2016). O Bacalhau: de Alimento de Penitência a lcone Português. *Argos - Revista Do Museu Marítimo de Ílhavo, 4*, 27–33.

Sobral, J. M., & Rodrigues, P. (2013). O "fiel amigo": O bacalhau e a identidade portuguesa. *Etnográfica (Lisboa)*, *17*(3), 619–649. doi:10.4000/etnografica.3252

Stevenson, N. (2019). Developing cultural understanding through story- telling. *Journal of Teaching in Travel & Tourism*, *19*(1), 8–21. doi:10.1080/15313220.2018.1560528

Teixeira Costa, P., & Veríssimo, M. (2020). The Sea by Tradition: From the Slogan to the Brand – Exploring a Successful Case of Territorial Strategy Anchored in the Sea and Tourism The Sea by Tradition. In M. Morais de Brito, A. Dias, & M. Patuleia (Eds.), *Managing, Marketing, and Maintaining Maritime and Coastal Tourism* (pp. 264–283). IGI Global. doi:10.4018/978-1-7998-1522-8.ch016

Teles, F., & Anselmo de Castro, E. (2015). *Plano Estratégico do Município de Ílhavo*. Academic Press.

Tussyadiah, I. P., & Fesenmaier, D. R. (2008). Marketing places through firstperson stories - An analysis of Pennsylvania Roadtripper Blog. *Journal of Travel & Tourism Marketing*, *25*(3–4), 299–311. doi:10.1080/10548400802508358

Tussyadiah, I. P., & Fesenmaier, D. R. (2009). Mediating Tourist Experiences. Access to Places via Shared Videos. *Annals of Tourism Research*, *36*(1), 24–40. doi:10.1016/j.annals.2008.10.001

Wang, N. (1999). Rethinking authenticity in tourism experience. *Annals of Tourism Research*, *26*(2), 349–370. doi:10.1016/S0160-7383(98)00103-0

Yin, R. (2003). *Case Study Research: Design and Methods* (3rd ed.). Sage Publications Ltd.

Zainal, Z. (2007). The Case Study as a Research Method. *Jurnal Kemanusiaan*, *9*, 1–6.

ADDITIONAL READING

Khanom, S., Moyle, B., Scott, N., & Kennelly, M. (2019). Host–guest authentication of intangible cultural heritage: A literature review and conceptual model. *Journal of Heritage Tourism*, *14*(5-6), 396–408. doi:10.1080/1743873X.2019.1574804

Kosfeld, M., Heinrichs, M., Zak, P. J., Fischbacher, U., & Fehr, E. (2005). Oxytocin increases trust in humans. *Nature*, *435*(7042), 673–676. doi:10.1038/nature03701 PMID:15931222

Kvítková, Z., & Petrů, Z. (2021). Approaches to Storytelling and Narrative Structures in Destination Marketing. *Tourism in South East Europe*, *6*, 425–438. doi:10.20867/tosee.06.28

Rizzolatti, G., & Craighero, L. (2004). The mirror-neuron system. *Annual Review of Neuroscience*, *27*(1), 169–192. doi:10.1146/annurev.neuro.27.070203.144230 PMID:15217330

Rizzolatti, G., Fadiga, L., Gallese, V., & Fogassi, L. (1996). Premotor cortex and the recognition of motor actions. *Brain Research. Cognitive Brain Research*, *3*(2), 131–141. doi:10.1016/0926-6410(95)00038-0 PMID:8713554

Zak, P. J. (2015). Why inspiring stories make us react: the neuroscience of narrative. *Cerebrum: The Dana Forum on Brain Science*.

KEY TERMS AND DEFINITIONS

Co-Creation of Tourist Experiences: The sum of psychological events a tourist goes through when actively contributing through physical and/or mental participation in activities and intervening with other subjects in the environment of the experience.

Cod Fishing With Line: Fishing carried out by a single man using a line and hooks from a small wooden boat (dóri).

Experience Economy: The sale of memorable experiences to customers.

Story World: Set of stories and activities linked to a central story or a particular character.

Transmedia Storytelling: The technique of telling a single story or story experience across multiple platforms and formats using current digital technologies.

Section 3

Storytelling, Information and Communication Technologies, and Social Media

Chapter 10
Cinematic Storytelling with Virtual Reality for Cultural Heritage Marketing and Accessibility

Alessandra Marasco
https://orcid.org/0000-0002-5028-4908
Institute for Research on Innovation and Services for Development, National Research Council, Italy

Francesca Nicolais
Università degli Studi Suor Orsola Benincasa, Italy

Guido Acampa
Lapej Communication, Italy

Barbara Balbi
Italian Ministry of Culture, SABAP Naples, Italy

ABSTRACT

This chapter addresses the potential of virtual reality (VR) as a storytelling medium for experiential marketing and the accessibility of cultural heritage sites through a case study of a cinematic VR storytelling aimed to enhance the visitor experience at the Archeological Park of Cuma (Italy). The immersive storytelling was used both as a persuasive tool in the pre-visit stage and as a core element of the on-site visit experience for improved accessibility. The VR storytelling was created through a user-centred design process involving users and in collaboration with heritage curators. The case study illustrates how the cinematic immersive narrative is used to engage viewers, developing a connection between two VR experiences that experiment different narratives in relation to their functions in the visitor experience. Theoretical, methodological, and managerial implications are discussed for the design of visitor-centric experiences and the development of experiential marketing strategies.

DOI: 10.4018/978-1-6684-3436-9.ch010

INTRODUCTION

Stories have always been at the core of cultural heritage interpretation, communication and experience design (Bedford, 2001; Nielsen, 2017; Moscardo, 2020). Storytelling in cultural heritage aims to make cultural contents more attractive for audiences and to facilitate the learning process, typically through the creation of a story/fiction (real or invented but reliable) set in a wider cultural and scientifically based context (Pietroni & Adami, 2014). The advancement in digital technologies provided unprecedented opportunities for telling stories to attract, engage and inspire cultural heritage visitors (Daley, 2021; Katifori et al., 2020; Palombini, 2017).

Research in this area devoted increasing attention to storytelling in the design of experiences mediated by Virtual Reality (VR) as a strategic approach to make heritage meaningful, memorable and accessible to different audiences (Bozzelli et al., 2019; Pagano et al., 2020; Škola et al., 2020). Triggering visitors' interest, empathy and imagination, storytelling represents a fundamental component in the design of memorable virtual heritage experiences (Roussou, 2008).

A growing number of successful cases illustrate the unprecedented and unmatched ability of VR to immerse visitors into stories and place them at the center of an experience (Gillam, 2017). The Oculus VR experience "Meeting Rembrandt: Master of Reality", developed in 2017 by ForceField in partnership with Rijksmuseum and Samsung Gear VR, transports visitors back to 1642 in Amsterdam to meet the painter at work in his house and discover details about his famous masterpiece The Night Watch. The 360° Viking Battle video produced by Black Dot Films for the National Geographic allows viewers to experience the life of Vikings and be immersed in the fray of an epic battle with over 600 hardcore reenactors armed with swords and axes. In 2019, the videogame company Ubisoft, creator of Assassin's Creed, with renowned expertise in re-creating historic sites, developed for the Arab World Institute (France) the VR experience "Age Old Cities: A Virtual Journey from Palmyra to Mosul" that transports visitors to some of humanity's oldest settlements destroyed or damaged by war. By the same company is the Discovery Tour series of dedicated educational games to explore and learn about the history and daily life of Ancient Greece, Ancient Egypt and the Viking Age. Successful examples also include VR productions premiered at films festivals, such as the live-action documentary film "Carriberrie" that celebrates Indigenous Australian song and dance through a journey across iconic Australian locations to experience first-person traditional and contemporary performances bringing together Aboriginal history, stories, music, and dance.

Notwithstanding the growing research and practice in this area, which is fueled by the availability of devices (including head-mounted displays - HMD) and distribution platforms, the potential of VR as a narrative medium in its own right (Aylett & Louchart, 2003) for promoting visitors' engagement and creating enhanced heritage experiences along the visitor journey is yet to be fully explored and exploited.

To contribute to advance knowledge in this field, this chapter addresses the use of VR as a narrative medium for experiential marketing and the accessibility of cultural heritage sites. Specifically, it focuses on cinematic storytelling in VR and its potential for the creation of immersive experiences of 'invisible' heritage, including sites that are not fully accessible by all potential visitors.

It presents a case study of a Cinematic VR (CVR) storytelling project created to enhance the visitors' experience of the Archeological Park of Cuma (Italy). This heritage site has been visited and appreciated since the times of the Grand Tour and is renowned for the cave of the prophetess Sibyl (Zevi et al., 2008); it is partially accessible to visitors with mobility disabilities, including the elderly and visitors with temporary disabilities. Building on the potential of immersive technologies and storytelling to enhance the visitor experience along the journey (Flavián et al., 2019; Moscardo, 2017, 2020) and integrate the pre-, during, and post-visit stages (Kuflik et al., 2015, Perry et al., 2017), the CVR storytelling was designed to promote visit intentions in the pre-experience stage and enhance the onsite experience supporting full access to the site and its relevant information.

The chapter describes the design process of the CVR experience, which was based on a user-centred design approach involving users and the collaboration with the curators of the archeological heritage site. It illustrates how the cinematic immersive storytelling is used to engage viewers, developing a connection between two immersive solutions that experiment different narratives in relation to their functions along the visitor journey. Theoretical, methodological and managerial implications are discussed for the design of visitor-centric experiences and the development of innovative experiential marketing strategies.

BACKGROUND

The last decades witnessed a growing trend in the applications of immersive reality technologies in cultural heritage and research has documented their role in the acquisition and dissemination of knowledge, exploration of and access to cultural heritage, and to improve the visitor experience at heritage sites (Bekele et al., 2018).

Further, studies on VR in cultural heritage tourism demonstrated its effectiveness as an *ex-situ* marketing tool in the pre-experience stage by enhancing intentions to visit heritage sites (Jung & tom Dieck, 2017; Lin et al., 2020; Marasco et al., 2018). In particular, Jung et al. (2017) highlighted the role of immersive VR applications in shifting the focus of marketing strategies to providing potential visitors the possibility to pre-experience the heritage site and thereby influence behavioural intention. Lin et al. (2020) provided evidence of the importance of emotional responses to VR experiences in heritage destination marketing.

Research addressed the potential of immersive VR technologies also for providing engaging visitor experiences of sites that are at risk of over-tourism, deterioration and destruction through alternative forms of access and consumption (Allal-Chérif, 2022; Bec et al., 2021). The ability of VR to simulate and transport users in imaginative and existing environments with a high level of multisensorial realism makes it an ideal technology to increase accessibility to cultural heritage, especially when physical access is limited or impossible (Bekele & Champion, 2019). Thus, research focused on the benefits of VR for the accessibility to heritage sites (Guttentag, 2010; Ozdemir, 2021; tom Dieck et al., 2019; Škola et al., 2020), which represents a significant aspect of heritage tourism development (Naniopoulos et al., 2015) and one of the greatest cultural heritage challenges (Council of Europe, 2017). The ability of VR to facilitate alternative access to heritage sites is particularly beneficial for visitors with disabilities, including elderly visitors, facing physical barriers that cannot be easily eliminated due to conservation requirements or prohibitive costs (Guttentag, 2010). Research has shown how VR can be usefully applied to provide realistic heritage experiences to visitors with disabilities improving physical and perceptual accessibility (Agostiano; 2016; Pérez et al., 2020; tom Dieck et al., 2019). Montagud et al. (2020) also highlighted the need and challenges to make immersive experiences in culture and heritage, including VR 360 videos, accessible to all for guaranteeing universal access and inclusion.

Growing research focuses on storytelling as an integral component of VR applications to convey cultural heritage information in a more effective way (Rizvic et al., 2020). For instance, the key strength of this technology – namely, the subjective 'realness' of immersive VR experiences, deriving from the place illusion (of being present in the simulated location) and the plausibility illusion that the depicted scenario is actually occurring (Slater, 2009) - is increasingly leveraged for storytelling in cultural heritage (Škola et al., 2020).

According to Roussou (2008), storytelling represents a crucial part of the design of virtual heritage experiences that can suspend the audience's disbelief - one of the key aspects of narrative engagement. Roussou (2001) pointed to the challenging synthesis between form (visual representation), story (narrative), and history (authenticated cultural content) for narrative development in the design of engaging and memorable virtual heritage experiences that combine historical accuracy, aesthetic pleasure, educational value and emotional involvement. In this regard, she also highlighted the importance of developing characters, preferably in a human form, that are not only believable but also unforgettable (Roussou, 2008). According to Pietroni and Adami (2014), the main challenges in the design of virtual applications for visitors' involvement and pleasant acquisition of cultural contents concern the creation of effective storytelling formats based on scientific contents, the alignment of contents with users' different profiles, the definition of interaction paradigms for enjoying digital contents without frustration. Argyriou et al. (2020) proposed a design methodology for 360-degree video-based VR cultural tours, articulated into two complementary layers: 1) the experience layer, related to the design of the narrative, the virtual scenes and actor role; 2) the interaction layer, related to the navigation in the virtual world and the development of gamified techniques for users' engagement. Based on this framework, they created and evaluated an immersive interactive cultural heritage virtual tour, providing a set of design guidelines for creating highly engaging experiences through the integration of branching storytelling, gamification techniques and motion effects.

In line with the relevance of digital storytelling for engaging visitors emotionally (Perry et al., 2017), studies addressed the potential of storytelling in virtual heritage experiences to elicit emotional responses as well as the issues involved in the measurement of visitors' emotions. In the project "Becoming Vincent", Calvi et al. (2015) designed an immersive experience with the Oculus Rift that allows tourists to discover the painter's heritage in an emotionally engaging way by becoming Vincent themselves and feeling his emotions. Little et al. (2020) addressed how advanced technologies allow to create 'windows into the past' introducing new emotional value to sites and monuments, thereby turning them from 'cold' into 'warm' for human experience. Rizvic et al. (2019) showed the range of positive and negative emotional responses to the experience of war heritage through VR, in line with an intended empathetic connection provided by the application. The works by Bozzelli et al. (2019) and Pagano et al. (2020) highlighted the influence of emotional involvement during the VR experience on attention, motivation and the memorability of information. In their ArkaeVision project, an immersive virtual exploration into the reconstructed Temple of Hera II (Paestum, Italy) was developed to discover the secret of the Hera cult beside the animated character of the priestess Ariadne, which was designed to build an emotional connection with users.

Stressing the need to better understand how emotions influence learning and the overall visitors' experience in the context of cultural tourism mediated by VR/AR, Han et al. (2019) proposed to use physiological measures, especially electroencephalography (EEG), in addition to post-experience self-reports for capturing the emotional dimension of experience. A mixed approach combining subjective (questionnaires) and objective (neurophysiological recording of the brain signals using EEG) evaluation methods was adopted by Škola et al. (2020) to assess the user experience of a storytelling 360° VR application for a cultural heritage submerged site. Their results highlighted positive users' rating in relation to engagement, emotions, and state of flow, suggesting the viability of using EEG for future research in this field.

Cinematic Storytelling with VR for Culture and Heritage

Research addressed the application of cinematic storytelling as a way to achieve a higher level of audience engagement and its intellectual and emotional connection with heritage (Demetrescu et al., 2016; Schiavottiello et al., 2016). The multidisciplinary discussion highlighted on one hand its role as an effective way of telling history, especially to the young audience, generating a completely new realm, on the other hand the need to carefully deal with its spectacular potential in the light of scientific interpretation (Palombini, 2017; Schiavottiello et al., 2016). Research focused on the opportunities, design challenges and effects of cinematic storytelling for the preservation and promotion of tangible and intangible heritage in a variety of applications and contexts (Table 1).

Table 1. Summary of previous studies

Authors (year)	Applications & findings
Demetrescu et al. (2016)	• Videomaking in virtual and mixed reality applications with real actors for the onsite fruition of archeological heritage sites (Italy). • New tools and methods shape innovative creative languages with high emotional involvement potential and lead to new types of jobs.
Dowling et al. (2018)	• The cinematic VR "Faoladh" filmed with real actors at the Irish National Heritage Park for experiencing a story based on an ancient legend (Ireland). • Traditional filmmaking techniques need to be adapted due to user ability to look in any direction.
Marasco et al. (2018)	• A cinematic VR trailer filmed with real actors to promote offsite the intention to visit a cultural heritage site (Italy). • The cinematic VR experience with HMD positively influences behavioral intentions to visit the cultural heritage site.
Marasco (2020)	• Online cinematic VR projects for various cultural heritage sites (worldwide). • CVR is used for creating different types of historyliving experiences that elicit a range of emotional responses.
Park et al. (2018)	• A video-based virtual tour, applied via film-induced tourism, developed for two heritage sites in an old downtown area (Korea). • The video-based experience is more effective than the basic virtual tour in improving attitude and visit intentions.
Pietroni et al. (2015)	• A novel approach combining different languages (VR, cinematographic paradigms, use of real actors, virtual set practices) for the onsite fruition of archeological heritage (Italy). • The proposed approach makes the archeological space alive and creates an experience that involves users at conceptual and emotional levels.
Rizvic et al. (2020)	• The VR film "Nine dissidents" on a story about banned intellectuals during the socialist regime (Bosnia and Herzegovina). • The film is set in a courtroom and tells the story in a satiric way. It is an effective approach for raising public attention on the topic.
Roussos & Bizri (1998)	• The narrative project "Mitologies" developed for the CAVE VR system based on various literary and artistic sources (USA). • The use of the cinematic approach in an artistic application of the medium allowed to capture the emotional essence of environments.
Seide & Slater (2020)	• A virtual cinematic heritage application for the preservation of a 'lost' historic film based on traditional folk culture (Singapore). • The cinematic experience immerses the viewer into the narrative world of the lost film as a 'walk-in movie' and is an effective approach for film heritage preservation.
Selmanovic et al. (2020)	• A 360° VR video storytelling application with HMD designed to preserve intangible heritage based on the case study of the Old Bridge diving tradition in Mostar (Bosnia and Herzegovina). • The experience supports immersion and presence and enables the creation of an intimate connection with intangible heritage.

In the pioneristic research by Roussos and Bizri (1998), the cinematic narrative form is used in the project "Mitologies" developed to explore the potential of VR as an artistic medium. The narrative experience is based on various themes, including the mythical journey of Theseus, and aimed to communicate the emotional essence of environments beyond their accurate representation through VR. Pietroni et al.

(2015) and Demetrescu et al. (2016) illustrated the development of innovative storytelling applications through the integration of videomaking techniques and cinematographic languages for engaging experiences of heritage, highlighting future perspectives in terms of the transformation of cultural and creative languages and development of new professions integrating heritage and audiovisual communication competencies (Demetrescu et al., 2016). Selmanović et al. (2020) investigated the use of immersive 360° video storytelling for preserving intangible heritage through a VR application developed for the bridge diving tradition from the Old Bridge in Mostar (Bosnia and Herzegovina). Their study highlighted the superior potential of this medium to elicit feelings of immersion and presence aligned with the physical reality as well as to enable viewers' empathic connection with the protagonist (the diver) of the tradition. The effectiveness of this medium to transfer the intangible aspects of heritage, including feelings and emotions, is stressed as a result of the user evaluation in the study. They also pointed to the narrative paradox challenge, relating the difficulties of applying traditional film language grammar as the viewer can freely turn around when the immersive content unfolds. The challenge to effectively guide the viewer's attention in CVR is addressed also by the study by Dowling et al. (2018) based on their VR film "Faoladh". This cinematic VR experience tells the journey of a Celt child whose village is attacked by Viking raiders and puts the viewers in the role of a wolf like guardian spirit and protector of the child from the woods of 9th century Ireland. According to the authors, spatial audio and actors' cues are effective practices to guide viewers' attention, but further research is needed to address the adaptations of cinematic practices in VR.

Rizvic et al. (2020) created the VR movie "Nine dissidents" for increasing public awareness of the story of Bosniak writers prosecuted during the socialist regime in Yugoslavia. The movie is set in a courtroom and tells the story in a satiric way. The authors described its directing, done so as to minimize the number of cuts for a better user experience. Seide and Slater (2020) pointed to the developments in cinematic VR for heritage as highly relevant in relation to film heritage preservation and recreated a scene of the lost Singaporean film "Pontianak" (1957) as a 'walk-in movie' experience that immerses the viewer in the centre of the scene for engaging both audiences familiar with the myth/film and those who are unaware of the story. Marasco (2020) explored the use of CVR for creating narrative experiences of cultural heritage, from the travel in the past to the immersion into a masterpiece to the encounter with the protagonists of cultural heritage, and highlighted their potential to elicit various emotional responses through an analysis of online comments.

Previous studies also explored the role of cinematic VR experiences in the marketing of cultural heritage sites. Park et al. (2018) demonstrated that video-based VR experiences, which were realized via film-induced tourism, positively influence attitudes and behavioral visit intention to a cultural heritage site compared to a basic virtual tour. The study by Marasco et al. (2018) showed the significant impact of a cinematic VR experience filmed as an immersive trailer with real actors on behavioral intentions to visit the featured cultural site. The study also highlighted the need to explore more deeply the factors underlying the emotional involvement of users in the immersive experience with HMD.

METHODS

Against this background, this study investigates how the power of VR as a cinematic storytelling medium can be leveraged for the experiential marketing and the accessibility of cultural heritage sites. This is achieved through a case study of a VR storytelling project developed for the Archeological Park of Cuma (Italy) for enhancing the heritage experience, from the pre-visit stage of the visitor journey.

The Archaeological Park of Cuma was founded in 1927 during the great excavation campaigns carried out in the early decades of the '900 under the direction of Amedeo Maiuri. The city of Cuma, probably the oldest Greek colony in Italy, had a very important economic and military role that continued throughout the Roman period and in particular during the Augustan age. The archeological area includes the cave of the prophetess Sibyl, the Temple of Apollo and the Temple of Jupiter on the summit of the Acropolis (Fig.1), but the site is particularly known for the Sibyl's cave around which history, myth and legend are intertwined. The site is partially accessible for visitors with mobility disabilities. The Sibyl's cave is fully accessible, but the paths to the temples of Apollo and Jupiter cannot be accessed by mobility-impaired visitors and can be difficult to access by the elderly and visitors with temporary disabilities.

Figure 1. The Archaeological Park of Cuma: Sibyl's cave (left) and Temple of Jupiter (right)

The study is based on the theoretical framework for experience design in tourism and specifically the fundamental approaches to designing tourism experiences (Tussydiah, 2014): the human-centered approach, the designing as iterative processes, and the holistic experience concept. Based on this framework, the approach adopted in this study emphasizes the in-depth understanding of the experience of visitors from their perspective at every stage of the design process through User-Centred Design (UCD) principles and methods, involving users from the beginning of the process. UCD is indeed supported as a preferred approach for digital innovation in cultural heritage (Beltrán et al., 2014; Perry et al., 2017; Southall et al., 2019). In the research, the UCD process was adopted to identify the experiential and functional requirements of potential users for the development of the VR solution in line with the principles of design for all.

Within this process, the narrative design, aimed to develop the concept and the design choices (setting, characters, theme, narrative, sounds), was carried out in close collaboration with the experts and curators of the heritage site. In particular, the storytelling was developed by a multidisciplinary team, including experts from archeology, communication, computer science, film directing, human-computer interaction, and cultural tourism. For the narrative development process, role of stories in tourism experience design was considered, and particularly in extending the temporal dimension of the tourism experience from the pre-visit experience (Moscardo, 2017, 2020). The design process of the cinematic immersive storytelling is described as follows.

Defining *Personas*, Scenarios and Users' Requirements

The UCD process was developed in line with the challenge to design a VR solution for enhancing the experience of partially visible cultural heritage sites and allowed to take into account the needs of potential users from the beginning. Its final aim was a VR solution informed by the *Design for All* approach (Arenghi & Agostiano, 2016) to support the widest possible audience. For this purpose, the project has adopted UCD principles (ISO 13407:1999 revised by 9241:2010), by developing from the beginning a co-design model to allow an effective involvement of users with particular attention to people with mobility disabilities. The UCD process developed a model to encompass all the interrelated dimensions of the experience of cultural heritage visitors mediated by a digital solution, including the hedonic dimensions and those more closely related to functional elements. The process was carried out in collaboration with three associations for people with disabilities and accessible tourism and is described in detail in Balbi and Marasco (2021).

The initial research phase for concept generation was conducted through several methods (survey, focus group, interviews) in order to obtain information on the behavior and needs of the target groups of potential users and allowed to identify the core elements of the visitor experience that could be improved by the VR solution, the characteristics, the needs, and the expectations of the main target groups, thus allowing to develop *personas* and scenarios, i.e. the tools necessary for the subsequent co-design phase. Based on previous studies (Guttentag, 2010; Michopoulou & Buhalis, 2013; Michopoulou et al., 2015), the core elements of the accessible visitor experience with VR were identified in relation to the whole journey (Marasco & Balbi, 2019), as shown in Fig. 2. This phase also addressed the needs of potential users in relation to the possible narrative modalities to be deployed in the VR solution, from a guided tour of the site to an approach based on storytelling.

Figure 2. VR role in the accessible visitor experience
Source: based on Marasco and Balbi, 2019

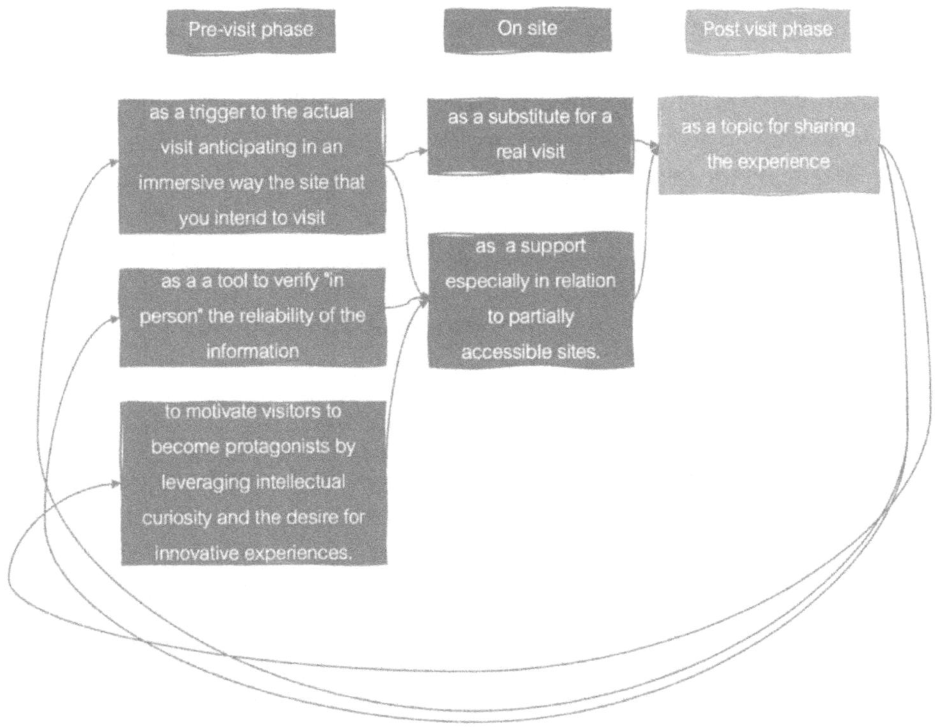

Through this research, it was possible to define four *personas* corresponding to the target groups through a real 'identity card' complete with accessibility needs, interest in cultural heritage and digital technologies in tourism, including VR. These *personas* were used in the subsequent co-design phase that adopted a scenario-based design approach to identify the experiential and functional requirements of the VR solution (Balbi & Marasco, 2021). To this end, four scenarios corresponding to the *personas* were designed to represent a familiar situation in which each *persona* was involved in organizing a cultural visit. In the co-design activity, the scenarios served as storyboards to pairs of users and designers to expand and continue the story imagining VR as a tool to support the experience. This phase provided insights for the storytelling with regard to the links between narrative modes, the technical features of the solution that support them, and intellectual or emotional involvement. For instance, the imagined modality of storytelling was found to functionally correspond to different users' requirements, relating in particular to:

- the suspension of disbelief satisfied from the user's point of view almost internal to the site or very close to the artifact;
- the enhanced experience of inaccessible areas by overcoming the barriers through a flight over the area to have a global vision;

- the authenticity of the experience through the use of diegetic and extradiegetic sound in a storytelling performed by a 'mentor' character not necessarily visible to visitors (e.g., voice over).

Importantly, the storytelling should be designed to support the visitor in the visit in a completely autonomous way. This requirement, which was experientially expressed with the desire to choose the type of visit and secondly the device, requires the absence of commands and/or controls; therefore, it corresponds to a story told without the need to activate points of view or paths with commands, but rather designed in a minimalistic way both from a narrative (cognitive autonomy) and gestural point of view (autonomy of use). In this sense, this is perhaps the requirement most closely linked to the actual concept of accessibility, which repeatedly emerged during the co-design activity with visitors with different mobility needs. The results obtained through the co-design helped to build a definitive scenario to support the designers of the CVR solution in creating the immersive experience, guided by a narrative mode that would satisfy the requirements identified together with the users.

Designing the Narrative

In line with the requirements that emerged in the UCD phase, the storytelling project involved the creation of a narrative for enhancing the heritage experience at the Archeological Park of Cuma, which was carried out in collaboration with the experts from the Ministry of Culture and site curators. The process started from the research around the Archaeological Park and the Sibyl. The oldest references to a cave of the Sibyl are found in a pseudo-Aristotelian text (De mirab. ause, LV-HI century BC) and in Lycophron (111 century BC), but the most famous evocation is certainly by Vergil in the sixth book of the Aeneid, who tells the story of Aeneas' journey and his arrival in Cuma, at the cave of the Sibyl, to consult the oracle and to know his own destiny and that of his descendants. This research brought to light a complex stratification of stories, names, facts and legends in which the myth is intertwined with the literature delivering for the Sibyl a character with a thousand shades and a thousand faces even in terms of imagery. The writing of the subject for an effective storytelling based on scientific knowledge was then oriented to give life to a character that could hold together all these facets and to create content that could be experienced without generating a sense of frustration, but rather fostering emotional engagement towards the story and the archaeological site itself.

The script process began by introducing some suggestions for the definition of the main character of the narrative, the Sibyl, and outlining the framework for building the narrative:

- she does not describe herself but refers to the descriptions that others make of herself.
- she does not want to be what she was, she wants to talk.
- she is one, but many.
- she may be fragile but every now and then she pokes the visitor... Too weak the modern man.
- if she is no longer a Sibyl, then what is she? (...there is time to figure it out).
- themes (to be introduced and then addressed): her in myth and history, the journey.

These considerations led also to the use, for some passages, of the only voice without any physical presence, suggested by the many literary sources that describe a Sibyl consumed in the body by an immortality that leaves unchanged only her voice. This choice, supported by the aid of spatial audio which allows a spatial movement of the voice around the visitor, is congenial to the mystery of the character.

The will was to represent, in some fragments, a stream of consciousness, the expression of a character that despite possessing many souls and much power, presents itself to us with a humanity that is in some ways fragile, perhaps tired of what it had been. Her stories talk about her, but also about history, places and legends linked to that territory. She who was once addressing emperors and great warriors, today feels like manifesting and telling herself to someone… no matter how worthy is that someone.

"…We were all something once.
I was many things in many places.
Once there seemed too little time.
Now it's all I have left.
I knew stories of magical creatures.
Now they are just that – stories!
Gone along with so many other things.
Magic was once the world's greatest force. It is no longer that.
The myths have disappeared and their children forgotten.
Today, men worship far more cruel earthly deities than ours.
But one thing today is the same as then.
It is impossible to take something that the earth cannot recover.
Everything goes back to it again".

Cinematic VR Experience Production

The VR experience was developed as full-fledged film production, including the writing of the subject, the drafting of the script and the creation of a detailed storyboard defining all the camera points to be filmed during the shooting phase. The director then filmed at the site with the help of a small crew and a professional cast. Live-shootage covered all the camera points defined in the storyboard that coincided with the viewer's/coprotagonist's point of view. A seagull flight, realized with a drone, was perfomed to allow viewers to take flight over the site and get a bird's eye view. The choice to never leave the viewer's point of view was aimed to generate maximum involvement of the viewer as the co-protagonist of the story and empathy with the character. The mysterious woman not only always addresses the spectator, but also involves the viewer in some personal reflections that refer to the history of the place.

THE CINEMATIC VR EXPERIENCE FOR THE ARCHEOLOGICAL PARK OF CUMA

Through this process, the storytelling "Traces of time" was conceived to create a connection between two CVR experiences that experiment different narratives in relation to their different purposes, notably influencing the intention to visit the Archeological Park of Cuma in the pre-visit offsite stage; and supporting visitors' full access to the site and to its relevant information during the visit.

The first CVR experience was designed as a 2-minutes trailer to be delivered through a web portal for accessible heritage tourism aimed to support decision-making and planning. It introduces the story with an engaging rhythm and mysterious implications, to involve the viewer with short and sharp hints motivating to visit the site. The initial sequence opens in the Sibyl's cave and the first lines "Do you need advice? Do you need advice?" also provide support for the viewer to absorb the surrounding environment and to get used to the 'new' reality. The voice 'rotates' around the viewer without ever allowing to identify a unique origin, but contributes to the level of presence by providing an awareness of the environment and guiding the user's attention towards desired points, assuming an almost interactive role and evolving into a hybrid between game audio and linear post-production.

After having looked at the site from above with a bird's eye view, the viewer returns to the cavern where the encounter with the mysterious voice, the carrier of so many stories, is interrupted by the entrance of a mysterious woman with the appearance of a tourist (Fig. 3); she triggers a contact with the viewer by asking an apparently obvious question: "You were here at one time ... Weren't you?"

Figure 3. A frame from the CVR trailer "Traces of time"

The second phase of the storytelling sees the entrance of this character, presented at the end of the trailer, with blurred contours: who could she be? A passer-by? A tourist? A guide? The Sibyl?

This immersive experience designed for the onsite visit, to be delivered through a mobile app, is presented as the casual meeting with an unknown character that becomes more and more intimate as the viewer is led to get to know better this 'source of knowledge'. The story starts from the terrace of the Acropolis (Fig. 4) with the interaction between the spectator and the mysterious woman seen in the cave at the end of the trailer. The woman accompanies the viewer on the visit to the site telling many stories, especially for the aspects related to time, a recurring topic in the words of the mysterious character.

"I was young! And all young minds harbour the deep longing for some sort of power that will make one's life... special!
But nobody is immune to time.
Those certainties waver with time.
Isn't this place wonderful? Layer upon layer of time,
and yet today you can barely see any trace of it."

The journey ends back in the cavern, perhaps because that place is the most precious to the woman ... a return to the mother earth where everything is born and everything is taken back.

Figure 4. Frames from the CVR experience "Traces of time"

The choice of introducing this character who actually activates a 'conversation' with the viewer (Fig. 5) serves a twofold purpose: to meet the requirements that emerged from the UCD process and to recognize the viewer as an entity in the scene and enhance sense of presence.

"Can you hear the sea now?
It's real! And it's saying something
Maybe we won't hear anything at all when it is all over.
I wonder when that day will come for me.
Come on, let's go!"

Figure 5. Frames from the onsite CVR experience: conversation with the character

In particular, studies on the so-called Swayze Effect (Burdette, 2016) show that the absence of an agency of the viewer actually creates an invisible wall between him and the virtual environment, and demonstrate the fact that ignoring the presence of the viewer in the virtual world, even though being in fact present 'here and now' (Škola et al., 2020), determines a lack of connection with the characters and especially loosens the link with the narrative. In order to hold these things together, i.e., the narrative, the environment, and the viewer, it is therefore necessary for the characters to somehow acknowledge the presence of the viewer.

This approach has a lot to do with another important element to consider when building a narrative for CVR: the sense of embodiment in VR. The term "sense of embodiment" (Kilteni et al., 2012) has been used in the context of VR to discern it from the general notion of embodiment. It consists of the components of the sense of self-location, the sense of agency, and the sense of body ownership. In the case of VR, the sense of owning, controlling, and being inside a biological body is complicated by the circumstance of also being immersed in a virtual environment or, as in the case of CVR, an environment that seems to be real. Recognizing the body of the viewer-explorer (Ruszev, 2021) and his sense of embodiment as a key element to expand the understanding of the narrative in VR then becomes a determining element. In VR, the spectator-explorer is immersed in a perceptually hybrid environment in which a double type of stimuli are perceived: those that come from the virtual environment and those that belong to the physical world where the experience is consumed.

As highlighted by previous studies (Dowling et al., 2018; Selmanović et al., 2020), a relevant issue in CVR experiences is that the viewer has the freedom to look around in a dimension much less guided by the director than in cinema and that though not being present with the body in the virtual world in a certain way inhabits it and lives its time and space. The body then becomes a sensitive storytelling device - a reason why narrative in VR must be viewed as an embodied, spatial and participatory process that unfolds through the viewer's exploration (Aylett & Louchard, 2003).

This work shows how narrative VR is open to new scenarios and new challenges in which the individual, the territory, heritage and tourism can be the drivers of new reflections and even new professions in line with Demetrescu et al. (2016), such as the Embodiment Director (Carpio & Birt, 2021) able to support directors to create truly immersive, educational, engaging and exciting cinematic experiences. The present tells us that heritage will be truly innovative.

"The truth is that the myth persists in you!
It expands in your imagination, like a pebble thrown into the pond of time.
This is why the search for the real cave has not ended
and probably never will be."

FUTURE RESEARCH DIRECTIONS

Future research will be aimed to provide further insights into the persuasive potential of the CVR experience as an innovative experiential marketing tool by empirically investigating its impact on viewers' visit intentions. To this end, the influence of narrative engagement elicited through the CVR trailer on behavioral intentions will be examined through an approach that combines self-report and physiological measures to better capture emotions elicited by the narrative, in line with recent research (Tuerlan et al., 2021).

A promising approach in this specific context is tracking the gaze of viewers as in narrative virtual reality the viewer is free from the fixed-frame point of view chosen by the director (Bender, 2019). This would contribute to fill a gap in research on the use of VR in marketing by better considering the effects of narrative components (Rach & Scott, 2020) and to provide useful insights to support experiential marketing strategies in heritage and tourism (Le et al., 2019).

Another avenue for research to contribute to advance knowledge on the role of immersive storytelling in cultural heritage and tourism is to examine the effectiveness of CVR for enhancing memorability as a key aspect of immersive heritage tourism experiences deserving further investigation (Bec et al., 2019; Han et al., 2019; Jiang et al., 2022). For instance, future research could explore how the characteristics of cinematic experiences in VR, including perceived realism of the virtual environment, audiovisual flow, spatial audio, sense of 'living' the film (Reyes, 2018), enhance the memorability of the heritage experience.

CONCLUSION

The research described in the case study provides insights into how innovative VR technologies can be used to create engaging narratives serving different functions in the heritage visitor experience, thereby contributing to the field of VR storytelling in cultural heritage and tourism. In particular, the perspective adopted in this study contributes to advance knowledge and practice in the area of heritage marketing and accessibility shifting the focus from artifacts/sites to the experience of visitors along the journey for devising innovative strategies to promote awareness, engagement and access to heritage. In this perspective, the persuasive potential of immersive VR storytelling is relevant for cultural heritage sustainability, especially by contributing to increase the visibility and valorisation of lesser-known heritage sites (Kontiza et al., 2020) and encouraging sustainable heritage tourism (Lin et al., 2020).

With specific regard to accessibility to heritage sites, the approach presented in the case study contributes to the design of technological 'compensatory solutions' as alternatives to direct access to sites in relation to the emotional experience associated with the public enjoyment of heritage in line with Agostiano (2016) and with an approach to accessibility that goes beyond physical accessibility (Deffner et al., 2015). The immersive storytelling "Traces of Time" was designed taking into account the need to stimulate emotional involvement through the cinematic narration along with providing access to accurate knowledge and realistic representation of the cultural heritage site.

In line with previous research on storytelling design in cultural heritage (Roussou, 2008) and tourism (Moscardo, 2020), a key issue addressed in the project was the design of an authentic credible and 'unforgettable' character to fuel viewers' immersion, emotional involvement and connection with heritage (Bozzelli et al., 2019; Selmanovic et al., 2018), reinterpreting insights of the past in a contemporary key.

From a methodological perspective, the study illustrates the application of UCD for the design of the immersive storytelling solution, from a perspective that integrates human-computer interaction, cultural heritage accessibility, tourism marketing and filmmaking for creating truly visitor-centric cultural experiences. This is achieved in the case study by involving potential users from the beginning of the design process as co-designers to better capture their experiential needs and requirements.

With regard to the managerial implications, it presents a case study of creative collaboration for innovation, where the potential of technology was leveraged to devise an innovative approach to marketing for reaching the broadest possible audience. As VR technologies usage continues to grow, cultural heritage institutions need to take into consideration their integration into experience design and marketing. The potential of these narrative experiences for innovative marketing can be leveraged through their integration into promotional platforms and communication channels that can potentially reach the greatest audience of potential visitors both online and onsite at destinations, for example at train stations, airports and other tourist spots.

ACKNOWLEDGMENT

This research was supported by Campania Region (POR FESR CAMPANIA 2014/2020) - Project #In-Heritage - Integrated Technology System for Accessing In-Visible Heritage.

The authors gratefully acknowledge the collaboration with Fabio Pagano, Filippo Demma, Marida Salvatori, Marzia Del Villano, Filippo Russo of the Parco Archeologico dei Campi Flegrei, Disabled Peoples' International (Comitato Territoriale della Campania), ItaliAccessibile, Unione Italiana Lotta alla Distrofia Muscolare (sezione di Caserta) and thank all the participants for their invaluable contribution to this research.

REFERENCES

Agostiano, M. (2016). Enjoyment of Cultural Heritage and "Compensatory Solutions" for an Alternative Accessibility. In Accessibility as a key enabling knowledge for enhancement of cultural heritage (pp. 131-147). Academic Press.

Allal-Chérif, O. (2022). Intelligent cathedrals: Using augmented reality, virtual reality, and artificial intelligence to provide an intense cultural, historical, and religious visitor experience. *Technological Forecasting and Social Change*, *178*, 121604. doi:10.1016/j.techfore.2022.121604

Arenghi, A., & Agostiano, M. (2016). Cultural Heritage and Disability: Can ICT Be the 'Missing Piece' to Face Cultural Heritage Accessibility Problems? In Smart Objects and Technologies for Social Good. Lecture Notes of the Institute for Computer Sciences, Social Informatics and Telecommunications Engineering (pp. 70-77). Springer.

Argyriou, L., Economou, D., & Bouki, V. (2020). Design methodology for 360° immersive video applications: The case study of a cultural heritage virtual tour. *Personal and Ubiquitous Computing*, *24*(6), 1–17. doi:10.100700779-020-01373-8

Aylett, R., & Louchart, S. (2003). Towards a Narrative Theory of Virtual Reality. *Virtual Reality (Waltham Cross)*, *7*(1), 2–9. doi:10.100710055-003-0114-9

Balbi, B., & Marasco, A. (2021). Co-designing Cultural Heritage Experiences for All with Virtual Reality: A Scenario-Based Design approach. *Umanistica Digitale*, (11), 71–98.

Bec, A., Moyle, B., Schaffer, V., & Timms, K. (2021). Virtual reality and mixed reality for second chance tourism. *Tourism Management*, *83*, 104256. doi:10.1016/j.tourman.2020.104256

Bec, A., Moyle, B., Timms, K., Schaffer, V., Skavronskaya, L., & Little, C. (2019). Management of immersive heritage tourism experiences: A conceptual model. *Tourism Management*, *72*, 117–120. doi:10.1016/j.tourman.2018.10.033

Bedford, L. (2001). Storytelling: The real work of museums. *Curator (New York, N.Y.)*, *44*(1), 27–34. doi:10.1111/j.2151-6952.2001.tb00027.x

Bekele, M. K., & Champion, E. (2019). A comparison of immersive realities and interaction methods: Cultural learning in virtual heritage. *Frontiers in Robotics and AI*, *6*, 91. doi:10.3389/frobt.2019.00091 PMID:33501106

Bekele, M. K., Pierdicca, R., Frontoni, E., Malinverni, E. S., & Gain, J. (2018). A survey of augmented, virtual, and mixed reality for cultural heritage. *Journal on Computing and Cultural Heritage*, *11*(2), 1–36. doi:10.1145/3145534

Beltrán, M. E., Ursa, Y., de los Rios, S., Cabrera-Umpiérrez, M. F., Arredondo, M. T., Páramo, M., Prados, B., & Pérez, L. M. (2014). Engaging people with cultural heritage: Users' perspective. In C. Stephanidis & M. Antona (Eds.), *Universal Access in Human-Computer Interaction. Universal Access to Information and Knowledge* (pp. 639–649). Springer. doi:10.1007/978-3-319-07440-5_58

Bender, S. (2019). Headset attentional synchrony: Tracking the gaze of viewers watching narrative virtual reality. *Media Practice and Education*, *20*(3), 277–296. doi:10.1080/25741136.2018.1464743

Bozzelli, G., Raia, A., Ricciardi, S., De Nino, M., Barile, N., Perrella, M., & Palombini, A. (2019). An integrated VR/AR framework for user-centric interactive experience of cultural heritage: The ArkaeVision project. *Digital Applications in Archaeology and Cultural Heritage*, *15*, e00124. doi:10.1016/j.daach.2019.e00124

Burdette, M. (2016). *The Swayze effect*. https://www.oculus.com/story-studio/blog/the-swayze-effect/

Calvi, L., Hover, M., Ouwens, F., & van Waalwijk, J. (2015). Visualising Vincent's life: An engaging experience into Van Gogh's heritage. Paper presented at DeSForM 2015, Aesthetics of Interaction: Dynamic, Multisensory, Wise. 9th International Conference on Design and Semantics of Form and Movement, 14-10; 2015, Milan, Italy.

Carpio, R., & Birt, J. (2021). The role of the Embodiment Director in virtual reality film production. *Creative Industries Journal*, 1–10.

Council of Europe (2017). *Recommendation of the Committee of Ministers to member States on the European Cultural Heritage Strategy for the 21st century*. Committee of Ministers of the Council of Europe CM/Rec(2017)1.

Daley, B. (2021). *Storytelling with digital culture is booming - find out why*. Europeana. https://pro.europeana.eu/post/storytelling-with-digital-culture-is-booming-find-out-why

Deffner, A., Psatha, E., Bogiantzidis, N., Mantas, N., Vlachaki, E., & Ntaflouka, P. (2015). Accessibility to Culture and Heritage: Designing for All. *Proceedings of the AESOP 2015 Conference Definite space—Fuzzy Responsibility*.

Demetrescu, E., Palombini, A., Pietroni, E., Forlani, M., & Rufa, C. (2016). Making movies: Nuove frontiere per la comunicazione audiovisiva del passato. *Archeologia e Calcolatori*, *8*, 203–210.

Dowling, D., Fearghail, C. O., Smolic, A., & Knorr, S. (2018). Faoladh: A case study in cinematic VR storytelling and production. In *International Conference on Interactive Digital Storytelling* (pp. 359-362). Springer. 10.1007/978-3-030-04028-4_42

Flavián, C., Ibáñez-Sánchez, S., & Orús, C. (2019). The impact of virtual, augmented and mixed reality technologies on the customer experience. *Journal of Business Research, 100*, 547–560. doi:10.1016/j.jbusres.2018.10.050

Gillam, S. (2017). *Spotlight VR/AR: Innovation in transformative storytelling*. MW17: MW 2017. https://mw17.mwconf.org/paper/spotlight-vrar-innovation-in-transformative-storytelling/

Guttentag, D. A. (2010). Virtual reality: Applications and implications for tourism. *Tourism Management, 31*(5), 637–651. doi:10.1016/j.tourman.2009.07.003

Han, D. I. D., Weber, J., Bastiaansen, M., Mitas, O., & Lub, X. (2019). Virtual and augmented reality technologies to enhance the visitor experience in cultural tourism. In *Augmented reality and virtual reality* (pp. 113–128). Springer, Cham. doi:10.1007/978-3-030-06246-0_9

ISO 9241-210: 2010 (2010). Ergonomics of human system interaction-Part 210: Human-centred design for interactive systems. International Standardization Organization (ISO).

Jiang, S., Moyle, B., Yung, R., Tao, L., & Scott, N. (2022). Augmented reality and the enhancement of memorable tourism experiences at heritage sites. *Current Issues in Tourism*, 1–16. doi:10.1080/13683500.2022.2026303

Jung, T., tom Dieck, M. C., Moorhouse, N., & tom Dieck, D. (2017). Tourists' experience of Virtual Reality applications. In *2017 IEEE International Conference on Consumer Electronics (ICCE)* (pp. 208-210). IEEE. 10.1109/ICCE.2017.7889287

Jung, T. H., & tom Dieck, M. C. (2017). Augmented reality, virtual reality and 3D printing for the co-creation of value for the visitor experience at cultural heritage places. *Journal of Place Management and Development, 10*(2), 140–151. doi:10.1108/JPMD-07-2016-0045

Katifori, A., Tsitou, F., Pichou, M., Kourtis, V., Papoulias, E., Ioannidis, Y., & Roussou, M. (2020). Exploring the potential of visually-rich animated digital storytelling for cultural heritage. In *Visual Computing for Cultural Heritage* (pp. 325–345). Springer. doi:10.1007/978-3-030-37191-3_17

Kilteni, K., Groten, R., & Slater, M. (2012). The Sense of Embodiment in Virtual Reality. *Presence (Cambridge, Mass.), 21*(4), 373–387. doi:10.1162/PRES_a_00124

Kontiza, K., Antoniou, A., Daif, A., Reboreda-Morillo, S., Bassani, M., González-Soutelo, S., Lykourentzou, I., Jones, C. E., Padfield, J., & López-Nores, M. (2020). On How Technology-Powered Storytelling Can Contribute to Cultural Heritage Sustainability across Multiple Venues—Evidence from the CrossCult H2020 Project. *Sustainability, 12*(4), 1666. doi:10.3390u12041666

Kuflik, T., Wecker, A. J., Lanir, J., & Stock, O. (2015). An integrative framework for extending the boundaries of the museum visit experience: Linking the pre, during and post visit phases. *Information Technology & Tourism, 15*(1), 17–47. doi:10.100740558-014-0018-4

Le, D., Scott, N., & Lohmann, G. (2019). Applying experiential marketing in selling tourism dreams. *Journal of Travel & Tourism Marketing, 36*(2), 220–235. doi:10.1080/10548408.2018.1526158

Lin, L. P. L., Huang, S. C. L., & Ho, Y. C. (2020). Could virtual reality effectively market slow travel in a heritage destination? *Tourism Management, 78*, 104027. doi:10.1016/j.tourman.2019.104027

Little, C., Bec, A., Moyle, B. D., & Patterson, D. (2020). Innovative methods for heritage tourism experiences: Creating windows into the past. *Journal of Heritage Tourism*, *15*(1), 1–13. doi:10.1080/1743873X.2018.1536709

Marasco, A. (2020). Beyond virtual cultural tourism: History-living experiences with cinematic virtual reality. *Tourism and Heritage Journal*, *2*, 1–16. doi:10.1344/THJ.2020.2.1

Marasco, A., & Balbi, B. (2019). Designing Accessible Experiences for Heritage Visitors through Virtual Reality. *e-Review of Tourism Research*, *17*(3), 426-443.

Marasco, A., Buonincontri, P., van Niekerk, M., Orlowski, M., & Okumus, F. (2018). Exploring the role of next-generation virtual technologies in destination marketing. *Journal of Destination Marketing & Management*, *9*, 138–148. doi:10.1016/j.jdmm.2017.12.002

Michopoulou, E., & Buhalis, D. (2013). Information provision for challenging markets: The case of the accessibility requiring market in the context of tourism. *Information & Management*, *50*(5), 229–239. doi:10.1016/j.im.2013.04.001

Michopoulou, E., Darcy, S., Ambrose, I., & Buhalis, D. (2015). Accessible tourism futures: The world we dream to live in and the opportunities we hope to have. *Journal of Tourism Futures*, *1*(3), 179–188. doi:10.1108/JTF-08-2015-0043

Montagud, M., Orero, P., & Matamala, A. (2020). Culture 4 all: Accessibility-enabled cultural experiences through immersive VR360 content. *Personal and Ubiquitous Computing*, *24*(6), 887–905. doi:10.100700779-019-01357-3

Moscardo, G. (2017). Stories as a tourist experience design tool. In D. Fesenmaier & Z. Xiang (Eds.), *Design science in tourism* (pp. 97–124). Springer International Publishing. doi:10.1007/978-3-319-42773-7_7

Moscardo, G. (2020). Stories and design in tourism. *Annals of Tourism Research*, *83*, 102950. doi:10.1016/j.annals.2020.102950

Naniopoulos, A., & Tsalis, P. (2015). A methodology for facing the accessibility of monuments developed and realised in Thessaloniki, Greece. *Journal of Tourism Futures*, *1*(3), 240–253. doi:10.1108/JTF-03-2015-0007

Nielsen, J. K. (2017). Museum communication and storytelling: Articulating understandings within the museum structure. *Museum Management and Curatorship*, *32*(5), 440–455. doi:10.1080/09647775.2017.1284019

Ozdemir, M. A. (2021). Virtual Reality (VR) and Augmented Reality (AR) technologies for accessibility and marketing in the tourism industry. In ICT Tools and Applications for Accessible Tourism (pp. 277-301). IGI Global.

Pagano, A., Palombini, A., Bozzelli, G., De Nino, M., Cerato, I., & Ricciardi, S. (2020). Arkaevision VR game: User experience research between real and virtual Paestum. *Applied Sciences (Basel, Switzerland)*, *10*(9), 3182. doi:10.3390/app10093182

Palombini, A. (2017). Storytelling and telling history. Towards a grammar of narratives for Cultural Heritage dissemination in the Digital Era. *Journal of Cultural Heritage*, *24*, 134–139. doi:10.1016/j.culher.2016.10.017

Park, H., Kim, J., Bang, S., & Woo, W. (2018). The effect of applying film-induced tourism to virtual reality tours of cultural heritage sites. In *3rd Digital Heritage International Congress (DigitalHERITAGE)* (pp. 1-4). IEEE. 10.1109/DigitalHeritage.2018.8810089

Pérez, E., Merchán, P., Merchán, M. J., & Salamanca, S. (2020). Virtual reality to foster social integration by allowing wheelchair users to tour complex archaeological sites realistically. *Remote Sensing*, *12*(3), 419. doi:10.3390/rs12030419

Perry, S., Roussou, M., Economou, M., Young, H., & Pujol, L. (2017). Moving beyond the virtual museum: Engaging visitors emotionally. In *2017 23rd International Conference on Virtual System & Multimedia (VSMM)* (pp. 1-8). IEEE.

Pietroni, E., & Adami, A. (2014). Interacting with virtual reconstructions in museums: The Etruscanning Project. *Journal on Computing and Cultural Heritage*, *7*(2), 1–29. doi:10.1145/2611375

Pietroni, E., Forlani, M., & Rufa, C. (2015). Livia's Villa Reloaded: An example of re-use and update of a pre-existing Virtual Museum, following a novel approach in storytelling inside virtual reality environments. 2015 Digital Heritage, 511-518.

Rach, M., & Scott, R. (2020). The Use of Virtual Reality in Marketing: Exploring the Need for Technology and Language Adaptation to Create High Quality Immersive Experiences. In *Marketing and Smart Technologies* (pp. 67–77). Springer. doi:10.1007/978-981-15-1564-4_8

Reyes, M. C. (2018). Measuring user experience on interactive fiction in cinematic virtual reality. In *International Conference on Interactive Digital Storytelling* (pp. 295-307). Springer. 10.1007/978-3-030-04028-4_33

Rizvic, S., Boskovic, D., Okanovic, V., Ivkovic-Kihic, I., & Sljivo, S. (2019). Virtual Reality Experience of Sarajevo War Heritage. In *EUROGRAPHICS Workshop on Graphics and Cultural Heritage* (pp. 1-6). Academic Press.

Rizvic, S., Djapo, N., Alispahic, F., Hadzihalilovic, B., Cengic, F. F., Imamovic, A., Okanovic, V., & Boskovic, D. (2017). Guidelines for interactive digital storytelling presentations of cultural heritage. In *9th International Conference on Virtual Worlds and Games for Serious Applications (VS-Games)* (pp. 253-259). IEEE. 10.1109/VS-GAMES.2017.8056610

Rizvic, S., Okanovic, V., & Boskovic, D. (2020). Digital storytelling. In *Visual computing for cultural heritage* (pp. 347–367). Springer. doi:10.1007/978-3-030-37191-3_18

Roussos, M., & Bizri, H. (1998). Mitologies: Medieval labyrinth narratives in virtual reality. In *International Conference on Virtual Worlds* (pp. 373-383). Springer. 10.1007/3-540-68686-X_36

Roussou, M. (2001). The interplay between form, story, and history: The use of narrative in cultural and educational virtual reality. In *International Conference on Virtual Storytelling* (pp. 181-190). Springer. 10.1007/3-540-45420-9_20

Roussou, M. (2008). The components of engagement in virtual heritage environments. In *New Heritage* (pp. 241–257). Routledge.

Ruszev, S. (2021). Tracing Embodied Narrative in VR Experiences. *Cinergie – Il Cinema E Le Altre Arti, 10*(19), 9–20.

Schiavottiello, N., Carpetudo, C., Correia, M., Viscomi, P., Cabral, R., Alarcón, S.P. (2016). CGI Storytelling in Archaeological and Cultural Heritage Public Interpretation: Scientific Facts or Hollywood Movies? *Entre Ciência E Cultura. Da Interdisciplinaridade a Transversalidade da Arqueologia*, 569-592.

Seide, B., & Slater, B. (2020). Virtual Cinematic Heritage for the Lost Singaporean Film Pontianak (1957). In *International Conference on Human-Computer Interaction* (pp. 396-414). Springer. 10.1007/978-3-030-50267-6_30

Selmanović, E., Rizvic, S., Harvey, C., Boskovic, D., Hulusic, V., Chahin, M., & Sljivo, S. (2020). Improving accessibility to intangible cultural heritage preservation using virtual reality. *Journal on Computing and Cultural Heritage, 13*(2), 1–19. doi:10.1145/3377143

Škola, F., Rizvić, S., Cozza, M., Barbieri, L., Bruno, F., Skarlatos, D., & Liarokapis, F. (2020). Virtual Reality with 360-Video Storytelling in Cultural Heritage: Study of Presence, Engagement, and Immersion. *Sensors (Basel), 20*(20), 5851. doi:10.339020205851 PMID:33081154

Slater, M. (2009). Place illusion and plausibility can lead to realistic behaviour in immersive virtual environments. *Philosophical Transactions of the Royal Society of London. Series B, Biological Sciences, 364*(1535), 3549–3557. doi:10.1098/rstb.2009.0138 PMID:19884149

Southall, H., Marmion, M., & Davies, A. (2019). Adapting Jake Knapp's design sprint approach for AR/VR applications in digital heritage. In *Augmented reality and virtual reality. The Power of AR and VR for Business, edited by M. Claudia tom Dieck and Timothy Jung* (pp. 59–70). Springer. doi:10.1007/978-3-030-06246-0_5

tom Dieck, M. C., Jung, T., & Michopoulou, E. (2019). Experiencing Virtual Reality in Heritage Attractions: Perceptions of Elderly Users. In Augmented Reality and Virtual Reality. The power of AR and VR for Business (pp. 89-98), Springer.

Tuerlan, T., Li, S., & Scott, N. (2021). Customer emotion research in hospitality and tourism: Conceptualization, measurements, antecedents and consequences. *International Journal of Contemporary Hospitality Management, 33*(8), 2741–2772. doi:10.1108/IJCHM-11-2020-1257

Tussyadiah, I. P. (2014). Toward a theoretical foundation for experience design in tourism. *Journal of Travel Research, 53*(5), 543–564. doi:10.1177/0047287513513172

Zevi, F., Demma, F., Nuzzo, E., Rescigno, C., & Valeri, C. (2008). *Museo Archeologico dei Campi Flegrei. Catalogo generale. Cuma.* Electa Napoli.

KEY TERMS AND DEFINITIONS

Cinematic Virtual Reality (CVR): While an established definition is still under development, the term is used to identify a type of VR experience that immerses users into fixed and predetermined narratives, including live-action film contents, unfolding in a linear way like in traditional movies within a 360-degree environment.

Digital Storytelling: The use of digital media for the production and diffusion of stories combining narrative with multimedia contents (including images, sound, video).

Experiential Marketing: An approach to marketing that focuses on customer experiences rather than on product functional characteristics and benefits, using emotional and experiential stimuli to influence decision-making processes.

Narrative Engagement: The multi-dimensional experience of becoming engaged in a narrative, including cognitive, affective and sensory dimensions.

User-Centred Design (UCD): An iterative design approach that puts users and their needs at the core of the design process. To create products that are satisfying, usable, and accessible to users, the design team in the UCD cycle includes end users at every stage through an increasing number and types of research and design methodologies.

Virtual Reality (VR): The range of technologies and devices (including head-mounted displays) that create the sense for users of being immersed in a computer-simulated environment different from the place where they are physically located.

Visitor Experience: The experience of visitors along the journey, from the pre-visit stage for decision-making and planning, to the visit stage with access to cultural resources and the post-visit stage for recalling memories, reflection and sharing.

Chapter 11
Digital Storytelling and 3D Technologies for Visitor Experience and Contested Heritage Preservation

Chris Little
Griffith University, Australia

Brent Moyle
Griffith University, Australia

Dale Patterson
https://orcid.org/0000-0002-6545-0905
Griffith University, Australia

Alexandra Bec
University of the Sunshine Coast, Australia

Liubov Skavronskaya
Griffith College, Australia

ABSTRACT

This chapter draws on the case study of aboriginal trackways at World Heritage-listed Willandra Lakes Region of New South Wales, Australia to explore how to re-create a unique, charismatic, and vulnerable heritage site by combining 3D technology and storytelling to assist with the virtual presentation of heritage. This chapter delves into the advantages and challenges of innovative 3D scanning methods available to accurately record fossilised Indigenous footprints, providing best practice guidelines on how to best deliver these outcomes in a form of engaging immersive visitor experience. Findings suggest that storytelling techniques improve intensity of virtual experience recreated using 3D scanning techniques through the process of narrative transportation, which elicits emotional arousal and improves emotional engagement with heritage. These findings contribute to the utilisation of 3D technologies and storytelling for AR/VR to create engaging immersive experiences for visitors.

DOI: 10.4018/978-1-6684-3436-9.ch011

INTRODUCTION

Exploring the past helps to make sense of the future, emphasising the need to preserve heritage. Innovative 3D technologies enable the creation of digital immersive experiences indistinguishable from the real world, providing immense potential for heritage preservation. Immersive technologies including virtual, augmented, and mixed reality represent the latest forms through which "historical narratives can be told, and heritage experienced" (Swords et al., 2021, p. 186).

Narration and storytelling have many forms and play a central role in how experiences are remembered and disseminated (Tivers & Rakić, 2012). The earliest forms of narration were usually oral, combined with gestures and expressions. Additionally, many ancient cultures used rock art and religious rituals for the purpose of storytelling (Yılmaz & Ciğerci, 2019). For instance, in Australian Indigenous culture painted symbols appear in stories on cave walls as a means of helping the storyteller to remember/document the story. The painted symbols are frequently accompanied by a combination of oral narrative, music, rock art and dance to assist telling the stories (Barton & Barton, 2014; Geia, Hayes & Usher, 2013). In Indigenous and many other cultures, storytelling is a critical tool for making, storing, and conveying historical legacy and culture (Barrett & Cocq, 2019).

Willandra Lakes Region is a unique landmark in the study of human evolution in Australia and represents an important legacy for Indigenous culture. Willandra Lakes Region is home to ancient Indigenous footprints known as the Willandra trackways in outback New South Wales at Mungo National Park. The footprints are the oldest ever found in Australia and have been placed on the World Heritage List by UNESCO (Macgregor, 2022; UNESCO World Heritage Centre, n.d.). In 2003, Indigenous Ranger Mary Pappin Jr discovered the early Aboriginal fossil trackways which had been unearthed due to erosion. Excavation and archaeological analysis revealed that the footprints are 20,000 years old. To minimise the risk of future damage of this Indigenous heritage, the location at Mungo National Park remains a secret. The footprints were also preserved under sand and shade cloth to stabilise against sand shifts and wind erosion (Little et al., 2018). The protective actions have limited access to the footprints and the stories they hold for the broader audience. However, growth of immersive technologies offers new opportunities for visitors to have memorable encounters with heritage, such as Willandra trackways, without the risk of physical damage for the vulnerable cultural artefacts.

Current literature suggests that incorporation of cutting-edge technologies into the design of visitor experiences has significant benefits for tourism stakeholders, communities, and tourism destinations. Incorporation of digital layers into the tourism experiences enhances competitiveness of tourism destinations, increases tourists' satisfaction, as well as encourages community pride and reaffirms a sense of identity (Murzyn-Kupisz, 2012). Furthermore, technologies allow interaction with vulnerable cultural heritage without causing damage, which is critical for preserving cultural heritage for future generations (Tscheu & Buhalis, 2016).

Despite there being significant benefit for utilising technologies for visitor experiences, limited studies have assessed the intricate connection between the use of 3D scanning and storytelling for immersive heritage tourism. Subsequently, this chapter explores the possibilities to re-create unique and vulnerable heritage sites. The Willandra trackways will be used as a case study for using innovative 3D technologies as a means of enabling memorable virtual experiences to share Indigenous stories and heritage.

HERITAGE PRESERVATION AND TOURISM

Heritage is a way of explaining and characterising such things as traditions, culture, and historical events, and are often displayed through landmarks, monuments, and objects (Little, Patterson, Moyle & Bec, 2018). Heritage has been defined in the relevant literature as both tangible materials and intangible cultural practices, with the definition continuing to evolve (Liu, Dupre & Jin, 2021). In fact, the definition of heritage has evolved and now encompass social, political, and cultural processes, as well as consumable experiences. Consequently, its meaning is continuously created, recreated, and validated for sustainable utilisation in the present and future (Aureli, & Del Baldo, 2022; Ashworth, 2014; Harrison, 2013).

Unfortunately, time, neglect, pollution, war, geopolitical conflicts, and other socio-ecological factors contribute to the disrepair of heritage sites, raising a need for innovative approaches to conservation. Furthermore, there are concerns surrounding visitors impact on the sustainability of heritage sites, as tourist visitation is both adding to and impeding the preservation of it (Schildgen, 2008). To address heritage preservation issues strategies have been introduced and implemented. Existing tourism literature suggests that traditional heritage preservation strategies for tourism sites include physical restoration, adaptation, and education approaches, which can be employed independently or in combination (Landorf, 2009).

Restoration approaches aim to preserve physical aspects of heritage in their original condition or maintain the current condition. However, restoration is typically a costly and ongoing process, often causing added financial considerations for the management of a site and limiting access for visitors (Nasser, 2003). Additionally, standalone restoration may not be appropriate for all heritage sites, as some have heightened fragility, or may be severely deteriorated (Alberts & Hazen, 2010). In turn, adaptation approaches to heritage site preservation involves designing tourism experiences that do not directly impact the site. Examples of adaptation approaches to heritage preservation include designing visitor paths which do not physically impede the site, and/or enforcing guidelines and restricting visitor behaviour or access. However, adaptation approaches have also incorporated physical restoration approaches to further preserve the site, particularly from natural deterioration (Deacon, 2006; Okech, 2010).

Educating visitors is another method of heritage site preservation which has been employed at sites where visitors are able to physically immerse themselves in the environment and historical monuments (Chen & Weiler, 2017). The main principle of the educational approach to heritage preservation is to communicate significance, as well as sensitivity of the site, and educate visitors on the required behavioural norms during the experience (Olsen, 2006). The underlying premise of this approach is to educate visitors on behavioural principles that will reduce their negative impact, thus contributing to preservation and sustainability (Okech, 2010).

Due to rapid advances in interactive technologies and industry 4.0, digital heritage preservation is becoming a more popular and accessible approach for heritage sites and tourism destinations to create memorable and engaging experience for visitors without physically experiencing the site (Hertzman, Anderson, & Rowley, 2008). Previous studies have identified digitisation as an effective way to preserve cultural objects and heritage (Shishido et al., 2017). However, there are challenges for using digital methods to preserve heritage, many of which are associated with converting the raw data to immersive technologies for tourism experience, such as virtual or augmented reality.

CHALLENGES ASSOCIATED WITH HERITAGE PRESERVATION: CONTESTATION AND DIGITALISATION

There are a number of issues associated with heritage preservations using cutting edge technologies, and the interpretation of the results for potential utilisation in tourism experience design. Disadvantages and challenges of traditional methods to heritage preservation has been widely covered in the existent literature, therefore in this chapter we discuss two challenges directly associated with digital heritage preservation and interpretation for tourism experiences. These two interconnected issues include the contested nature of heritage artefacts, as well as the transfer process of heritage artefacts to digital mediums. Both of these issues have significant impact on how visitors can experience heritage in the present and future, as well as how digital experiences can be used for conservation and educational purposes.

The contested nature of some heritage artefacts is a growing topic in the current tourism and heritage preservation literature. Ingerpuu (2018) suggests that a conflict inevitably presents itself in every heritage site through aspects of ownership, authenticity, interpretation, and its multiple uses due to the deemed intrinsic nature of heritage. As 'all the heritage is someone's heritage and then logically not someone else's' (Ingerpuu, 2018, p.21).

In a broader sense, heritage is a subject of two types of contestations. The first one is involving different social groups claiming the ownership of the same heritage, but with different interpretations. The second one occurred when heritage was interpreted and used differently by various divisions within one group (Olsen & Timothy, 2002). Liu, Dupre & Jin (2021) report the increasing number of stakeholders from different sectors trying to manipulate heritage to fulfil multiple functions. The contestation also stimulates debates on the nature, definition, and functions of heritage, and how it should be used, as it accentuates the contestation between continuity and change in heritage management (Liu, Dupre & Jin, 2021). One of the relevant examples of Indigenous heritage contestation associated with the Willandra Lakes region is the Mungo Man and Mungo Lady story, and the subsequent tension between the two groups. The archaeological and anthropological communities present conflicting approaches regarding the handling and treatment of skeletal remains and archaeological sites (Gibson, 2015).

Surprisingly, heritage digitalisation can provide a potential solution to some issues associated with heritage ownership, authenticity, and interpretation. For instance, digital models of the contested heritage sites and artefacts can be utilised to create multiple immersive experiences with different interpretations and alternate scenarios (Bec et al., 2019). Recent studies confirmed that the contested perspectives on heritage are also important parts of cultural legacy that can form the basis of a memorable tourism experience for visitors (Naef & Ploner, 2016). It has been confirmed that a contested heritage object as well as a surrounding story "makes it something of interest to an audience, as they want to see it and judge for themselves" (Little et al., p.4)

Non-linear storytelling, where multiple versions of historical events are portrayed, has successfully been adopted in a variety of contexts (Berg, 2006). One of the most notable examples are the alternative versions of reality presented in Quentin Tarantino's movies such as "Once upon a time in Hollywood" and "Inglourious Bastards" (Beckman, 2015). The same principal of counterfactual versions of reality that are presented in Tarantino movies can be used for creating multiple immersive digital tourism experiences using the same 3D models of contested heritage artefacts to create different interpretations of historical events, while addressing heritage contestations issues.

However, to address the issues associated with contested nature of heritage artefacts and to create alternative realities, artefacts need to be transferred into digitals mediums, such as 3D models. Accurate

3D models of heritage artefacts are essential for creating immersive digital tourism experiences for visitors, including alternate realities. Little and Patterson (2016) suggest that there are several difficulties associated with digitalisation of cultural heritage for tourism as only a small amount of the digitally preserved data is being transferred. Photogrammetry and 3D laser scanning are two of the most popular methods being used within the tourism industry for heritage digitalisation and the creation of immersive visitor experiences (Poux et al., 2020). The next section focuses on 3D technologies, including 3D scanning, and its potential for enhancing visitor experiences while achieving contested heritage preservation.

3D TECHNOLOGIES AND TOURISM HERITAGE PRESERVATION

The development and expansion of 3D technologies has led to the development of new experiences such as 3D printing, 3D television, 3D audio, 3D modelling and 3D gaming etc. (Shishido et al., 2017). This creates an opportunity for increasingly accurate methods of 3D data capture and measurement, which is critical for heritage preservation.

As result of advancements in 3D technology, applications in a social context are also expanding, leading to a growing expectation towards 3D interaction or the co-creation of experiences (Buonincontri & Marasco, 2017). Due to the diversity of applications, 3D scanning is receiving increased attention from the scientific community, with the recognition that 3D digitisation techniques can be applied to cultural heritage and contribute to conservation (Shishido et al., 2017). Immersive technologies have been embedded within the tourism industry since the 1990s (Lutz & Weintke, 1999). 3D reality capture methods have been applied to devise tourism experiences which create a window into the past and preserve the cultural heritage of sites (Little et al., 2020).

3D scanning is the non-contact process of digital reproduction of the physical shape of an object. The digital 3D model created from the scans accurately defines the size, shape, and colour of a real-world surface/object (Poux et al., 2020). In turn, a 3D scanner is a device which analyses an object to collect distance point measurements to then accurately recreate and transform the surfaces of the scanned objects or environments into a digital 3D model (Little et al., 2018). The 3D scanning technology is now becoming an integral tool for analysis and presentation of information for heritage preservation as it enables a clearer depiction of the past (Pieraccini, Guidi & Atzeni, 2001). While 3D scanning is widely used for the preservation of heritage, the potential of this rapidly advancing technology for the purpose of visitors' engagement is yet to be fully identified, with techniques such as storytelling underexplored in the literature. Existing projects utilising 3D technologies for heritage preservation predominantly focus on raw data capture aiming to enhance interpretation, analysis, delivery, and display of heritage artefacts (Little et al., 2020). However, a successful transition of 3D models into an immersive digital experience requires further attention and represents a prominent area for future academic inquiry (Macgregor, 2022).

3D scanning techniques are critical for the provision of cultural understanding in a tourism context and beyond, as it offers visitors the opportunity to interact with 3D scanned items in immersive realities, while also ensuring heritage preservation (Bec et al., 2019; Macgregor, 2022). Therefore, 3D reality capture of heritage items is a non-invasive method that not only can preserve and protect cultural artefacts and vulnerable heritage sites, but also with a potential of communicating the stories of past civilizations to future generations (Taçon & Baker, 2019). However, studies focused on the development of the technology, need to connect with advances in literature on the creation of memorable tourism experiences, through analysis of in-depth case studies. The next section discusses application of 3D scanning as a method for preserving the ancient aboriginal trackways site in the Willandra Lakes World Heritage Area in New South Wales, Australia, and explores how to best deliver these outcomes as a form of immersive experiences.

WILLANDRA TRACKWAYS: INTRODUCTION

The Willandra Lakes is located in the south-west New South Wales region of Australia, with a part of the area forming the Lake Mungo National Park. Willandra Lakes Region is a unique landmark in the study of human evolution in Australia and represents an important legacy for Indigenous culture as it is home to ancient Indigenous footprints known as the Willandra trackways. In ancient times, the system of five lakes was the centre of a rich landscape providing a living for Indigenous communities.

The Willandra trackways are the oldest footprints ever found in Australia and have been placed on the World Heritage List by UNESCO (Macgregor, 2022; UNESCO World Heritage Centre, n.d.). In 2003, Indigenous Ranger Mary Pappin Jr discovered the early Aboriginal fossil trackways which had been unearthed due to erosion. Excavation and archaeological analysis revealed that the footprints are 20,000 years old and represents the largest ice-age (late Pleistocene) human trackways on the planet. In total the Willandra Lakes site consists with more than 500 footprints recorded within the 22 trackways. The trackways are a part of Mungo National Park which attracts 50,000 visitors a year, which created a challenge to ensure this fragile site short-term and long-term preservation for future visitors who would like to explore the famous trackways (Macgregor, 2022)

To minimise the risk of future damage of this Indigenous heritage, the location of the footprints at Mungo National Park remains a secret. The footprints were also preserved under sand and shade cloth to stabilise against sand shifts and wind erosion (Little et al., 2018). The protective actions have limited access to the footprints and the stories they hold for the broader audience. Fortunately, to further preserve this remarkable evidence of early Aboriginal life, authorities recognised the need for preservation using innovative 3D scanning techniques before reburying the footprints. Subsequently, prior to covering and reburying the footprints in the sand, the world heritage site was 3D laser scanned for the first time in 2006. The 3D data was initially used to monitor deterioration of individual footprints. In 2007, the footprints were reburied with only a select few identified to excavate for further 3D scanning periodically as part of the monitoring program (Little et al., 2018). However, the dynamic nature of the landscape generates challenges for the site preservation and more footprints are periodically exposed. The site provides a powerful link with the ancestors of the current Indigenous community of the region and offers an opportunity to unearth Willandra stories for the broader public. Thus, for the purpose of heritage preservation, there was a need to create more detailed 3D scans of all the footprints (Macgregor, 2022).

Although the Willandra trackways area is known for the high levels of archaeological, cultural, spiritual, and scientific significance, the region has numerous other sites and items of interest. This provides visitors with endless possibilities for engaging with past Aboriginal society. For Aboriginal community, this window into the past develops a very personal connection to their ancestors and ways of life (Little et al., 2020). All the same, it is not purely the data of the sites and artefacts that are critical to preserve, but also the stories that have been passed down through the Indigenous communities who live there, such as a story of a one-legged man or a giant kangaroo called Bookamurra (Douglas, 2010). Archaeological analysis of the trackway captures a moment in time and a record of how a family crossed the landscape. Possible versions of the story could include a hunter throwing a spear and missing a kangaroo, a child cutting the path of the adults as they walk behind, a mother carrying a child and shifting the weight from one hip to the other, a one-legged man, possibly assisted by a walking stick (Macgregor, 2022). The protective actions have limited access to the footprints and the remarkable stories they hold for the broader audience. However current growth of immersive technologies offers new opportunities for visitors to have memorable encounters with trackways, without physical damage of the vulnerable cultural artefacts. The following section presents 3D data capture workflow, outlines the stages of the 3D scanning process, and documents technical challenges associated with converting data into deliverable outcomes.

3D DATA CAPTURE WORKFLOW

There are four main steps in the 3D scanning workflow process: capture, processing, analysis, and presentation. In 2016, an inquiry from the Aboriginal Advisory Group (AAG) and Willandra Lakes Region World Heritage Area Executive Officer was made to Mr Chris Little from Griffith University to determine whether the previously made 3D scan data files can be converted for creating interpretations. Data was to be used for web sites, touch screens, physical reproductions, and virtual reality experiences, as well as 3D printing. Files also needed to be geospatially rectified to create an immersive visitor experience of the full Willandra trackway complex.

3D footprint scans were able to be converted to a useable format. A workflow was successfully developed to convert trackway data for export, enabling the converted files to be accessible to others for further analysis and presentation and import into virtual environments. The developed workflow also translates trackway information to its correct geospatial position from existing survey control points taken at the time of scanning, which is critical for potential virtual re-creation of the Willandra trackway.

The second stage of the 3D scanning workflow process took place at Willandra site. Through sustained engagement with the Aboriginal community, NSW National Parks, and Wildlife Service, a range of possible deliverables were targeted relating to the 3D scanning of the Willandra trackways. These included 3D printed replicas of trackway site, establishing a web-based viewer for analysis and educational purposes, developing an interactive virtual reality visitor experience for mobile devices to bring to life the stories surrounding the trackways, and a potential to generate augmented reality hotspots to give interesting information by touching mobile device screens. However, 3D technology has greatly improved since the earliest 3D scans of the trackways, which made combining and geospatially referencing old scans to new scans difficult. Issues with converting these 3D scans into new 3D models presented a number of challenges, including limited control points, inconsistent or no scans in some areas, no detailed record of workflow or process, and no layout information.

Additionally, the dynamic nature of the landscape generates more challenges for the site, with more footprints being periodically exposed. While this enabled further 3D scans of the newly exposed footprints to be undertaken, it intensified the challenges. When registering 3D scan data together, over time, every piece of trackway starts to look the same as the last and stitching multiple 3D scans starts to feel like joining pieces of a jigsaw puzzle, therefore initial planning is an integral step in the 3D data capture process. To simplify post-processing, new 3D scans were done using a measured grid pattern and a sufficient number of small markers which made similar points of overlap easier to find. These new 3D scans then needed to be referenced to the original 3D scan data. To do so required intelligent algorithmic formulas to find common points between old and new data sets. This was only possible due to advancements in post-processing software tools.

The outputs of 3D scanning processes are very high-quality 3D data. However, a further research and analysis of the footprints and the surrounding area is required to understand Aboriginal culture and experiences passed down through generations at the Willandra site. A combination of forensically accurate 3D data with recording of the stories and narratives allows us to potentially place this data in a VR/AR environment and to experience the heritage of Willandra Lake and trackways in a new in-depth level. This is a different level of digital information accuracy to that required for presentation of the outlined final deliverables of the 3D scanned data. For example, the VR/AR systems may not require high accuracy of the scanned object, but the forensic analysis does require extremely high accuracy to enable a reconstruction of the scenes from the past. In the case of Willandra, 3D scanned data accuracy is critical in capturing most historically significant items, not only because of extremely limited access to the footprints but also the 3D scans of the Willandra trackways will be vaulted, recorded, and documented for heritage preservation and future research as part of Australian cultural history.

Combined with research into early Aboriginal culture and the stories shared by the Indigenous community of the Willandra Lakes and surrounding areas, such 3D scanned data with other historical information present immense potential to recreate history in a form of immersive visitor experience through AR/VR technologies which brings stories back to life. For the 3D captured data to be effectively used to communicate historical events for visitor experiences, the immersive digital product needs to elicit emotional engagement. The following sections will focus on exploring how digital storytelling and techniques such as narrative transportation can assist in communicating heritage and engage visitors at a more profound level for a memorable experience.

DIGITAL STORYTELLING AND VISITOR EXPERIENCES

Recent research on human-computer interaction suggests that storytelling is the next stage of development for immersive technologies (Wiederhold, 2018). The value of storytelling for immersive technologies contributes to the intensity and memorability of an experience. This is increased through the level of emotional engagement and persuasion effect that storytelling can offer (Lien & Chen, 2013).

Studies on the tourism experience suggest that narration and storytelling "lies at the very heart of tourism as experiences" (Lund, Cohen & Scarles, 2018, p.10). Existent literature defines storytelling as a process of sharing knowledge and experiences through narrative to communicate lessons, complex ideas, concepts, and causal connections (Sole & Wilson, 2002). The importance of storytelling refers to its ability to be a powerful mode of human expression that helps to understand possible futures and make sense of the past (Hinchman & Hinchman, 1997). Humans are likely to remember stories better than facts (Lundqvist et al., 2013), which is a more effective approach to heritage preservation and cultural transmission.

The evolution of cutting-edge technologies facilitates the development of digital storytelling as an advanced tool to engage visitors at a profound level (Van Limburg, 2009). Digital storytelling involves constructing stories using a range of technologies or data including voice, images, 3D digital models, and other sources focused on a central 'theme' with an identified 'message' (Tang, 2016). Digital storytelling has been identified as a prominent combination of narrative and technology, advancing the understanding of a variety of social phenomena, including different aspects of human experience (Rossiter & Garcia, 2010). This research aims to explore the potential of 3D technologies and storytelling to create engaging immersive visitor experiences through the process of narrative transportation.

NARRATIVE TRANSPORTATION

Narrative transportation is the process of "an integrative melding of attention, imagery, and emotion, focused on story events" (Green, Brock & Kaufman, 2004, p. 248). This definition includes three functions of narrative transportation: attention arousal, imagination stimulation (mental imagery), and emotional resonance or empathy (Green & Brock, 2002). Thus, narrative transportation occurs whenever the story receiver or visitor experiences a feeling of entering a world evoked by the narrative because of empathy for the story characters and imagination of the story plot (Van Laer et al., 2014).

An alternative definition of narrative transportation refers to the sensation of being transported during the process of storytelling, the capability of which is to regulate and enhance the effects of narrative persuasion (Wong, Lee & Lee, 2015). The persuasive power of narrative is determined by the number of people that are transported by a well-developed story (Green & Brock, 2002; Hall & Bracken, 2011). Consequently, one of the main aspects of effective narrative is the emotional responses evoked during transportation into a narrative world using different mediums and techniques (Green, Brock & Kaufman, 2004).

Narrative transportation is a form of experiential response to narratives and often described in the literature as being similar to concepts such as cognitive absorption, flow, and immersion. However, cognitive absorption refers to a tendency to be immersed in life experiences, while narrative transportation is an engrossing temporary experience (Sestir & Green, 2010). In turn, flow is a more general construct (i.e., people can experience flow in a variety of activities), whereas narrative transportation involves empathy and mental imagery, which do not occur in flow experiences (Bracken, 2006). Immersion is primarily an experiential response to visual elements of images, whereas narrative transportation relies on a story with a plot and characters - features that are not necessarily present in immersion (Van Laer et al., 2014). However, principles from narrative transportation are yet to be integrated into digital heritage tourism experiences, with limited work focused on the operationalisation and transformation into immersive tourism experiences.

A 'pilot' 3D phone-based application or 'app' of the immersive experiences for the Willandra Trackway was constructed. Stories were collected from Indigenous elders, ensuring the accuracy of the content to be developed for visitors. Following this, their stories were transcribed and digitalised for further analysis and integration into the virtual environment. Storytelling 'touchpoints' were identified through a process of consultation with key stakeholder groups. Once 'touchpoints' were identified, further consultation with stakeholder groups identified techniques from narrative transportation which could be utilised to increase visitor engagement with the scan data. Preliminary analysis identified four core goals of stakeholder groups as: 1) to elicit empathy for Indigenous people, 2) to improve visitors' emotional engagement, 3) to feel connected to the past, and 4) to create deep memories through 'experiencing' the Willandra trackway which visitors cannot physically access due to site closures. Future work will explore how 'cold' digital 3D scan data can be transformed into 'warm' sites through deep emotional engagement articulated by the principles underpinning narrative transportation. At the time of this book chapter, the piloted immersive experiences of the Willandra have been developed and are currently undergoing an evaluative component.

CONCLUSION

Creating immersive heritage-led experiences that enable visitors to engage with history and heritage in meaningful and memorable ways remains challenging. One key issue with the creation of immersive experiences is maintaining accuracy within the presentation of information, including balancing the contested information with known accurate facts (Mura, Tavakoli, & Sharif, 2017). Current literature highlights a need to balance historical facts and heritage artefacts with elements of fiction or storytelling to achieve memorability of the immersive visitor experience, while maintaining preservation (Bec et al., 2019). Furthermore, it is important to maintain authentic replications of artefacts while using immersive technologies, which can be achieved by acknowledging spiritual and emotional significance of artefacts through rich narrative and storytelling techniques (Alberts & Hazen, 2010). In a case of Willandra, which has significant cultural and spiritual value for the Aboriginal community, maintaining 3D data and information accuracy is critical for further documentation, preservation, and virtual re-creation of this remarkable part of Australian history.

The Willandra trackway has been actively analysed and researched over an extended period of time. However, until recently the focus of the previous studies was on archaeological, educational and conservation efforts, rather than creation of memorable visitor experiences and other tourism products. Further exploration of innovative ways to engage with Aboriginal contested heritage, specifically the Willandra trackways, has been further explored in this study. This study focused on the possibilities to re-create the unique and vulnerable heritage site using a combination of 3D technologies and storytelling to assist with the virtual presentation of heritage for visitors. This study was also undertaken as a part of ongoing conservation and research initiatives surrounding Willandra Lake heritage. This chapter has explored how heritage preservation can be achieved through the creation of immersive virtual experiences using a combination of innovative 3D scanning methods, storytelling techniques, and theories from modern cognitive psychology. The 3D data capture workflow was presented, outlining the stages of the 3D scanning process and documenting technical challenges associated with converting data into deliverable outcomes for education, conservation, and tourism purposes. It has been found that these innovative 3D scanning techniques provide very high-quality 3D data compared with earlier methods for 3D data capture of the site and can improve traditional methods of heritage preservation. The challenges associated with converting earlier 3D scans into new 3D models for the purpose of integrating with AR/VR workflows emphasise the importance of initial planning as an integral step in the 3D data capture process, specifically keeping with a detailed record of the workflow necessary for further virtual experience re-creation.

Although accuracy is necessary for both 3D scanned data and other historical information, including stories and narratives, a high level of realism is also required for this remarkable piece of Australian Indigenous history to be represented as an immersive and memorable visitor experience. This study also emphasises the importance of deep emotional engagement in virtual presentation or re-creation of heritage though the utilisation of psychological theories, and explains how an integrative melding of attention, flow, imagery, and emotion elicits empathy through the process of narrative transportation.

FUTURE RESEARCH DIRECTIONS

To further expand this research and the application of storytelling within the digital experiences created for contested heritage artefacts, future studies should extend to explore the creation of alternative realities and Metaverse. Another potential area for future research includes improving emotional engagement for visitors by applying other theories of emotions from modern cognitive psychology and neuroscience to a virtual immersive environment using 3D scanned data. These new emotional connections developed to culturally significant objects and stories will provide visitors a new level of understanding and attachment to the past.

REFERENCES

Alberts, H. C., & Hazen, H. D. (2010). Maintaining authenticity and integrity at cultural world heritage sites. *Geographical Review*, *100*(1), 56–73. doi:10.1111/j.1931-0846.2010.00006.x

Ashworth, G. J. (2014). Heritage and economic development: Selling the unsellable. *Heritage & Society*, *7*(1), 3–17. doi:10.1179/2159032X14Z.00000000015

Aureli, S., & Del Baldo, M. (2022). Stakeholders' consciousness of cultural heritage and the reconciliation of different needs for sustainable development. *Journal of Cultural Heritage Management and Sustainable Development*. Advance online publication. doi:10.1108/JCHMSD-10-2020-0156

Barrett, J., & Cocq, C. (2019). Indigenous storytelling and language learning: digital media as a vehicle for cultural transmission and language acquisition. In *Perspectives on Indigenous writing and literacies* (pp. 89–112). Brill.

Barton, R., & Barton, G. (2014). Storytelling as an arts literacy: Use of narrative structure in Aboriginal arts practice and performance. In *Literacy in the Arts* (pp. 251–268). Springer. doi:10.1007/978-3-319-04846-8_15

Bec, A., Moyle, B., Timms, K., Schaffer, V., Skavronskaya, L., & Little, C. (2019). Management of immersive heritage tourism experiences: A conceptual model. *Tourism Management*, *72*, 117–120. doi:10.1016/j.tourman.2018.10.033

Beckman, F. (2015). Ambivalent Screens: Quentin Tarantino and the Power of Vision. *Film Philosophy*, *19*(1), 85–104. doi:10.3366/film.2015.0006

Berg, C. R. (2006). A taxonomy of alternative plots in recent films: Classifying the" Tarantino effect. *Film Criticism*, *31*(1/2), 5–61.

Bracken, C. C. (2006). Perceived source credibility of local television news: The impact of television form and presence. *Journal of Broadcasting & Electronic Media*, *50*(4), 723–741. doi:10.120715506878jobem5004_9

Buonincontri, P., & Marasco, A. (2017). Enhancing cultural heritage experiences with smart technologies: An integrated experiential framework. *European Journal of Tourism Research*, *17*, 83–101. doi:10.54055/ejtr.v17i.295

Chen, H., & Weiler, B. (2017). Heritage interpretation as a tool for sustainable management of national parks in China. *CAUTHE 2017: Time for Big ideas? Re-thinking the field for tomorrow*, 278-287.

Deacon, H. (2006). The 2003 Intangible Heritage Convention: keynote address. *South African Museums Association Bulletin, 32*(1), 2-9.

Douglas, K. (2010). *Pictures of time beneath: science, heritage, and the uses of the deep past.* CSIRO Publishing. doi:10.1071/9780643100251

Geia, L. K., Hayes, B., & Usher, K. (2013). Yarning/Aboriginal storytelling: Towards an understanding of an Indigenous perspective and its implications for research practice. *Contemporary Nurse, 46*(1), 13–17. doi:10.5172/conu.2013.46.1.13 PMID:24716757

Gibson, J. (2015). Message from Mungo. *Australian Historical Studies, 46*(2), 308–310. doi:10.1080/1031461X.2015.1044159

Green, M. C., & Brock, T. C. (2002). In the mind's eye: Transportation-imagery model of narrative persuasion. In M. C. Green, J. J. Strange, & T. C. Brock (Eds.), *Narrative impact: Social and cognitive foundations* (pp. 315–341). Lawrence Erlbaum Associates Publishers.

Green, M. C., Brock, T. C., & Kaufman, G. F. (2004). Understanding media enjoyment: The role of transportation into narrative worlds. *Communication Theory, 14*(4), 311–327. doi:10.1111/j.1468-2885.2004.tb00317.x

Hall, A. E., & Bracken, C. C. (2011). I really liked that movie - testing the relationship between trait empathy, transportation, perceived realism, and movie enjoyment. *Journal of Media Psychology, 23*, 90–99. doi:10.1027/1864-1105/a000036

Harrison, D. (2004a). Introduction: Contested narratives in the domain of World heritage. *Current Issues in Tourism, 7*(4-5), 281–290. doi:10.1080/13683500408667985

Hertzman, E., Anderson, D., & Rowley, S. (2008). Edutainment heritage tourist attractions: A portrait of visitors' experiences at Storyeum. *Museum Management and Curatorship, 23*(2), 155–175. doi:10.1080/09647770802012227

Hinchman, L. P., & Hinchman, S. (Eds.). (1997). *Memory, identity, community: The idea of narrative in the human sciences.* Suny Press.

Ingerpuu, L. (2018). Socialist architecture as today's dissonant heritage: Administrative buildings of collective farms in Estonia. *International Journal of Heritage Studies, 24*(9), 954–968. doi:10.1080/13527258.2018.1428664

Landorf, C. (2009). Managing for sustainable tourism: A review of six cultural World Heritage Sites. *Journal of Sustainable Tourism, 17*(1), 53–70. doi:10.1080/09669580802159719

Lien, N. H., & Chen, Y. L. (2013). Narrative ads: The effect of argument strength and story format. *Journal of Business Research, 66*(4), 516–522. doi:10.1016/j.jbusres.2011.12.016

Little, C., Bec, A., Moyle, B. D., & Patterson, D. (2020). Innovative methods for heritage tourism experiences: Creating windows into the past. *Journal of Heritage Tourism, 15*(1), 1–13. doi:10.1080/1743873X.2018.1536709

Little, C., Patterson, D., Mitchell, L., Pappin, D., & Rosendahl, D. (2018, May). Willandra Lakes Fossil Human Trackway: Presentation and Protection. In *Proceedings of the 3r International Conference on Best Practices in World Heritage: Integral Actions Menorca, Spain* (pp. 4 - 7). Academic Press.

Little, C., Patterson, D., Moyle, B., & Bec, A. (2018, January). Every footprint tells a story: 3D scanning of heritage artifacts as an interactive experience. In *Proceedings of the Australasian Computer Science Week Multiconference* (pp. 1-8). 10.1145/3167918.3167974

Liu, Y., Dupre, K., & Jin, X. (2021). A systematic review of literature on contested heritage. *Current Issues in Tourism, 24*(4), 442–465. doi:10.1080/13683500.2020.1774516

Lund, N. F., Cohen, S. A., & Scarles, C. (2018). The power of social media storytelling in destination branding. *Journal of Destination Marketing & Management, 8*, 271–280. doi:10.1016/j.jdmm.2017.05.003

Lundqvist, A., Liljander, V., Gummerus, J., & Van Riel, A. (2013). The impact of storytelling on the consumer brand experience: The case of a firm-originated story. *Journal of Brand Management, 20*(4), 283–297. doi:10.1057/bm.2012.15

Lutz, B., & Weintke, M. (1999). Virtual Dunhuang Art Cave: A cave within a CAVE. *Computer Graphics Forum, 18*(3), 257–264. doi:10.1111/1467-8659.00346

Macgregor, C. (2022). Preserving the ancient human trackways site in the Willandra lakes world heritage area. *Studies in Conservation, 67*(sup1), 150–155. Advance online publication. doi:10.1080/00393630.2022.2050616

Mura, P., Tavakoli, R., & Sharif, S. P. (2017). 'Authentic but not too much': Exploring perceptions of authenticity of virtual tourism. *Information Technology & Tourism, 17*(2), 145–159. doi:10.100740558-016-0059-y

Murzyn-Kupisz, M. (2012). Cultural, economic and social sustainability of heritage tourism: Issues and challenges. *Economic and Environmental Studies, 12*(2), 113–133.

Naef, P., & Ploner, J. (2016). Tourism, conflict and contested heritage in former Yugoslavia. *Journal of Tourism and Cultural Change, 14*(3), 181–188. doi:10.1080/14766825.2016.1180802

Nasser, N. (2003). Planning for urban heritage places: Reconciling conservation, tourism, and sustainable development. *Journal of Planning Literature, 17*(4), 467–479. doi:10.1177/0885412203017004001

Okech, R. N. (2010). Socio-cultural impacts of tourism on World Heritage sites: Communities' perspective of Lamu (Kenya) and Zanzibar Islands. *Asia Pacific Journal of Tourism Research, 15*(3), 339–351. doi:10.1080/10941665.2010.503624

Olsen, D. H., & Timothy, D. J. (2002). Contested religious heritage: Differing views of Mormon heritage. *Tourism Recreation Research, 27*(2), 7–15. doi:10.1080/02508281.2002.11081215

Pieraccini, M., Guidi, G., & Atzeni, C. (2001). 3D digitizing of cultural heritage. *Journal of Cultural Heritage*, *2*(1), 63–70. doi:10.1016/S1296-2074(01)01108-6

Poux, F., Valembois, Q., Mattes, C., Kobbelt, L., & Billen, R. (2020). Initial user-centered design of a virtual reality heritage system: Applications for digital tourism. *Remote Sensing, 12*(16), 1-32.

Rossiter, M., & Garcia, P. A. (2010). Digital storytelling: A new player on the narrative field. *New Directions for Adult and Continuing Education*, *126*(126), 37–48. doi:10.1002/ace.370

Sestir, M., & Green, M. C. (2010). You are who you watch: Identification and transportation effects on temporary self-concept. *Social Influence*, *5*(4), 272–288. doi:10.1080/15534510.2010.490672

Shishido, H., Ito, Y., Kawamura, Y., Matsui, T., Morishima, A., & Kitahara, I. (2017, December). Proactive preservation of world heritage by crowdsourcing and 3D reconstruction technology. In *2017 IEEE International Conference on Big Data (Big Data)* (pp. 4426-4428). IEEE.

Sole, D., & Wilson, D. G. (2002). Storytelling in organizations: The power and traps of using stories to share knowledge in organizations. LILA, Harvard, Graduate School of Education, 1-12.

Swords, J., Nally, C., Rogage, K., Watson, R., Charlton, J., & Kirk, D. (2021). Colliding epistemologies, productive tensions and usable pasts in the generation of heritage-led immersive experiences. *International Journal of Heritage Studies*, *27*(2), 186–199. doi:10.1080/13527258.2020.1780462

Taçon, P. S., & Baker, S. (2019). New and emerging challenges to heritage and well-being: A critical review. *Heritage*, *2*(2), 1300–1315. doi:10.3390/heritage2020084

Tang, S. (2016). Digital storytelling approach in a multimedia feature writing course. *Journal of Language Teaching and Research*, *7*(3), 572–578. doi:10.17507/jltr.0703.19

Throsby, D. (2009). Tourism, heritage and cultural sustainability: Three 'golden rules. In L. F. Girard & P. Nijkamp (Eds.), *Cultural tourism and sustainable local development* (pp. 13–29). Routledge.

Tivers, J., & Rakić, T. (Eds.). (2012). *Narratives of travel and tourism*. Ashgate.

Tscheu, F., & Buhalis, D. (2016). Augmented reality at cultural heritage sites. In *Information and Communication Technologies in Tourism* (pp. 607–619). Springer.

UNESCO World Heritage Centre. (n.d.). *Willandra Lakes Region*. Retrieved September 8, 2021, from https://whc.unesco.org/en/list/167/video

Van Laer, T., De Ruyter, K., Visconti, L. M., & Wetzels, M. (2014). The extended transportation-imagery model: A meta-analysis of the antecedents and consequences of consumers' narrative transportation. *The Journal of Consumer Research*, *40*(5), 797–817. doi:10.1086/673383

Van Limburg, B. (2009). The anatomy of a digital storytelling system: The architecture. *Information Technology & Tourism*, *11*(1), 31–42. doi:10.3727/109830509788714604

Wiederhold, B. K. (2018). The next level of virtual reality isn't technology—It's storytelling. *Cyberpsychology, Behavior, and Social Networking*, *21*(11), 671–672. doi:10.1089/cyber.2018.29129.bkw PMID:30421994

Wong, J. Y., Lee, S. J., & Lee, W. H. (2015). 'Does it really affect me?' Tourism destination narratives, destination image, and the intention to visit: Examining the moderating effect of narrative transportation. *International Journal of Tourism Research, 18*(4), 458–468.

Yılmaz, R., & Ciğerci, F. M. (2019). A brief history of storytelling: From primitive dance to digital narration. In *Handbook of research on transmedia storytelling and narrative strategies* (pp. 1–14). IGI Global. doi:10.4018/978-1-5225-5357-1.ch001

ADDITIONAL READING

Bec, A., Moyle, B., Schaffer, V., & Timms, K. (2019). Virtual and mixed reality for second chance tourism. *Tourism Management*, 83.

Hassan, A. T., & Fritsch, D. (2019). Integration of laser scanning and photogrammetry in 3D/4D cultural heritage preservation – A review. *International Journal of Applied Science and Technology*, *9*(4).

Skublewska-Paszkowska, M., Milosz, M., Powroznik, P., & Lukasik, E. (2022). 3D technologies for intangible cultural heritage preservation – literature review for selected databases. *Heritage Science*, *10*(1), 50. doi:10.118640494-021-00633-x PMID:35003750

Webb, S., Cupper, M. L., & Robins, R. (2006). Pleistocene human footprints from the Willandra Lakes, south-eastern Australia. *Journal of Human Evolution*, *50*(4), 405–413. doi:10.1016/j.jhevol.2005.10.002 PMID:16343597

Webb, S., Cupper, M. L., & Robins, R. (2006). Pleistocene human footprints from the Willandra Lakes, south-eastern Australia. *Journal of Human Evolution*, *50*(4), 405–413. doi:10.1016/j.jhevol.2005.10.002 PMID:16343597

KEY TERMS AND DEFINITIONS

3D Technologies: Computer graphics that use three-dimensional representation of geometric data.

Contested Heritage: Heritage that has multiple conflicting viewpoints, interpretations, and/or perspectives.

Data Capture Workflow: The process of gathering information from the real world using x, y, and z coordinates, and making it digital.

Chapter 12
Show Me a Story:
Social Media-Based User-Generated Videos and Destination Brand Engagement

Michael William Lever
Fairleigh Dickinson University, Canada

Statia Elliot
University of Guelph, Canada

ABSTRACT

User-generated videos (UGVs) are stories, and stories reflect how we experience the world. This chapter explores the story-like nature of UGVs by applying narrative analysis to deconstruct a tourism destination's Instagram Reels, revealing their influence on user engagement. Posted comments for each of 20 Reels are coded based on their representation of cognitive, affective, and behavioral dimensions of engagement as they relate to the story elements of plot, character, and verisimilitude. Results indicate that the most engaging Reels feature all story elements, unique locations, and personalized stories. These fulsome video stories are most likely to provoke cognitive engagement, whereas UGVs with fewer elements provoke emotional engagement. This study uniquely connects destination engagement and narrative elements in user-generated storytelling videos to guide destination social media marketing effectiveness.

INTRODUCTION

The use of story elements in advertising is a powerful way to convey a marketing message (Brechman & Purvis, 2015; Kılıç & Yolbulan Okan, 2020). This holds true in a consumer-to-consumer context, as evidenced by the continuing proliferation of consumer-created marketing, a form of organic promotion in which consumers create and share user-generated content (UGC) (Muñiz & Schau, 2007), often through the use of stories shared on social media. Most recently, platforms such as *Instagram* emphasize these story-based elements through video-based UGC. In *Instagram*'s case, these videos are referred to as 'Reels', 15-second videos that can be stitched together to create multi-clip content that is then shared with the general public (Instagram, 2020). Despite this increase in user-generated videos (UGV) and

DOI: 10.4018/978-1-6684-3436-9.ch012

the comprehensive examination of how consumer-created marketing impacts user engagement in the marketing field, less research has explored this phenomenon in a tourism context. Therefore, the aim of this study is to explore the increasing impact of UGV on destination brand engagement. The research supports the vital role of destination resilience during crises such as the devastating impacts of COVID-19 on the tourism industry by revealing how UGV can encourage visitation to a destination, supporting destination marketing organizations' (DMOs) pathways to recovery.

DMOs have responded to the advancement of media technologies by shifting marketing efforts from website development and online marketing to social media marketing (Pan, MacLaurin & Crotts, 2007). Taking advantage of social media's decentralization, openness, and greater utility, DMOs continue to explore applications to host message boards and virtual communities as effective means to market their destinations, mixing messages either produced and/or distributed by consumers and/or other organizations. At the forefront, UGVs can be considered visual stories, told by subjects to an audience through self-made productions, with the intent to convey a meaning (Pace, 2008). Online communities provide the forums for stories to be shared (Muniz & O'Guinn, 2001), and increasingly these stories relate to the consumption of goods, services, brands, and destinations. The popularity of UGVs is recognized by destinations and travelers alike, and research is emerging to substantiate their effectiveness (Zhang et al., 2021).

Advertising research sheds much light on the popularity of UGVs. Structuring advertisements as stories, with a plot and a brand as the hero, can make the message more persuasive than an analytical illustration of product features (Escalas, 2007). Self-referencing theory is also relevant to UGV's effectiveness. When consumers see advertisements that they can refer to themselves, they elicit more positive attitudes (Debevec & Iyer, 1988). Another relevant concept is authenticity, historically seen as positively correlated to traveler attitudes (Cohen, 1988; Hughes, 1995).

Simon Anholt (2007), a popular destination brander, describes place branding strategy as coming up with a brand story for a geographic region that is inspiring and magnetic but also credible. For destinations that can create a visual brand story, there is evidence that viewers will become absorbed (i.e., narratively transported) in the story and persuaded by the information (Escalas, 2007). Narrative transportation is different from traditional elaboration-based persuasion models that focus on cognitive response processes (Petty, Cacioppo, and Schumann, 1983). With narrative transportation, persuasion is more of an affective response to being immersed in a story (Escalas, 2007), signaling the potential of UGVs to persuade through narrative transportation.

Given the story-like nature of UGVs, this study adopts a qualitative consumer research method of narrative analysis, which involves the analysis of data in which significant amounts of narrative are present, including case studies, social media, and direct/indirect observations. Using narrative concepts such as plot, character, structure, and language (Pace, 2008), UGVs can be deconstructed and their meanings interpreted. As artifacts of actual destination experiences, UGVs may reveal the most salient trip experiences - those the traveler chooses to share - a visual glimpse of behavior that was only interpretable in the not-so-distant past via consumer survey.

BACKGROUND

User-Generated Content

One of the first to study UGVs, Tussyadiah and Fesenmaier (2009) analyze 120 UGVs of New York City, finding that UGVs provide pleasure to viewers by stimulating fantasies, daydreams, and past travel memories. Beyond pleasure, UGC is a valuable source of information during the critical decision-making pre-travel stage (Zeng and Gerritsen, 204; Xiang and Gretzel, 2010) and is posted by those who desire to share their experiences with others (Nezakati et al., 2015). UGC has been found to directly influence tourist expectations and to have an indirect effect on tourist satisfaction (Kaosiri et al., 2019). For millennials, a cohort that is particularly engaged with social media, seeing the positive experience of peers can trigger destination visit intentions (Liu et al., 2019), supported by Expedia research which indicates that 87% of millennials attribute their destination choice to social media (Expedia and The Center for Generational Kinetics, 2018).

Studies that specifically consider UGV content have found that its influence on users' perception and intended behavior depends on the technical quality of the video (Hautz et al., 2014), more so than whether the video was posted by an agency or an individual user. Yet, in the case of low quality UGVs, users have significantly stronger positive effect on source trustworthiness and expertise than agencies (Hautz et al., 2014), a finding that suggests UGVs as a cost-effective media. Huertas (2018) interviewed users to show how videos can communicate authenticity, and proximity, generate tourism experiences, and make a destination more attractive by creating an enticing destination image.

It is indeed a powerful aspect of social media that both the poster and viewer find pleasure in the experience. Oliveira et al. (2020) found perceived enjoyment to be the most essential motive for travelers to share their experiences online. DMOs are anxious to harness this power. Zhang et al. (2021) study UGC effects and provide destinations with an approach to facilitate the co-productive role of UGC. Their comprehensive study sheds much light on UGCs potential to generate tourist arrivals. This study builds upon past research to consider the actual form of UGVs, the features and style, and their influence on viewer engagement. Thus, a cross-disciplinary review of social media, tourism, and technology literature was undertaken to identify an appropriate method to evaluate UGVs.

Narrative Analysis

Narrative analysis, while historically common in fields of psychology, sociology, and the humanities, has been gaining ground in consumer and market research (Shankar & Goulding, 2001). It is used to examine the flow of first-person accounts of experiences in the stories they tell, with the objective of understanding their meaning (Gubrium & Holstein, 2009). Narrative analysis denotes a set of methods of inquiry to study representations of realities that can be written, oral or visual (Mura and Sharif, 2017). Given that tourism is such a rich field in terms of its visualizations and descriptions, it is somewhat surprising that narrative analysis has been relatively neglected by tourism scholars. As more researchers turn to critical analysis and qualitative approaches, narrative analysis will undoubtedly gain ground. When employed in the context of a destination, it can reveal a very different narrative than what is offered by the official brand (Cassel and De Bernardi, 2021), bringing forth new and meaningful perspectives.

UGVs are stories, and stories reflect how we experience the world. People's stories are an important lens through which we can explore social reality (Mura and Sharif, 2017). Like stories, narratives follow common organizing principles and plots, characters and trajectories (Dessart & Pitardi, 2019). Most commonly, narrative analysis is used to analyze themes. Here, the authors take a different approach, using the organizing principles of a good narrative to analyze UGVs as stories.

Destination Brand Engagement

Consumer brand engagement has been studied extensively in the nascent marketing and branding literature (Hollebeek, 2011; Leckie, Nyadzayo & Johnson, 2016; Sprott, Czellar, & Spangenberg, 2009). In a consumer brand context, engagement has come to replace the traditional concept of involvement in explaining consumers' deep connection to a given brand (Hollebeek, Glynn and Brodie, 2014). Much of this research centers around the role of social media given the opportunity it creates for deeper interactive brand relationships (Malhotra, Malhotra & See, 2013).

Despite its popularity in a consumer branding context, destination brand engagement research remains limited, though recent studies have begun to popularize its usage in the tourism field. For instance, Chen et al. (2020) observed a direct positive relationship between destination brand engagement and destination brand loyalty. Similarly, time frame and posting seasonality were strong predictors of engagement within a tourist-based Facebook group (Villamediana, Kuster, & Vila (2019). Destination brand engagement has also been found to mediate the relationship between destination brand authenticity and destination brand self-congruence (Chen et al., 2020). There have also been environmental benefits of being engaged with a destination brand, as intention and behaviors related to environmental conservation strengthened as engagement increased (Saleem, Li, & Afzal, 2021).

Given that tourist-generated communication is a more reliable source of information than is traditional (i.e., supply-created) communication (Huerta-Álvarez, Cambra-Fierro, & Fuentes-Blasco, 2020), DMOs must understand who their engaged brand promoters are within their social media channels. Therefore, this chapter considers the important role of destination brand engagement as it relates to storytelling within Instagram.

NARRATIVE ANALYSIS OF USER-GENERATED VIDEOS

There remains a limited breadth of research exploring engagement using storytelling content (Dessert and Petardi, 2019). This gap is even more pronounced in the context of tourism research in particular. The few tourism studies that examine storytelling intentions and behavior have considered co-creation (Mathisen & Chen, 2018), destination brand communication (Pachucki, Grohs, and Scholl-Grissemann, 2021), and tourist satisfaction (Cater et al., 2021) but surprisingly not the connection between story and engagement. While Elliot (2009) used narrative analysis to deconstruct UGVs using four dimensions of narratives – plot (point of the story), characters (players and setting), structure (pattern of beginning, middle, end), language (verbal expression) and overarching theme, they did not consider the impact of UGVs on user engagement. Based on an analysis of the inclusion or exclusion of literary dimensions, Elliot (2009) concluded that "good" stories were more complete and would be preferred by viewers yet, they did not measure engagement to make this connection. To the author's knowledge, there are no studies to date which apply narrative analysis to the deconstruction and evaluation of UGV's impact on user engagement. Therefore, this chapter provides this view.

Data Collection

Destination Canada (DC) was selected as the focal organization for the stories to be analyzed. DC is Canada's national DMO, known for its innovative use of electronic marketing. DC was, in fact, one of the first DMOs to incorporate UGVs into their commercial tourism marketing campaign (CTC, 2008). To explore engagement within social media, *Instagram*'s Reels using the #ExploreCanada hashtag (initiated by DC) were selected as the study's object of interest. The videos were then analyzed using a narrative analysis approach to examine an individual's first-hand experiences shared in stories to understand their influence.

Data Analysis

Storytelling Narrative Concepts

The analysis was informed by the key storyteller narrative concepts outlined in the extended transportation-imagery model developed by Van Laer et al. (2014) and depicted in Figure 1.

Figure 1. Extended transportation-imagery model (Van Laer et al., 2014)

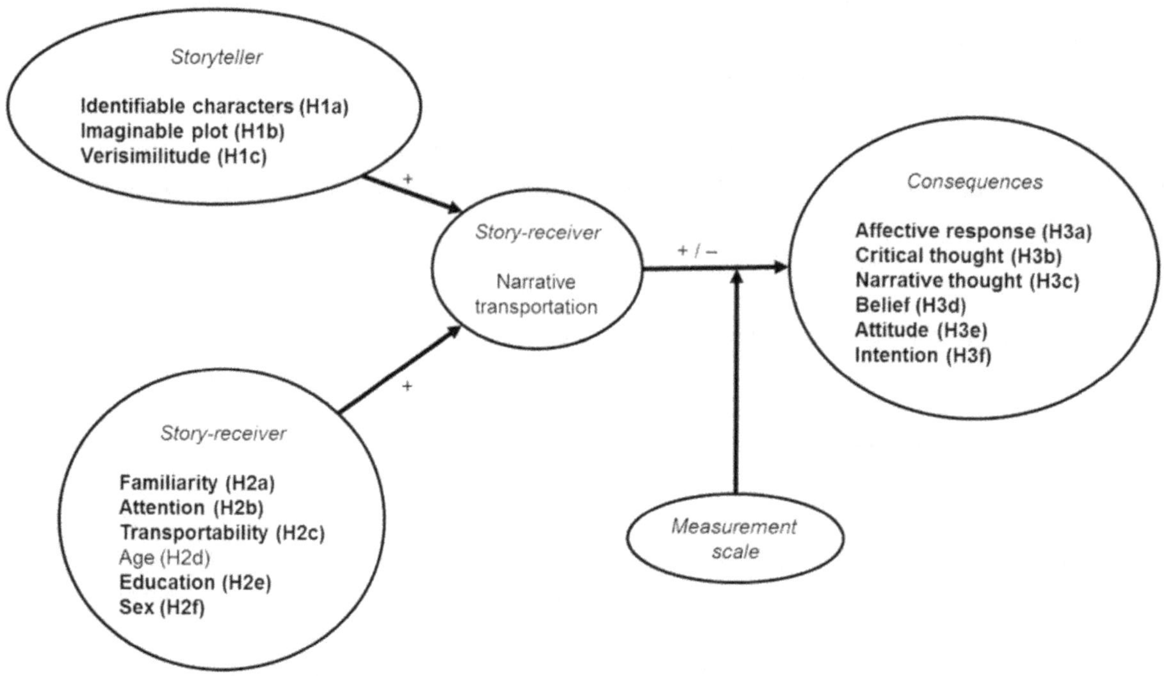

Focusing on the storytelling component of the model, Van Laer et al. (2014) conducted a meta-analysis of past research to identify three storyteller antecedents that influence the strength of the impact of a story on the story-receiver: identifiable characters, imaginable plot, and verisimilitude. Defined as "invented personas the story receiver clearly pinpoints from the storyteller's use of context-derived assumptions", the characters represent the key players and settings that convey the story's beliefs and emotions, allows the viewer to empathize with the character, and builds a deeper connection to the story being told (Van Laer et al., 2014, p.802). The plot represents the point of the story – a temporal sequence of events that play out to guide the viewer from the beginning to the end in an understandable progression. A compelling plot can provoke strong mental imagery beyond the visuals depicted in the story itself. Finally, verisimilitude represents the likelihood of events in the story coming true (often referred to as the story's 'lifelikeness') (Dessart & Pitardi, 2019). To analyze UGVs using these storytelling elements, the authors located 20 Reels uploaded to *Instagram* using the 'ExploreCanada' hashtag between June 28th and July 6th, 2021.

Cognitive, Emotional, and Behavioral Dimensions of Engagement

Recognizing that stories can impact users' perceptions, feelings, and actions (Dessart & Pitardi, 2019), engagement was further analyzed within each narrative element in terms of cognitive, emotional, and behavioral dimensions (Hollebeek & Macky, 2019). Cognitive dimensions are those thoughts and mental elaborations that require one to stay focused (i.e., attention) and remain immersed (i.e., absorption) with the story being told. Emotional dimensions are those expressed through feelings such as enjoyment (pleasure derived) or enthusiasm (intrinsic level of excitement involved in the act of viewing the story). Finally, behavioral dimensions focus on the effort or energy to engage with the story, often demonstrated through actions such as sharing one's content or information with the storytellers, using the content as a learning opportunity, or recommending the storytelling content to others.

RESULTS

The 20 Reels are summarized below in Table 1, including the posts' number of likes and comments, date posted, featured location in the Reel, characters depicted, and a summary of the story's plot and climax.

Table 1. Summary of Instagram reels used for analysis

ID	Likes	Comments	Date posted	Featured location	Characters	Story plot and climax
1	5029	46	03-Jul-21	Peyto Lake, Alberta	Photographer	A first-person perspective of the photographer entering the water at a knee-deep level to frame the perfect shot (climax) of the rocky mountain landscape.
2	2998	101	02-Jul-21	Beaver Creek, Yukon	Residents of Beaver Creek	A walking tour of her hometown of Beaver Creek, showing off the various sights and businesses, including the only gas station in town and the limited-service post office. The video emphasizes the remoteness and tranquility of her location. The video ends (climax) with her revealing the creek that she visits at the end of every day.
3	2749	54	02-Jul-21	Okanagan Highlands, British Columbia	Young woman	A slow pull-away shot of a Hobbit-inspired Airbnb. While the entire house is revealed, a young woman emerges from the entrance with a cup of coffee in hand, taking in the breathtaking scenery (climax).
4	1375	81	01-Jul-21	Parliament Hill, Ottawa, Ontario	CF Snowbirds	Dramatic view of the CF Snowbird jets flying overtop of the Parliament Hill buildings during Canada Day celebrations before forming together and flying off in a small group (climax).
5	1036	24	29-Jun-21	Vancouver, British Columbia	Young woman	In an elegant black dress, a young woman walks along her hotel's terrace with a dramatic sunset in the background. She reaches a small table and picks up a glass of wine which she then turns to face the water and surrounding views (climax).
6	1026	92	06-Jul-21	Yoho National Park, British Columbia	A photographer and her subject	Rushing blue water from between two large boulders in the foreground. Suddenly, the camera pans up to reveal a photographer and her subject atop the two boulders (climax) with views of large green trees and a mountain range in the background.
7	1009	11	03-Jul-21	Various locations from Sidney to Gabriola Island on the British Columbia coast	A family consisting of two adults, an infant, and a toddler	This post highlights this family's experience moving into a 26-foot sailboat and taking it for a 10-day tour of the BC coastline.
8	890	17	05-Jul-21	Vancouver, British Columbia	A young woman and various others in her social group	The reel opens with a young woman indicating that this video is a recap of her month of June, followed by rapidly changing scenes of various activities, including baseball, hiking, camping, watersports, log splitting, and more. Each scene is very quick and gives a sense of a fun-filled experience (climax).
9	785	21	02-Jul-21	Berg Lake, British Columbia	First-person view (no characters)	A first-person view of a walk through a rustic-looking cabin leads the viewer through the front door and reveals a majestic mountain range with turquoise blue water (climax) in front.
10	749	37	02-Jul-21	Banff, Alberta	A young couple	A young girl and her partner travel to various locations around Banff National Park, including a canoe ride, views of the mountains and lakes, and hiking. A final shot of a blue lake with the mountain behind it wraps up the various scenes depicted (climax).
11	259	59	06-Jul-21	Georgian Bay, Ontario	A group of young women	A group of friends travels together through Georgian Bay. One of them sits atop a steep cliff edge overlooking the bay. This is followed by several views of the area. The caption in the video reads, "That's in Ontario? Dam !?! Looks like Bahamas". The final scene shows all three women raising their arms in victory at having explored the area (climax).
12	245	27	02-Jul-21	Alberta, Canada	A young woman	With a backdrop of mountain ranges and a green field, a young woman does a slow-motion spin with a big smile throughout (climax).
13	239	17	29-Jun-21	Alberta, Canada	A young woman	As a young woman steps out onto a solitary rock overlooking a dramatic mountain range across a blue lake, a sudden rapid succession of quick video clips of her various sights across multiple parks, including sitting in a hammock atop a cliff edge, paddling a canoe, seeing a bear up close, and so on are played (climax).

Continued on following page

Table 1. Continued

ID	Likes	Comments	Date posted	Featured location	Characters	Story plot and climax
14	194	19	02-Jul-21	Kingston, Ontario	First-person view but with a man's legs showing	A man crosses some rocks to reveal a beautiful sunset over a relatively calm lake (climax).
15	184	31	28-Jun-21	British Columbia	Various young men and women	A quick scene of a car traveling down an empty road is followed by various scenes changing at a medium pace of views of British Columbia's lakes and trees. Many scenes include one of the Reel's characters interacting with the environment, including swinging on a rope swing into the water, riding a zipline, or jumping off a cliff's edge into the water. The final scene shows a peaceful view of a sun setting over a beach with the waves calmly lapping (climax).
16	184	13	03-Jul-21	Alberta, Canada	First-person view (no characters)	The camera slowly pans across a raging waterfall that travels into a bottomless abyss below (climax).
17	159	15	04-Jul-21	Revelstoke, British Columbia	A woman and a hang-gliding expert	Two people begin running forward with a parachute following behind them as they launch into the sky (climax) and into the far distance as a snow-capped mountain and lake can be seen in the background.
18	151	17	29-Jun-21	British Columbia	Family traveling with kids of various ages	The opening scene shows a rushing stream and includes a monologue that eventually they're going to die, and no one will remember them, "so f**k it", followed by a rapid succession of still images depicting their various experiences (climax).
19	143	12	29-Jun-21	Banff, Alberta	Mom with young daughters	The family walks down a path alongside short cliff walls as a stream of water rushes next to them (climax).
20	113	10	04-Jul-21	Moraine Lake, Alberta	Photographer	A photographer walks along the edge of a tall cliff edge before bringing his camera up to his eye and taking a photo.

The comments for each Reel were coded based on their representation of cognitive, affective, or behavioral dimensions as they relate to the three-story elements (story plot, characters, and verisimilitude). The results reveal that posts that most closely mirror the elements of effective storytelling have greater levels of engagement in terms of the number of likes and number and types of comments. Additionally, the dimensions of engagement are uniquely woven through each story element. As previously mentioned, cognitive engagement focuses on a consumers' thoughts and mental elaborations. When considered in relation to a story's plot, this may translate into a deeper immersion with the story, more attention paid to it, or a reflection on the story's message. Examples from the analysis demonstrate how this may translate to users' comments on a social media post, such as the relationship between cognitive engagement and a story's plot revealing immersion with the story by indicating a strong immersion with the storyteller's content without yet being fulfilled (e.g., "*Can't get enough of this* 🐺"). Similarly, examples of how a story's characters connected to cognitive engagement include the relation of users to the characters' own experiences ("*You remind me of myself*", and verisimilitude by being able to imagine oneself actually visiting the place being depicted ("*This reminds me of my own time there; How can I get there myself*").

Next, emotional engagement demonstrates one's summative and enduring levels of emotion. When the audience is surprised or excited about the story, they may feel an emotional connection in relation to its plot (*"This is beautiful! Wow!* 😊 *"*). This connection could also apply to the characters rather than the story itself (*"You look so happy here! You're so amazing!"*). Lastly, the behavioral engagement dimension consists of a consumers' energy, effort, and time spent on an engagement object. For example, users may repeatedly watch a video or approve of its message (*"I've already watched this a hundred times"*). Or, users may choose to promote the brand or story content with others, mainly in the form of tagging, as an example of behavioral verisimilitude.

In terms of user engagement, the posts revealed considerable variability across visual elements and significant landmarks. Geographically, Western Canada was prominent, specifically the provinces of British Columbia and Alberta, which may reflect their more dramatic mountain and lake views compared to other Canadian regions. Interestingly, the posts featured more women than men. The number of likes ranged from 113 to just over 5,000, and the number of comments on each post ranged from 10 to 101. There was a large difference in the average number of likes between the top and bottom 10 (top: 1,877, bottom: 179). Many of the bottom ten posts had concise descriptions, whereas top posts were more descriptive and invited higher levels of participation from the audience –suggesting the importance of crafting a story and associating it with the message and video. Notably effective were cases that asked for engagement from the audience (e.g., "*let me know if you have any questions about the community in the comments section*" from the Reel with the second-highest engagement).

Distinguishing features of the story elements at the top and bottom posts from our data set, it was observed that the top posts focused more on unique locations and walk-through type videos, like a tour of one's hometown with highlights presented along the way (Reel #2). Other examples of unique destinations included a view of a Hobbit-hole-inspired Airbnb (Reel #3) and the Canadian Forces Snowbird jets flying over Parliament Hill (Reel #3). By contrast, the bottom posts predominantly featured places that, while visually appealing, were not as unique, such as a blue lake or a large mountain (Reel #13). This suggests that unique locations and personalized stories encourage more meaningful user engagement with the content.

Finally, as a further indication of engagement, the authors found a higher average number of comments on top posts (51 comments per post) than on bottom posts (18 comments per post). When coded to capture the engagement dimensions and storytelling elements, it was observed that stories with more potent storytelling elements appeared to have more cognitive comments about the story's plot. That is, the audience more frequently described their immersion with the story, gave it more of their attention, and seemed to reflect more deeply on the message. Users would also exhibit greater behavioral verisimilitude by sharing the content with others and promoting the story content. Relations between story elements and engagement dimensions are summarized in Table 2.

Table 2. Relationship between storytelling elements and engagement dimensions

Branded Story Elements	Engagement Dimension	Relationship to storytelling elements	Sample User Comment
Story Plot	Cognitive	More prevalent when visual storytelling elements are strong	*Can't get enough of this* 😻
	Emotional	Prevalent regardless of the strength of visual storytelling elements	*Peyto is like the land before time - what a beautiful place.*
	Behavioral	Uncommon for both strong and weak visual stories	*Love this tour! Would love to see more of your small community* 😍
Characters	Emotional	More prevalent when visual storytelling elements are weak	*I love your pics & videos. But this is definitely the best of all your account* 👏🔥👏😍
	Cognitive	Slightly more prevalent when visual storytelling elements are strong	*I spent a summer in Dawson city Yukon, one of my favourite summers of my life.*
Verisimilitude	Behavioral	Slightly more prevalent when visual storytelling elements are strong	*Awesome! They kept flying over our place and I would keep missing them! Tagging @s.m.carriere @wtl @vale8080*
	Cognitive	Slightly more prevalent when visual storytelling elements are weak	*I remember being there on Parliament Hill on Canada Day. It was wonderful. I can't wait to do that again. Thanks for sharing! #ottawa #canadaday #canadaday*

Interestingly, users focused on the characters for posts with fewer storytelling elements, particularly by emotionally engaging with them through comments. Rather than discussing features of the destinations, users would make comments like "*You look great here!*" or "*You look so happy!*". This suggests that when the story is less clear, users will focus instead on the people in the story, overshadowing the promotion of the setting. Users' comments on posts with few storytelling elements were slightly more prone to cognitive verisimilitude, questioning the reality of the place and the Reels' overall believability concerning their desire to visit (or memory of a past trip), perhaps a reflection of the fantastical element of stories that are not always meant to be an exact mirror of reality.

DISCUSSION

This study contributes to theory by connecting elements of narrative analysis captured in user-generated storytelling videos to destination engagement. UGVs represent co-constructed artifacts of places that, like the process of branding, are made meaningful through their visualization place (Cassel and De Bernardi, 2021), unfolding like a story to be told and retold, interpreted, and reinterpreted. Given that the content is shared by unpaid fans of the brand (i.e., consumer-created marketing), this research provides an understanding of how traveler experiences engage the brand's consumers in ways not typically captured through traditional business-to-consumer advertising. Further, the narrative analysis methodology within social media builds on existing tourism communication research in a unique video storytelling context. In terms of its practical implications, the findings support the potential for DMOs to further adapt to the changing social media landscape by acknowledging the key role of UGC in inspiring visitors' engagement, and specifically how effective user-generated video stories can act as a complement to a DMOs existing social media strategy, particularly during times of great uncertainty.

FUTURE RESEARCH DIRECTIONS

Generally, this research provides a starting point for tourism researchers to adopt narrative analysis techniques to better understand the link between destination brand engagement and social media-based UGV. Looking ahead, there are three key trends in this research area that may be of particular interest to those who study social media-informed destination brand engagement.

First, research is emerging that compares engagement with narrative transportation. This chapter focused mainly on identifying where linkages between cognitive, affective, and behavioral dimensions of engagement and story elements (i.e., plot, characters, and verisimilitudes) exist. However, whether the Reels examined here had the effect of causing one to temporarily lose awareness of their physical surroundings and emotionally react to a narrative through the creation of vivid mental story imagery is unclear. Since engagement comes from experiencing and can manifest as stimulation and intrinsic enjoyment of a brand (Kim, Lloyd, & Cervellon, 2016), it is valuable to understand how it could lead to one being transported into a story's elements more meaningfully. The second emergent research trend regarding this chapter's topic is the role of engagement beyond a destination's brand. Since engagement can occur across various roles and networks (Brodie et al., 2013), it is increasingly worthwhile to consider the impact of UGV on, say, regional or private tourism operators. For example, one of the Reels in this study focused on a tour of the user's personal experiences living in Beaver Creek (Reel #2). In her video, she showcased several local attractions and businesses that may also be interested in understanding how her story impacted them directly. This trend would provide unique opportunities for DMOs to collaborate with others in their shared network. Third, there is a growing importance in understanding the complex relationship between social tone and social presence (Pachucki & Scholl-Grissemann, 2021). When more social terms (e.g., "friends") are used in the UGV, along with the use of first-person pronouns (e.g., "I", "we", etc.), what are the impacts on the viewer's perceptions of that video's trustworthiness, familiarity, and so on? While this study shed light on the importance of the characters within the story in terms of establishing a bond between the viewer and the poster of the content, there is a growing need for research that shows at what point this bond changes from being strong and effectual to weak and inauthentic.

In terms of replicating this research in future study contexts, the authors have summarized their research analysis process in Figure 2.

Figure 2. Conceptual model for analyzing social media-based destination brand engagement using narrative analysis

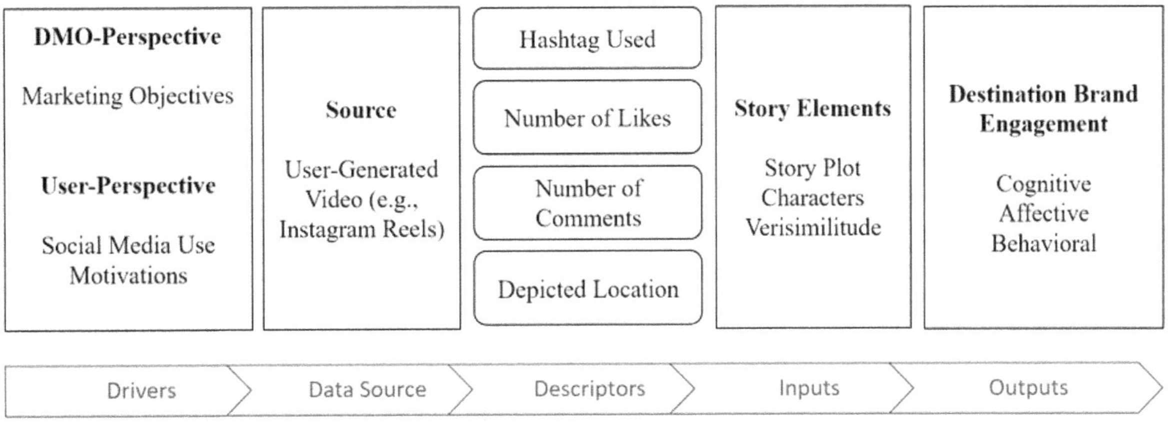

First, the process involves outlining the specific drivers of the data being explored. Researchers must identify, for instance, what objectives have been identified by the subject DMO and what the motivations are of those who are posting content and its relevant meaning. Misunderstanding the motivations of social media users can present challenges in accurately interpreting the data that populates social media pages. Similarly, failing to identify the core objectives of the DMO may lead to challenges in summarizing the findings and justifying the benefits to the organization. Next, the data source must be identified which aligns with those objectives and motivations. Regardless of the source, the content must be user-generated (i.e., not developed directly or indirectly by the DMO or a paid stakeholder). Once identified, researchers may begin the coding process, focusing first on descriptive elements such as the number of likes and comments on a hashtag of interest and noting the featured attractions and locations within the destination. Once this information has been collected, a deeper level of coding may begin, focusing on relating the story elements (plot, characters, and verisimilitudeCOVID-19 had) to destination brand engagement dimensions (cognitive, affective, and behavioral).

This process of narrative analysis presents a variety of future research opportunities to better understand the important role of UGV. For instance, this chapter focuses on the implications for DMOs in terms of their marketing impact, but future researchers may consider how to best measure and predict social media users' motivations based on the type of content shared. This would be of value when considered from the perspective of destination resilience, as significant upheavals to the tourism industry, such as the impact of COVID-19, influence the motivations of those who choose to share their travel experiences. Additional data sources go beyond the specific context of *Instagram* Reels explored here, such as TikTok, YouTube, and Facebook Live. Each has its strengths and weaknesses as a research platform but can be equally as valuable as a source of information depending on the research questions.

CONCLUSION

For travellers, new media has provided a sophisticated means of sharing travel stories. For researchers, UGV calls for much refinement of specific methods of research for such a new medium (Pace, 2008).

Narrative analysis is a useful method of analyzing the meaning and effectiveness of UGVs and is suggested as one approach for tourism market researchers to study social media content. The richness of narrative analysis is its ability to capture and interpret the subtle innuendos and situational particulars of travel stories and visuals. Narrative analysis of UGV within social media will help inform visual storytelling as a research approach by focusing on the prolific use of social media as a platform to create, share, and discuss timely video-based content. By mining this rich UGC data, destination marketers can more accurately examine users' story-based narratives to determine what makes their destination attractive, thereby revealing a more authentic account of their destination's strengths directly from the perspective of their market. For this gain, however, the trade off is a loss in generalizability of findings. As a methodology, narrative analysis is considered an open process of collective knowledge (Riessman, 1993). Additional applications of narrative analysis to social media content in a travel context will help refine methodology and deepen our understanding of the role of UGVs in tourism communication.

ACKNOWLEDGMENT

The authors would like to acknowledge Destination Canada for its support and encouragement during the data collection phase.

This research received no specific grant from any funding agency in the public, commercial, or not-for-profit sectors.

REFERENCES

Anholt, S. (2007). *Competitive Identity: Brand Management for Nations, Cities and Regions*. Palgrave Macmillan. doi:10.1057/9780230627727

Brechman, J. M., & Purvis, S. C. (2015). Narrative, transportation and advertising. *International Journal of Advertising*, *34*(2), 366–381. doi:10.1080/02650487.2014.994803

Brodie, R. J., Ilic, A., Juric, B., & Hollebeek, L. (2013). Consumer engagement in a virtual brand community: An exploratory analysis. *Journal of Business Research*, *66*(1), 105–114. doi:10.1016/j.jbusres.2011.07.029

Canadian Tourism Commission. (2008). CTC Debuts Edgy New Broadcast Spots, a Key Part of its 2009 Global Campaign. CTC Media Centre.

Cassel, S. H., & De Bernardi, C. (2021). Visual representations of Indigenous tourism places in social media. *Tourism, Culture & Communication*, *21*(2), 95–108. doi:10.3727/109830421X16191799471980

Cater, C., Albayrak, T., Caber, M., & Taylor, S. (2021). Flow, satisfaction and storytelling: A causal relationship? Evidence from scuba diving in Turkey. *Current Issues in Tourism*, *24*(12), 1749–1767. doi:10.1080/13683500.2020.1803221

Cohen, E. (1988). Authenticity and Commodification in Tourism. *Annals of Tourism Research*, *15*(3), 371–386. doi:10.1016/0160-7383(88)90028-X

Debevec, K., & Iyer, E. (1988). Self-referencing as a Mediator of the Effectiveness of Sex-Role Portrayal in Advertising. *Psychology and Marketing*, *5*(1), 71–85. doi:10.1002/mar.4220050106

Dessart, L., & Pitardi, V. (2019). How stories generate consumer engagement: An exploratory study. *Journal of Business Research*, *104*, 183–195. doi:10.1016/j.jbusres.2019.06.045

Elliot, S. (2009). User-Generated Videos in Tourism Destination Marketing: Using Narrative Analysis to Deconstruct Video Travel Stories. *TTRA International Conference*.

Escalas, J. E. (2007). Self-referencing and Persuasion: Narrative transportation Versus Analytical Elaboration. *The Journal of Consumer Research*, *33*(4), 421–429. doi:10.1086/510216

Expedia and The Center for Generational Kinetics. (2018). *Generations on the Move*. https://viwfinder.expedia.com/wp-content/uploads/2018/01/Expedia-Generations-on-the-move.pdf

Gubrium, J. F., & Holstein, J. A. (2009). *Analyzing Narrative Reality*. Sage Publications, Inc. doi:10.4135/9781452234854

Hautz, J., Fuller, J., Hutter, K., & Thurridl, C. (2014). Let users generate your video ads? The impact of video source and quality on consumers' perceptions and intended behaviors. *Journal of Interactive Marketing*, *28*(1), 1–15. doi:10.1016/j.intmar.2013.06.003

Hollebeek, L. D., & Macky, K. (2019). Digital content marketing's role in fostering consumer engagement, trust, and value. *Journal of Interactive Marketing*, *45*, 27–41. doi:10.1016/j.intmar.2018.07.003

Huertas, A. (2018). How live videos and stories in social media influence tourist opinions and behaviour. *Information Technology & Tourism*, *19*(1-4), 1–28. doi:10.100740558-018-0112-0

Hughes, G. (1995). Authenticity in Tourism. *Annals of Tourism Research*, *22*(4), 781–803. doi:10.1016/0160-7383(95)00020-X

Instagram. (2020). *Introducing Instagram Reels*. https://about.instagram.com/blog/announcements/introducing-instagram-reels-announcement

Kaosiri, Y. N., Fiol, L. J. C., Tena, M. A. M., Artola, R. M. R., & Garcia, J. S. (2019). User-generated content sources in social media: A new approach to explore tourist satisfaction. *Journal of Travel Research*, *58*(2), 253–25. doi:10.1177/0047287517746014

Kılıç, F., & Yolbulan Okan, E. (2020). Storytelling and narrative tools in award-winning advertisements in Turkey: An interdisciplinary approach. *Journal of Marketing Communications*, 1–16.

Kim, J. E., Lloyd, S., & Cervellon, M. C. (2016). Narrative-transportation storylines in luxury brand advertising: Motivating consumer engagement. *Journal of Business Research*, *69*(1), 304–313. doi:10.1016/j.jbusres.2015.08.002

Liu, H., Wu, L., & Li, X. (2019). Socia media envy: How experience sharing ons ocial networking site drives millennials' aspirational tourism consumption. *Journal of Travel Research*, *58*(3), 355–369. doi:10.1177/0047287518761615

Mathisen, L. I. N. E., and Chen, J. S. (2014). Storytelling in a co-creation perspective. *Creating experience value in tourism*, 157-168.

Muñiz, A. M. Jr, & Schau, H. J. (2007). Vigilante marketing and consumer-created communications. *Journal of Advertising, 36*(3), 35–50. doi:10.2753/JOA0091-3367360303

Mura, P., & Sharif, S. P. (2017). Narrative analysis in tourism: A critical review. *Scandinavian Journal of Hospitality and Tourism, 17*(2), 194–207. doi:10.1080/15022250.2016.1227276

Nezakati, H., Amidi, A., Jusoh, Y. Y., Moghadas, S., Aziz, Y. A., & Sohrabinezhadtalemi, R. (2015). Review of social media potential on knowledge sharing and collaboration in tourism industry. *Procedia: Social and Behavioral Sciences, 172*, 120–125. doi:10.1016/j.sbspro.2015.01.344

Oliveira, T., Araujo, B., & Tam, C. (2020). Why do people share their travel experiences on social media? *Tourism Management, 78*, 1–14. doi:10.1016/j.tourman.2019.104041 PMID:32322615

Pace, S. (2008). YouTube: An Opportunity for Consumer Narrative Analysis? *Qualitative Market Research, 11*(2), 213–226. doi:10.1108/13522750810864459

Pachucki, C., Grohs, R., & Scholl-Grissemann, U. (2021). No story without a storyteller: The impact of the storyteller as a narrative element in online destination marketing. *Journal of Travel Research*. doi:10.1177/00472875211046052

Pan, B., MacLaurin, T., & Crotts, J. C. (2007). Travel Blogs and the Implications for Destination Marketing. *Journal of Travel Research, 46*(August), 35–45. doi:10.1177/0047287507302378

Petty, R., Cacioppo, J., & Schumann, D. (1983). Central and Peripheral Routes to Advertising Effectiveness: The Moderating Role of Involvement. *The Journal of Consumer Research, 10*(Sept), 135–146. doi:10.1086/208954

Riessman, C. (1993). *Narrative Analysis*. Sage Publications, Inc.

Shankar, A., & Goulding, C. (2001). Interpretive Consumer Research: Two More Contributions to Theory and Practice. *Qualitative Market Research, 4*(1), 7–16.

Shankar, A., & Goulding, C. (2001). Interpretive Consumer Research: Two More Contributions to Theory and Ppractice. *Qualitative Market Research, 4*(1), 7–16. doi:10.1108/13522750110364523

Tussyadiah, I., & Fesenmaier, D. R. (2009). Mediating Tourist Experiences: Access to Places via Shared Videos. *Annals of Tourism Research, 36*(1), 24–40. doi:10.1016/j.annals.2008.10.001

Van Laer, T., De Ruyter, K., Visconti, L. M., & Wetzels, M. (2014). The extended transportation-imagery model: A meta-analysis of the antecedents and consequences of consumers' narrative transportation. *The Journal of Consumer Research, 40*(5), 797–817. doi:10.1086/673383

Xiang, Z., & Gretzel, U. (2010). Role of social media in online information search. *Tourism Management, 31*(2), 179–188. doi:10.1016/j.tourman.2009.02.016

Zeng, B., & Gerritsen, R. (2014). What do we know about social media in tourism? A review. *Tourism Management Perspectives, 10*, 27–36. doi:10.1016/j.tmp.2014.01.001

Zhang, Y., Gao, J., Cole, S., & Ricci, P. (2021). How the spread of user-generated contents (UGS) shapes international tourism distribution: Using agent-based modeling to inform strategic UGC marketing. *Journal of Travel Research, 60*(7), 469–491. doi:10.1177/0047287520951639

KEY TERMS AND DEFINITIONS

Destination Brand Engagement: An individual's personal connection to a place as manifested through their cognitive, affective, and behavioral responses to that place as a tourist destination.

Destination Marketing Organizations (DMO): A firm that is tasked with promoting a destination in an effort to increase its attractiveness to potential visitors and/or businesses.

Destination Resilience: The intrinsic ability of a tourist destination to manage and recover from various degrees of external stressors.

Narrative Analysis: A qualitative research approach involving the analysis of data in which significant amounts of narrative are present, including case studies, social media, and direct/indirect observations.

Narrative Transportation: The ability to cause one to temporarily lose awareness of their physical surroundings and emotionally react to a narrative through the creation of vivid mental story imagery.

Social Media: The broad term for those websites and applications where users' content (e.g., photos, text, videos, reviews, etc.) may be created, shared, and discussed.

User-Generated Content (UGC): Any content created by people, rather than by brands.

User-Generated Videos (UGV): Any video-based content created by people, rather than by brands. Video-based content typically include video blog entries and/or video-based social media uploads (e.g., Reels, Stories, etc.).

Visual Storytelling: An approach to storytelling in which the message is conveyed primarily using visual media (i.e., videos and/or photos).

Chapter 13
Social Media as Storytelling Destination Communication Channels:
The Case of Fátima Sanctuary

Mariana Inácio Marques
https://orcid.org/0000-0003-1898-0740
GOVCOPP, Instituto Superior de Gestão, Portugal

João Caldeira Heitor
https://orcid.org/0000-0002-3369-6126
CiTUR, Instituto Superior de Gestão, Portugal

ABSTRACT

This chapter reports a study on a religious tourism destination, Fátima, Portugal, with a focus on the contribution of social media to destination communication. The authors analysed the Facebook pages of four entities related to Fátima tourism to examine the number and type of posts and analyse to which extent these entities are sharing destination information or stories and encouraging visitors to tell their own stories about their visit to Fátima. The authors conducted a quantitative and qualitative analysis to Facebook posts during May, August, and October 2021 when Fátima receives the highest number of pilgrims.

INTRODUCTION

Destination marketing is increasingly difficult in a context of high importance of social media for internet users. People use internet to plan their trips, find out prices as well as to search for information about what they want to do at the destination (Buhalis & Foerste, 2015). However, social media is also used to share stories and experiences, not just to share raw data and information. As personal stories help selling destinations, visitors engage in co-creation (Lund, Cohen & Scarles, 2018). Also, in the digital era,

DOI: 10.4018/978-1-6684-3436-9.ch013

user-generated content (UGC) is becoming popular. This refers to content created by end users, when sharing online content, such as videos, opinions, comments, photos or even a story, thus influencing tourist decision (Moscardo, 2021).

In terms of destination governance, digital media use is seen as a way to encourage travellers to speak about their trips and share their experiences and stories. Local governments stimulate this sharing as it facilitates communication and engagement among stakeholders, involving them in joint place identity creation (Bassano et al., 2019). Creative storytelling brings benefits to tourism pilgrimage routes, mostly because it allows to improve tourists' and locals' awareness regarding pilgrimage routes and their contribution to local development (Soria & Molendowska-Ruiz, 2017).

ICT (Information and Communication Technology) is being used as a channel to promote the religious experience and related activities. Due to digital technologies, temple sounds from all parts of the world can be heard. According to Borrelli (2010), religious discourse is moving from the altar to the mediatic spaces.

Fátima is an international destination recognized as a sacred place immersed in a spiritual atmosphere, located in the Ourém Municipality in the centre region of Portugal. *Fátima* is both a location and a sanctuary which is visited by people from all over the world. In 2017, Pope Francis visited *Fátima* sanctuary for the celebration of the Holy Mary apparition's centenary and the event increased the region's visibility and subsequently tourism statistics escalated. According to the *Fátima Sanctuary Management Board*, in 2017, 374.586 pilgrims from over 100 countries visited the sanctuary. *Fátima* is a destination totally related to pilgrimage. Every year, especially in May, August and October, the region receives thousands of visitors from the whole world regions.

In this chapter, an analysis of *Facebook* pages from entities related to *Fátima* tourism is conducted comprising the months of May, August and October of 2021, aiming to uncover the amount of *Facebook* posts and the type of content published. Authors will look at content shared with a particular focus on stories about the destination and the experience of visiting *Fátima*. Accordingly, the guiding question for this research is: "to what extent is *Facebook* being used as a storytelling tool for developing religious tourism and the religious experience in Fátima"

The chapter is divided in four sections. In the first one, a theoretical discussion about storytelling and religious tourism is presented. In the second section, *Fátima* is addressed and characterized. The third section described the methodology adopted in the study and the last section includes discussion of main results and suggestions for future challenges in research and destination managers.

BACKGROUND

Places are more than geographic locations for visitors, they include landscape, history, culture, heritage, gastronomy, relationships, and other dimensions that elicit story generation. Several techniques are used to tell a story about a place. There is storytelling based on oral tradition, but also storytelling using print, video, graphic and digital media (Smith, 2015). According to Pompílio (2019), trips and stories are related, as trips create stories and conversely these stories can become a motivational source to travel. Storytelling is an important communication tool that captures tourists attention. Being so, it is important to involve the tourist in place-making, by allowing them to feel that they are part of the destination and can contribute to build the story about the place.

Storytelling

Storytelling as the power to wake up the human senses, sharing emotional, spiritual and symbolical values (Frias et al., 2017). Storytelling can be defined as the process of creating a story and it is increasingly important in place and tourism planning. However, stakeholders involved in destination management and planning should be aware of what storytelling is and how stories can be developed that attract visitors, building bridges between them and the region and locals (Hartman et al., 2019).

Stories are important to tourism development because they help to create emotional arguments that appeal to the tourist even more than rational ones. These stories can be planned, provided, and managed before the tourist goes on a trip (pre-experience), during the experience or after the travel experience. Local communities can help building stories, using resources to develop destinations, mostly by using emotional engagement (Moscardo, 2020). Moreover, having tourists emotionally attached to the destination can be accomplished by involving them in creating stories about what they saw and experienced at the destination. The *experience* concept is key to succeed in visitors' emotional attachment to the place. The better the experience, the more likely is the tourist to share stories about the place visited (Su, Cheng, & Swanson, 2020).

In current tourism practice, stories are used to create fictional worlds that resonate in the tourist's imagination through telling a story about the destination that goes beyond images. Moscardo (2021) proposes a model that integrates the local community in the tourism system, thus arguing that this way residents are allowed the opportunity to tell and share their stories.

Lund et al.'s (2018) study on destination branding discusses the role of social media storytelling. The authors recommend DMOs to implement five lines of action using digital technologies in the process of branding the destination:

i. DMOs must break the barriers between the online and the offline worlds. Users want to feel that they have a real impact when they interact online (e.g. using the *Like* button, writing a comment or sharing a post);
ii. DMOs must recognize the impact of collaborative storytelling using social media, being proactive and reactive. Sometimes, they need to interact and create more dialogue with online users;
iii. DMOs must create and maintain an emotional and *persona* connection with users, listening and maintaining dialogue, continually co-creating stories with them;
iv. DMOs must understand social media as allies and invest time on it;
v. DMOs must think as individuals storytellers and design a plan to know how they must perform on social media, to interact with online users.

It is of critical importance that DMOs can recognize the power of the digital in storytelling and how to use it strategically to marketing places. Digital storytelling in the tourism sector is considered to reinforce communication with the markets, to involve stakeholders and improve value co-creation (Bassano et al., 2019). Due to the emotional component, storytelling stimulates tourist motivation and has impact on future visit intention. Hence, it is a differentiating element on destinations' strategic plans (Gonçalves, Seabra & Silva, 2018).

Videos are also used to promote destinations using storytelling and application on social media facilitates a continuous monitoring of visitor co-creation and reaction (Moin, Hosany & O'Brien, 2020).

However, storytelling must adapt to tourists' needs to allow them to co-create and sustain a positive relationship with them (Youssef, Leicht, & Marougiu, 2019).

Storytelling is a way of sharing knowledge and experiences, communicating facts and creating connections (Camprubi and Planas, 2020). But as tourists are bombarded with marketing communication, stories are seen as truthful. Kim and Youn (2017) argue that storytelling influences emotional attachment with the destination, long term memory, and future visit intention. Memories, emotional feelings, and positive behavioural intention increase if tourist are exposed to some kind of audio-visual stimuli.

Social media changed the way tourists see and share information about their travel. They provide an opportunity to personalize communication and directly engage with others (Ferreira, Alves & Quico, 2014). By using social media to share travel-related content, tourists are co-creating and helping to construct the destination brand. In the digital era led by social media communication, it is urgent and mandatory that DMOs monitor and manage digital content, reducing the risk of value co-destruction (Lund, Scarles & Cohen, 2020).

Technology plays an important role in tourism experience, mainly by use of digital and interactive platforms and tools that appeal to tourist participation, making them increasingly involved in the creation of stories about the destination. These digital tools create tourist engagement and simultaneously provide information about local heritage (Dionisio & Nisi, 2021). Social networks now complement traditional destination marketing and *YouTube* is the most influential platform to share video contents (Alegro & Turnsek, 2020). Li and Liu (2020) emphasise the influence of micro-movies on the tourist predisposition to visit places. According to these authors, a micro-movie can use storytelling to promote a touristic product, but also to activate the tourist's brain and be able to elicit emotions, creating empathy about the place. In sum, in order to better meet tourists' needs, online content and social media must become a dimension included in destination strategies (Pachucki, Grohs & Scholl-Grissemann, 2021).

Impact of Social Media on Destination Storytelling

Storytelling stimulates dialogue between audiences online, because they can share stories using the same hashtags, playing an active role in cultural and local promotion. This social media use can be applied to religious tourism (Soria & Molendowska-Ruiz, 2017). DMOs must pay attention to what is happening on social media considering that this interactive space can provide important guidance to their development strategies. Also, destinations may adapt their branding to this digital environment and social media is essential to spread stories about the destination (Lund et al., 2018). Social media platforms allow peer-to-peer communication and tourists provide content about destination, creating several stories about the places they visit. Social media is also important to destination because this social environment creates virtual communities around stories about destination (Tiago, Couto, Faria, & Borges-Tiago, 2018) contributing to destination storytelling and place-making (Conti & Cassel, 2020). Pachucki et al. (2021) argue that when storytelling is a first-person narrative, engagement on social media increases. These narrative serve to share their experiences. However, social media can also be a tool to destruct destination image. Thus, DMOs are urged to be aware of negative content on social media (Lund et al., 2020).

The Use of New Technologies in Destination Storytelling

Storytelling has been used by tourism stakeholders as a way to communicate with tourists and facilitating them memorable experiences, a role played by tourist guides that tell stories to visitors about places and attractions. Currently, new devices are being used based on new technologies, such audio-guides and interactive mobile devices, which allow an interactive storytelling experience (Kontiza et al., 2020).

The use of new technologies has been helping destination storytelling in the context of the Covid-19 pandemic (Karzen & Demonja, 2020), emphasising the critical importance of the digital tools. Other tools, like chatbots, are able to sustain a logical conversation with tourists and helping to satisfy specific needs, and interaction created between the destination and the visitor through chatbots may accommodate the emergence of stories. However, chatbots lack the human dimension (Casillo et al., 2020). Video assembly for online sharing of content are also a destination storytelling tool. There is evidence that advances in multimedia technologies are allowing the increase of online videos. Tourists understand videos as a creative way to promote places, with the potential to become viral (Alegro & Turnsek, 2020).

Religious Tourism and Fátima

All religions communicate through symbols, myths and rites. In this regard, it is important to remember Lindstrom (2013) evoking the role of evangelism in the history of religion. In addition to satisfying basic needs, like food, physiological, emotional, or cognitive, humans need guidance when addressing existential questions and dilemmas (Heitor, 2020). With the "religious motivation as the main cause that leads the individual to leave home, to make trips moved by faith, fed by belief, looking for answers and inner reflections" (Heitor, 2020, p. 27), pilgrimage is considered "a universal phenomenon of all religions", a social phenomenon that expresses "a time of strong spirituality, a passage from the ephemeral to the real, from death to life, from man to divinity" (Dantas, 2007, p. 93). The development of many "tourist destinations around sacred sites and buildings, traversed by pilgrims" (Heitor, 2020, p. 33) is explained by this specific motivation.

Churches, chapels, sanctuaries, monasteries, cemeteries, and other locations dedicated in a special way to prayer are considered sacred places (Wisniewsk, 2018). Different elements can contribute to the recognition of a space as a sacred one. Those elements include: (1) the supernatural character of the space due to an apparition, a miracle, or (2) the historic events that took place there.

Religious tourism in one of the oldest types of tourism. Through the ages, people have visited religious places with different motivations. It has been acknowledged that people may visit religious attractions without the spiritual motivation. However, pilgrimage is a strong driver of religious tourism (Tsai, 2021). Pilgrimage describes "a phenomenon of a spiritual nature, representing an approximation of Man to God, allowing the baptismal renewal of the pilgrim, by exposing his weaknesses, by challenging his physical strengths, justifying the sacrifice", in a process of reconciliation with the divine (Heitor, 2020, p. 28). According to Coleman (2021), pilgrimage as an experience is composed of place, movement, and motivation.

During the pilgrimage journey, the pilgrim uses his senses and records images of the places he has passed, the ceremonies he attended, the images and religious objects he seen. The pilgrim breaks with everyday life, on a journey (s)he materializes towards the sacred place, a sanctuary, aiming to participate in great events, fulfil a promise, deepen the faith or simply to get to know the place. Although the pilgrimage incorporates a religious component to the highest degree, there are people who visit pilgrimage centres to discover the heritage or live the experience of the spirituality of the place, to the detriment of traditional practices (Heitor, 2020). Indeed, pilgrimage can fulfil human needs, especially those related to the quest of finding spirituality, so a journey starts with the motivation of faith and hope (Wisniewsk, 2018).

According to Liu and Minton (2018), religion is a force in society, with religious values and references increasingly incorporated into marketing efforts. Religion is conceptualised as referring to the values and beliefs that guide the sense of purpose and meaning in life, often rooted in the belief in a God or other divine being. Religious tourism occurs through "the existence of sacred sites, sanctuaries, religious centres of worship and spiritual reference with recognized built heritage" (Heitor, 2019, p. 60), which are visited by people with religious and spiritual, cultural and architectural motivations. *Fátima* conveys "a particular religious and spiritual legacy" (Heitor, 2019, p. 60), amplified by the architectural wealth of the *Basílica of Nossa Senhora do Rosário* and the *Basílica of the Santíssima Trindade*, encouraging the performance of tourist activities.

Fátima Sanctuary is the effect to the request of Nossa Senhora do Rosário, assumedly made on August 13, 1917, and expressly communicated in the apparition of October 13 of that year to the peasant children, Lúcia, Francisco and Jacinta: "I want to tell you that you must build a chapel here in my honour, I am the Lady of the Rosary (First Memory of Sister Lúcia)". The chapel was built in 1919 on then same place where the 1917's apparitions where reported, in Cova da Iria, and, since then the Sanctuary area has been developed as a result to the significant influx of pilgrims (*Fátima* Sanctuary, 2022).

There is a "search for new religious identities" (Heitor, 2020, p. 29) that takes place in religious centres and shrines, with people taking part in major manifestations of faith, such as those that take place in *Fátima*, during the *Procession of the Candles*, on May 12 and October 12 of each year, with an intense emotional charge on both the believer and the non-believer. In 2019, the centre region of Portugal registered an increase in tourism statistics and *Turismo Centro de Portugal*, the public office in charge of tourism policies for the region, won some national and international awards. The film "Are you ready?" won nine awards and was placed the 8th most awarded film at the international film festivals that took place in 2019, among more than 3000 tourism films produced. In 2019, at least 6.3 million people visited the Sanctuary. However due to the Covid-19 pandemic, in the first 10 months of 2020, only approximately 1.5 million people visited the Sanctuary, 4.8 million less than in the same period of 2019, a drop of almost 24% in visitor numbers. In 2021, there were still some limitations to travel caused by the pandemic, so the number of pilgrims was still below the usual figures.

Despite the pandemic context, religious tourism is recognized as strategic and a priority to the region's promotion strategy. The Regional Plan for Tourism Development for the Centre Region of Portugal (2020-2030) set five strategic pillars for development, and one of them addresses Spiritual and Religious Tourism. This plan highlights that religious tourism is important to the region and for that reason it is incorporated on the plan's strategy.

Religious Tourism and ICT

ICT is an important tool to contribute to the dissemination of tangible and intangible heritage knowledge and must include local communities and tourists. And currently, social media are important to the religious experience too. Despite this, for many people, religion and ICT are not compatible realities. For some, religious practice must take place in a calm and focused atmosphere induced by prayer, in contrast with ICT devices that merely add "noise" to it. However, nowadays, ICTs help people in religious activities. By using the Internet, people can watch religious services or ceremonies, read holy books and religious texts without leaving home. Furthermore, people can take pictures during activities or while inside sacred buildings or places, and share them on social media with as personal faith testimonial (De Ascaniis *et al*, 2018).

Ramos and Rodrigues (2017) presented the Religious Tourism Experience Model (RTEM) representing the technological solutions that can be incorporated into religious tourism. The digital and technological solutions proposed allow an aggregation of experiences (religious, cultural and touristic), leading to the enhancement of heritage, helping defining and creating a cultural identity for the region. ICT tools considered in this model are mobile devices, augmented reality, gamification, social networks and business intelligence.

According to Zlatanov and Popesku (2021), religious tourism continues to be seen as a traditional experience. However, in the digital era, the promotion of religious places cannot be done in a traditional way. According to Wisniewsk (2018), *Fátima* Sanctuary *Facebook* account contains information that includes reflections about God's word, information about religious services and activities, and communicates pilgrims' testimonies, providing evidence of the combination of religious tourism practice, management and ICT use.

METHODOLOGY

This study adopted a mixed-methodology approach, using both quantitative and qualitative analysis. Authors analysed *Facebook* posts from organisations related to *Fátima* tourism during the months of May, August and October of 2021 and examined posts statistics and content type published. This research method was used previously by Tiago et al. (2018), which studied social media metrics, such as fan numbers, likes, posts, shares and content posts, of cruise companies *Facebook* pages.

The tourism-related organisations included in this study were: ACISO (Ourém-*Fátima* Business Association) CIM Médio Tejo (Intermunicipal Tourism Community from Media Tejo) *Fátima* Sanctuary, and Ourém Municipality. All these organisations maintain a *Facebook* page with updated content about the destination and are described below:

(1) ACISO: Ourém *Fátima* Business Association represents individuals or businesses who carry out or represent any business activity in the Ourém municipality. Their *Facebook* page was created on October 14 2011 and is managed by a Direction board. The focus of the page is not *Fátima* Sanctuary, however it disseminates also religious events. During data collection, the ACISO *Facebook* page registered 4278 likes and 4658 followers.

(2) CIM Médio Tejo (Intermunicipal Tourism Community from Media Tejo): this organisation aims to promote the planning and management of the economic, social and environmental development strategy for the territory covered by the municipalities of Abrantes, Alcanena, Constância, Entroncamento, Ferreira do Zêzere, Mação, Ourém, Sardoal, Sertã, Tomar, Torres Novas, Vila de Rei and Vila Nova da Barquinha. The work developed by CIM supports the enhancement of regional identity, by combining and optimizing public services, stimulating the cooperation of local and regional actors, and encouraging business competitiveness. Their *Facebook* page was created on February 10 2014 and is managed by the Support Unit for the Intermunicipal Executive Secretariat. During data collection, the page registered 5732 likes and 6160 followers. Again, CIM's *Facebook* page doesn't concentrate on *Fátima* Sanctuary, providing general information on topics of interest to the local community, and posts related to the location of *Fátima* and religious events are not prominent.

(3) Ourém Municipality: this is the administrative body that carries out activities through the exercise of legally defined competences in the areas of consultation, planning, investment, management, licensing, administrative control, and inspection. The organisation is responsible for managing and promoting the integrated and sustainable development of the territory and community, improving the quality of life and well-being of residents, boosting the local economy, competitiveness and investment, protecting culture and local. The *Facebook* page was created on November 4 2014 and is managed by an internal communication office. During data collection, this *Facebook* page counted 26271 likes and 28422 followers.

(4) Sanctuary of Our Lady of the Rosary of *Fátima*, best known as *Fátima* Sanctuary, is a catholic structure that manages, studies and disseminates the message of the *Fátima* religious events, originating from the Our Lady Apparitions, as well as the spaces and structures surrounding it. The *Facebook* page was created on May 28 2010 and is managed by an internal Communication Office. The core of the Sanctuary's *Facebook* page is to promote the religious events at the site. During data collection the Sanctuary *Facebook* page registered 1.107154 likes and 1.310065 followers.

In this study, authors analysed the number of likes and followers, posts, shares, and comments. Additionally, they examined what kind of content organisations communicate about *Fátima*'s history or experiences. Analysis included posts containing the words "Fátima" or "Fátima Sanctuary", "pilgrim" or "pilgrimage".

The specific months of May, August and October were chosen because they correspond to the dates of Virgin Mary's apparitions, and thus the times of the year that *Fátima* receives the highest numbers of visitors.

RESULTS

Table 1 shows data on likes and followers from the analysed organisations' *Facebook* pages. *Fátima* Sanctuary's *Facebook* page is, by far, the most popular with the highest number of 'likes' and followers.

Table 1. Organisations' likes and followers on Facebook pages

Entity	Likes	Followers
ACISO	4278	4658
CIM Média Tejo	5732	6160
Ourém Municipality	26271	28422
Fatima Sanctuary	1.107154	1.310065

(Source: Authors' own elaboration)

During the analysed months, taking into account the four *Facebook* pages, the total amount of posts was 817, however only 138 were related to *Fátima*. Table 2 presents global posts count. Results show the *Facebook* page with more posts on *Fátima* was the Sanctuary (102). Ourém Municipality's *Facebook* page ranks second (22), then ACISO (10) and the last position is occupied by CIM Media Tejo (4).

Table 2. Global posts perspective

ACISO												Total Posts on the considered months	Total Posts Related to the theme
May 2021			August 2021			October 2021			Related	Not related	Total		
Related	Not related	Total	Related	Not related	Total	Related	Not related	Total					
4	9	13	1	5	6	5	14	19	**10**	**28**	**38**		
CIM Media Tejo													
May 2021			August 2021			October 2021			Related	Not related	Total		
Related	Not related	Total	Related	Not related	Total	Related	Not related	Total					
2	78	80	0	53	53	2	50	52	**4**	**181**	**185**		
Fátima Sanctuary													
May 2021			August 2021			October 2021			Related	Not related	Total		
Related	Not related	Total	Related	Not related	Total	Related	Not related	Total					
32	0	32	26	0	26	44	2	46	**102**	**2**	**104**		
Ourém Municipality													
May 2021			August 2021			October 2021			Related	Not related	Total		
Related	Not related	Total	Related	Not related	Total	Related	Not related	Total					
11	106	117	2	181	183	9	181	190	**22**	**468**	**490**	**817**	**138**

(Source: Authors' own elaboration)

The authors also analysed the type of publication, only on related posts and the results could be seen on Table 3.

Table 3 presents the type of publications on *Facebook* pages: posts with text and image are the most common publication (65,2%). Video content (and links to websites occupy the second and the third positions with, respectively, 20.3% and 14,5% of total publications.

Table 3. Types of publication with posts related to Fátima location and religious events

Entities	October 2021				August 2021				May 2021				Global Count				Total
	Video	Just Text	Just link for website	Text and image	Video	Just Text	Just link for website	Text and image	Video	Just Text	Just link for website	Text and image	Video	Just Text	Just link for website	Text and image	
ACISO	0	0	0	5	0	0	0	1	0	0	0	4					
CIM Media Tejo	1	0	0	1	0	0	0	0	1	0	0	1					
Fátima Sanctuary	10	0	8	26	3	0	4	19	10	0	5	17	28	0	20	90	138
Ourém Municipality	1	0	2	6	0	0	0	2	2	0	1	8					
Total per month	**12**	**0**	**10**	**38**	**3**	**0**	**4**	**22**	**13**	**0**	**6**	**30**					

(Source: Authors' own elaboration)

According to Moin et al. (2020), Kim and Youn (2017), Lund et al., (2020), Alegro and Turnsek (2020), and Li and Liu (2020), videos are important to storytelling helping in the creation of the destination brand, and sustaining the relationship between the visitor and the place. Table 4 synthesises data on video posts linked to *Fátima*:

Table 4. Video posts results

	Total	Reactions	Facebook Sharing	Comments
Posts with videos	28	145290	33550	18681

(Source: Authors' own elaboration)

As seen on Table 4, 28 videos elicited a considerable number of reactions. Crossing data from Tables 4 and 5 (total reactions, comments and *Facebook* sharing of posts on *Fátima* Sanctuary), reactions to videos amount to 145 290, representing 26,6% of total number. In turn, video comments are 39,3% from the total and sharing of *Facebook* posts 44,3%.

Table 5. Engagement analysis

Reaction								Total of reactions	Comments	Facebook sharing
ACISO	100	3	0	0	0	0	0	103	1	7
CIM Media Tejo	12	1	0	0	0	0	0	13	0	1
Fátima Sanctuary	400690	139977	3889	46	145	170	52	544969	47451	75466
Ourém Municipality	2049	86	1	2	2	1	0	2141	102	282
Total	**402851**	**140067**	**3890**	**48**	**147**	**171**	**52**	**547226**	**47554**	**75756**

(Source: Authors' own elaboration)

During the period under analysis, post count showed a total of 547226 reactions, 47554 comments and 75756 sharing. From Table 5 it is observed that *Fátima* Sanctuary *Facebook* page is, by far, the one eliciting more reactions, comments, and shares.

Calculation of post engagement rate was considered important to understand interaction with each analysed organisation. According to Social Bakers (2011), to assess the brand post engagement rate on *Facebook,* the formula *(Likes + Comments + Shares) / followers x 100* is used. Table 6 presents results based on this formula:

Table 6. Engagement rate assessment

Entity	Formula	Engagement Rate
ACISO	(103 +1 + 7) / 4658 x 100	2,4%
CIM Media Tejo	(13 + 0 + 1) / 6160 x 100	0,2%
Fatima Sanctuary	(544969+47451+75466) / 1 310065 x 100	51%
Ourém Municipality	(2141 + 102+ 282) / 28422 x 100	8,9%

(Source: Authors' own elaboration)

On *Facebook*, an engagement rate superior to 1% is considered a good one and bellow 0,5%, engagement is considered negative. According to Table 6 results, CIM Media Tejo is the only organisation showing negative engagement. *Fátima* Sanctuary *Facebook* page stands out with the highest engagement rate (51%).

A content analysis conducted to the first 50 comments posted on *Facebook* pages aimed at exploring the type of comment and feed. The Bright Local study (2015) indicates that 73% of online consumers need to read just 1 to 6 online comments to make their own opinion about the theme.

Results from the analysis provide evidence that no interaction occurred between organisations and users, and comments amount to weak positive responses to posted content (congratulations and applause). One particular publication on a film preview named "Fátima" posted by Ourém Municipality elicited negative reactions from *Facebook* users. Despite this, no official response from the municipality was observed. Comments on *Fátima* Sanctuary hint at the religious worldview, with users using religious words (e.g. "Amen" or "praying for help to Our Lady") as a proxy for prayer.

SOLUTIONS AND RECOMMENDATIONS

From this study, authors recommend that the analysed organisations should take action to ensure a better management of digital content and visitor interactions on social networks they manage. Content type is mostly informative and in most cases concern with creating stories or managing relationships with users that comment the posts is low. *Fátima* Sanctuary *Facebook* page is a standalone example among the region's organisations, making an effort to create stories and appeal to interaction with visitors. Storytelling about *Fátima* location and Sanctuary should be emphasised through greater use of digital tools, e.g. interactive videos stimulating users' participation.

FUTURE RESEARCH DIRECTIONS AND LIMITATIONS

Research on destination storytelling could develop into complementary studies focusing on other social media platforms, such as *YouTube* and *Instagram*. Also, a mixed-methods approach, by combining qualitative and quantitative data, could uncover different conclusions from the ones achieved in this research.

CONCLUSION

Fátima is a place impregnated with an atmosphere of mystery and the supernatural, which is visited by millions of people every year. Religious tourism is one of the oldest types of tourism and involves not only religion, but also spirituality, culture, and architecture, among other endogenous factors that stimulate individuals to visit that sacred site. Concerning the religious experience, Shinde (2021) emphasised multidimensionality as a result of the combination of the touristic dimension (related to heritage, recreation and leisure), the spiritual dimension (related to philosophy, mediation and engagement), and the religious dimension (related to traditions and rites). Management of the religious experience in *Fátima* should present evidence of consideration of these three dimensions.

ICTs are increasingly important tools in the promotion of religious events, sites and experiences. *Fátima* was studied as a specific case of religious tourism, but research on the relation between religious sites, their management, the visitor experience, and ICTs is still scarce. Future studies are welcome that explore the advantages of satisfying religious tourists using digital tools. Social media can make a difference in the building of an emotional bound with tourists, and also in what concerns understanding of their motivations. Religious sites must make an effort to manage social media platforms, in search of better interaction with visitors by promoting a closer relationship with them (Caber et al., 2021).

In this study, *Facebook* pages from four organisations related to *Fátima* were analysed. The aim was to examine to what extent this social network is being used to share stories about the site and the religious events and managing interaction with visitors through ICT. Findings highlight the distinctive relevance of *Fátima* Sanctuary, which achieved a high number of reactions, comments and content sharing. Other organisations' *Facebook* pages don't contribute to enhance the destination connection to the religious themes, site, activities, and experience. In general, posts perform poorly in terms of reactions, comments or sharing. It can be concluded that up to the present there has not been enough effort to work storytelling associated to the religious experience in *Fátima* site using social media and ICT. Moreover, video posts are considered richer than text-only posts (Sabate, Berbegal-Mirabent, Cañabate, & Lebherz, 2014). Accordingly, as this study has shown, there is an opportunity to explore and create content which engages users at a deeper level by combining different sensory stimuli.

Concerning this study's research question "to what extent is *Facebook* being used as a storytelling tool for developing religious tourism and the religious experience in *Fátima*?", the answer is *no*. Metrics from the analysed organisations demonstrate that they aren't making use of *Facebook* to communicate stories about the destination or to encourage visitors to create their own stories about the *Fátima* experience.

REFERENCES

Alegro, T., & Turnsek, M. (2020). Striving to Be Different but Becoming the Same: Creativity and Destination Brands' Promotional Videos. *Sustainability*, *13*(1), 139. doi:10.3390u13010139

Bassano, C., Barile, S., Picciochi, P., Spohrer, J., Iandolo, F., & Fisk, R. (2019). Storytelling about places: Tourism marketing in the digital age. *Cities (London, England)*, *87*, 10–20. doi:10.1016/j.cities.2018.12.025

BrightLocal. (2015). *Local Consumer Review Survey*. https://searchengineland.com/87-percent-customers-wont-consider-low-ratings-228607

Buhalis, D., & Foerste, M. (2015). SoCoMo marketing for travel and tourism: Empowering co-creation of value. *Journal of Destination Marketing & Management*, *4*(3), 151–161. doi:10.1016/j.jdmm.2015.04.001

Caber, M., Drori, N., Albayrak, T., & Herstein, R. (2021). *Social media usage behaviors of religious tourists: The cases of the Vatican, Mecca, and Jerusalem*. John Wiley & Sons Ltd. DOI: doi:10.1002/jtr.2444816

Camprubí, R. & Planas, C. (2020). El Storytelling en la marca de destinos turísticos: el caso de Girona. *Cuadernos de Turismo*, *46*, 269-289. doi:10.6018/turismo.451841

Casillo, M., Clarizia, F., D'Aniello, G., De Santo, M., Lombardi, M., & Santaniello, D. (2020). CHAT-Bot: A cultural heritage aware teller-bot for supporting touristic experiences. *Pattern Recognition Letters*, *131*, 234–243. doi:10.1016/j.patrec.2020.01.003

Coleman, S. (2021). Pilgrimage. In R. A. Segal & N. P. Roubekas (Eds.), *The Wiley Blackwell Companion to the Study of Religion*. John Wiley & Sons. doi:10.1002/9781119092797.ch26

Conti, E., & Cassel, S. (2020). Liminality in nature-based tourism experiences as mediated through social media. *Tourism Geographies*, *22*(2), 413–432. doi:10.1080/14616688.2019.1648544

Dantas, V. (2007). O futuro dos santuários no contexto da mobilidade humana. In Santuário de Fátima (Ed.), Atas das Jornadas Aljustrel e Valinhos – o outro pulmão de Fátima (pp. 83-100). Fátima: Santuário de Fátima.

De Ascaniis, S., Mutangala, M., & Cantoni, L. (2018). ICTs in the tourism experience at religious heritage sites: A review of the literature and an investigation of pilgrims' experiences at the sanctuary of Loreto (Italy). *Church. Communication and Culture*, *3*(3), 310–334. doi:10.1080/23753234.2018.1544835

Dionisio, M., & Nini, V. (2021). Leveraging Transmedia storytelling to engage tourists in the understanding of the destination's local heritage. *Multimedia Tools and Applications*, *80*(26-27), 34813–34841. Advance online publication. doi:10.100711042-021-10949-2

Ferreira, S., Alves, A., & Quico, C. (2014). The role of social media: The TravelPlot Porto case study. *Revista Turismo & Desenvolvimento*, (21/22), 151–160. doi:10.34624/rtd.v2i21/22.11407

Frias, C., Silva, C., & Seabra, C. (2017). Storytelling: Motivações, autenticidade e envolvimento dos Turistas Culturais. *Revista Turismo & Desenvolvimento*, *27/28*, 1355–1366. doi:10.34624/rtd.v1i27/28.9977

Gonçalves, J., Seabra, C., & Silva, C. (2018). Histórias de cultura. O poder do Storytelling em destinos de Turismo Cultural. *Caderno de Geografia*, (37), 113–120. doi:10.14195/0871-1623_37_9

Hartman, S., Parra, C., & Roo, G. (2019). Framing strategic storytelling in the context of transition management to stimulate tourism destination development. *Tourism Management*, *75*, 90–98. doi:10.1016/j.tourman.2019.04.014

Heitor, J. (2020). *A Imagem Peregrina de Nossa Senhora de Fátima*. Cordel D'Prata.

Heitor, J. C. (2019). *A imagem peregrina de Nossa Senhora de Fátima. Uma marca na promoção do turismo religioso de Fátima?* Tese doutoramento em Turismo. Universidade de Lisboa, Instituto de Geografia e Ordenamento do Território. http://hdl.handle.net/10451/39519

Karzen, M. & Demonja, D. (2020). Importance of Storytelling: how to create more resilient cultural heritage? *Nova prisutnost, 18*(3), 653-668. doi:10.31192/np.18.3.15

Kim, J., & Youn, H. (2017). How to Design and Deliver Stories about Tourism Destinations. *Journal of Travel Research*, *56*(6), 808–820. doi:10.1177/0047287516666720

Kontiza, K., Antoniou, A., Daif, A., Reboreda-Morillo, S., Bassani, M., González-Soutelo, S., Lykourentzou, I., Jones, C., Padfield, J., & López-Nores, M. (2020). On How Technology-Powered Storytelling Can Contribute to Cultural Heritage Sustainability across Multiple Venues—Evidence from the CrossCult H2020 Project. *Sustainability*, *12*(4), 1666. doi:10.3390u12041666

Li, C.-H., & Liu, C.-C. (2020). The effects of empathy and persuasion of storytelling via tourism micro-movies on travel willingness. *Asia Pacific Journal of Tourism Research*, *25*(4), 382–392. doi:10.1080/10941665.2020.1712443

Lindstrom, M. (2013). *Brand Sense. Os Segredos Sensoriais que nos levam a comprar*. Gestãoplus Edições.

Liu, L., & Minton, A. (2018). Faith-filled brands: The interplay of religious branding and brand engagement in the self-concept. *Journal of Retailing and Consumer Services*, *44*, 305–314. doi:10.1016/j.jretconser.2018.07.022

Lund, N., Cohen, S., & Scarles, C. (2018). The power of social media storytelling in destination branding. *Journal of Destination Marketing & Management*, *8*, 271–280. doi:10.1016/j.jdmm.2017.05.003

Lund, N., Scarles, C., & Cohen, S. (2020). The Brand Value Continuum: Countering Co-destruction of Destination Branding in Social Media through Storytelling. *Journal of Travel Research*, *59*(8), 1506–1521. doi:10.1177/0047287519887234

Moin, S., Honasy, S., & O'Brien, J. (2020). Storytelling in destination brands' promotional videos. *Tourism Management Perspectives*, *34*, 100639. doi:10.1016/j.tmp.2020.100639

Moscardo, G. (2020). Stories and design in tourism. *Annals of Tourism Research*, *83*, 102950. doi:10.1016/j.annals.2020.102950

Moscardo, G. (2021). The story turn in tourism: Forces and futures. *Journal of Tourism Futures*, *7*(2), 168-173.

Pachucki, C., Grohs, R., & Scholl-Grissemann, U. (2021). No Story Without a Storyteller: The Impact of the Storyteller as a Narrative Element in Online Destination Marketing. *Journal of Travel Research*, 1–16. doi:10.1177/00472875211046052

Pompílio, M. (2019). *O storytelling como recurso na promoção do turismo em Fortaleza*. Master Dissertation, Universidade Estadual do Ceará.

Ramos, C. & Rodrigues, J. (2017). Turismo e Tecnologia: Qualificando a Experiência do Turista Religioso. *Revista Rosa dos Ventos – Turismo e Hospitalidade*, *9*(1), 32-48. doi:10.18226/21789061.v9i1p032

Sabate, F., Berbegal-Mirabent, J., Cañabate, A., & Lebherz, P. (2014). Factors influencing popularity of branded content in Facebook fan pages. *European Management Journal*, *32*(6), 1001–1011. doi:10.1016/j.emj.2014.05.001

Sanctuary, F. (2022). *Identidade e Missão do Santuário de Fátima*. https://www.fatima.pt/pt/pages/identidade-e-missao

Shinde, K. (2021). Sacred Sites, Rituals, and Performances in the Ecosystem of Religious Tourism. *Religions*, *12*(7), 523. doi:10.3390/rel12070523

Smith, S. (2015). A sense of place: Place, culture and tourism. *Tourism Recreation Research*, *40*(2), 220–233. doi:10.1080/02508281.2015.1049814

Soria, A., & Molendowska-Ruiz, J. (2017). Creative Storytelling for European Pilgrimage route's audience development. In Book of Abstracts, Cultural Management and Governance for European Pilgrimage routes and religious tourism. Academic Press.

Su, L., Cheng, J., & Swanson, S. (2020). The impact of tourism activity type on emotion and storytelling: The moderating roles of travel companion presence and relative ability. *Tourism Management*, *81*, 104138. doi:10.1016/j.tourman.2020.104138

Tiago, F., Couto, J., Faria, S., & Borges-Tiago, S. (2018). Cruise tourism: Social media content and network structures. *Tourism Review*, *73*(4), 433–447. doi:10.1108/TR-10-2017-0155

Tsai, H. (2021). Exploring the motivation-based typology of religious tourists: A study of Welcome Royal Lord Festival. *Journal of Destination Marketing & Management*, *21*, 100623. doi:10.1016/j.jdmm.2021.100623

Wisniewsk, L. (2018). Religious tourism in Christian sanctuaries - the implications of mixed interests for the communication of the faith. *Church, Communication and Culture*, *3*(3), 199–220. doi:10.1080/23753234.2018.1537674

Youssef, K., Leicht, T., & Marongiu, L. (2019). Storytelling in the context of destination marketing: An analysis of conceptualisations and impact measurement. *Journal of Strategic Marketing*, *27*(8), 696–713. doi:10.1080/0965254X.2018.1464498

Zlatanov, S., & Popesku, J. (2021). Promotion of Religious Tourism Sites on Facebook by NTOs. *International Journal of Religious Tourism and Pilgrimage*, *9*(3), 70–81. doi:10.21427/5gct-vb53

ADDITIONAL READING

Hay, N., Chien, P., & Ruhanen, L. (2022). Tell me your story: Branding destinations through resident's (place) stories. *Journal of Vacation Marketing*, *28*(3), 319–334. doi:10.1177/13567667211060567

Lovell, J., & Hitchmought, S. (2020). Stimulated authenticity: Storytelling and mythic space on the hyper-frontier in Buffalo Bill's Wild West and Westworld. *Tourist Studies*, *20*(4), 409–428. doi:10.1177/1468797620937912

Olsson, A., Therkelsen, A., & Mossberg, L. (2016). Making an effort for free-volunteer's roles in destination-based storytelling. *Current Issues in Tourism*, *19*(7), 659–679. doi:10.1080/13683500.2013.784242

So, K., King, C., Sparks, B., & Wang, Y. (2016). The role of customer engagement in building consumer loyalty to tourism brands. *Journal of Travel Research*, *55*(1), 64–78. doi:10.1177/0047287514541008

KEY TERMS AND DEFINITIONS

Co-Creation: Product or service design process in which there are several inputs from all stakeholders.

Destination Management Organization (DMO): Coordinated and thoughtful planning of all elements that make up a tourism destination.

Information and Communication Technology (ICT): Includes any technological communication device.

Social Media (SM): All the interactive technologies used to promote people and business.

User-Generated Content (UGC): All types of content, such as images, videos, text, and audio, that has been posted by users on any online platforms.

Compilation of References

Aaker, J. L. (1997). Dimensions of brand personality. *JMR, Journal of Marketing Research, 34*(3), 347–356. doi:10.1177/002224379703400304

Ababneh A. (2018) Tour guides and heritage interpretation: guides' interpretation of the past at the archaeological site of Jarash, Jordan. *Journal of Heritage Tourism, 13*(3), 257-272. doi:10.1080/1743873X.2017.1321003

Abbott, H. P. (2012). *The Cambridge introduction to narrative.* Cambridge University Press.

Adaval, R., & Wyer, R. S. Jr. (1998). The Role of narratives in consumer information processing. *Journal of Consumer Psychology, 7*(3), 207–245. doi:10.120715327663jcp0703_01

Agostiano, M. (2016). Enjoyment of Cultural Heritage and "Compensatory Solutions" for an Alternative Accessibility. In Accessibility as a key enabling knowledge for enhancement of cultural heritage (pp. 131-147). Academic Press.

Aikawa, N. (2004). An historical overview of the preparation of the UNESCO international convention for the safeguarding of intangible heritage. *Museum International, 56*(1-2), 137–149. doi:10.1111/j.1350-0775.2004.00468.x

Ajzen, I. (1987). Attitudes, traits, and actions: Dispositional prediction of behavior in personality and social psychology. In L. Berkowitz (Ed.), Advances in Experimental Social Psychology (Vol. 20, pp. 1–63). Academic Press. doi:10.1016/S0065-2601(08)60411-6

Ajzen, I. (1991). The theory of planned behavior. *Organizational Behavior and Human Decision Processes, 50*(2), 179–221. doi:10.1016/0749-5978(91)90020-T

Ajzen, I. (2001). Nature and operation of attitudes. *Annual Review of Psychology, 52*(1), 27–58. doi:10.1146/annurev.psych.52.1.27 PMID:11148298

Ajzen, I., & Fishbein, M. (1977). Attitude-behavior relations: A theoretical analysis and review of empirical research. *Psychological Bulletin, 84*(5), 888–918. doi:10.1037/0033-2909.84.5.888

Ajzen, I., & Fishbein, M. (2000). Attitudes and the attitude-behavior relation: Reasoned and automatic processes. *European Review of Social Psychology, 11*(1), 1–33. doi:10.1080/14792779943000116

Akgün, A. E., Keskin, H., Ayar, H., & Erdoğan, E. (2015). The influence of storytelling approach in travel writings on readers' empathy and travel intentions. *Procedia: Social and Behavioral Sciences, 207*, 577–586. doi:10.1016/j.sbspro.2015.10.129

Akurgal, E. (2020). *Anadolu Uygarlıkları.* Phoenix Yayınevi.

Alberts, H. C., & Hazen, H. D. (2010). Maintaining authenticity and integrity at cultural world heritage sites. *Geographical Review, 100*(1), 56–73. doi:10.1111/j.1931-0846.2010.00006.x

Alegro, T., & Turnsek, M. (2020). Striving to Be Different but Becoming the Same: Creativity and Destination Brands' Promotional Videos. *Sustainability*, *13*(1), 139. doi:10.3390u13010139

Allal-Chérif, O. (2022). Intelligent cathedrals: Using augmented reality, virtual reality, and artificial intelligence to provide an intense cultural, historical, and religious visitor experience. *Technological Forecasting and Social Change*, *178*, 121604. doi:10.1016/j.techfore.2022.121604

Almeida, C. da C. (2004). O marketing das cidades. *Gestao e Desenvolvimento*, *12*(12), 9–45. doi:10.7559/gestaoedesenvolvimento.2004.104

Almeida, S. (2022). Stories. In *Encyclopedia of tourism management and marketing*. Edward Elgar Publishing. doi:10.4337/9781800377486.stories

Alvarez, M. D., Go, F. M., & Yüksel, A. (Eds.). (2016). *Heritage tourism destinations: Preservation, communication and development*. CABI. doi:10.1079/9781780646770.0000

American Marketing Association. (n.d.). *Definition of Brand*. Available at: https://www.ama.org/the-definition-of-marketing-what-is-marketing/

Ammarell, G. (1999). *Bugis navigation*. Yale University Southeast Asia Studies.

Anholt, S. (2004). Editor's foreword to the first issue. *Place Branding and Public Diplomacy*, *1*(1), 4–11. doi:10.1057/palgrave.pb.5990001

Anholt, S. (2006). The Anholt-GMI City Brands Index. How the world sees the world's cities. *Place Branding*, *2*(1), 18–31. doi:10.1057/palgrave.pb.5990042

Anholt, S. (2007). *Competitive Identity: Brand Management for Nations, Cities and Regions*. Palgrave Macmillan. doi:10.1057/9780230627727

Aragonez, T., & Caetano Alves, G. (2012). Marketing territorial: o futuro das cidades sustentáveis e de sucesso sucesso. *Tourism and Management Studies International Conference*, *1*, 316–319.

Araújo, M. J. (2021). "Portugal at Table" - Food Heritage, Identity and Tradition: The tourist gaze. *Journal of Tourism & Development*, *35*, 243–257. https://dialnet.unirioja.es/servlet/articulo?codigo=7891681

Arenghi, A., & Agostiano, M. (2016). Cultural Heritage and Disability: Can ICT Be the 'Missing Piece' to Face Cultural Heritage Accessibility Problems? In Smart Objects and Technologies for Social Good. Lecture Notes of the Institute for Computer Sciences, Social Informatics and Telecommunications Engineering (pp. 70-77). Springer.

Argyriou, L., Economou, D., & Bouki, V. (2020). Design methodology for 360° immersive video applications: The case study of a cultural heritage virtual tour. *Personal and Ubiquitous Computing*, *24*(6), 1–17. doi:10.100700779-020-01373-8

Aristóteles. (2018). *Poética* (A. M. Valente, Trans.). Fundação Calouste Gulbenkian.

Arnold, D. (1992). *The handbook of brand management*. Century Business: The Economist Books.

Arslan, F. (2009). *Legends of Hagia Sophia (determine-examine)* [Unpublished doctoral dissertation]. İstanbul University, İstanbul, Turkey.

Ashworth, G. J. (2014). Heritage and economic development: Selling the unsellable. *Heritage & Society*, *7*(1), 3–17. doi:10.1179/2159032X14Z.00000000015

Ashworth, G. J., Graham, B. J., & Tunbridge, J. E. (2007). *Pluralising pasts: Heritage, identity and place in multicultural societies*. Pluto Press.

Åstrøm, J. K. (2019). Why theming? Identifying the purposes of theming in tourism. *Journal of Quality Assurance in Hospitality & Tourism*, 1–22. doi:10.1080/1528008X.2019.1658147

Aureli, S., & Del Baldo, M. (2022). Stakeholders' consciousness of cultural heritage and the reconciliation of different needs for sustainable development. *Journal of Cultural Heritage Management and Sustainable Development*. Advance online publication. doi:10.1108/JCHMSD-10-2020-0156

Aveiro, V. (2009). *Histórias Desconhecidas dos Grandes Trabalhadores do Mar*. Editorial Futura.

Aveiro, V. (2014). *Ecos do Grande Norte*. Âncora Editora.

Aveiro, V. (2016). *Nómadas do Oceano*. Âncora Editora.

Aylett, R., & Louchart, S. (2003). Towards a Narrative Theory of Virtual Reality. *Virtual Reality (Waltham Cross)*, *7*(1), 2–9. doi:10.100710055-003-0114-9

Bakas, F. E., Duxbury, N., Silva, S., & Castro, T. G. (2020). Connecting to place through creative tourism. In S. M. Cabeça, A. R.Gonçalves, J. F.Marques, M. Tavares (Eds.), Creative tourism dynamics: connecting travellers, communities, cultures and places (pp. 119-139). Grácio Editor.

Baker, C. (2016). Creative choices and fan practices in the transformation of theme park space. *Transformative Works and Cultures, 22*.

Baker, C. (2018b). The use of themed entertainment design in museums and heritage sites. *The Museum Review*, *3*(1).

Baker, C. (2018a). Summary of exploring a three-dimensional narrative medium: The theme park as "de Sprookjessprokkelaar," the gatherer and teller of stories. *Journal of Themed Experience and Attractions Studies*, *1*, 1–9.

Baker, C. (2018c). Universal's Wizarding World of Harry Potter: A primer in contemporary media concepts. In A. Firestone & L. A. Clark (Eds.), *Harry Potter and convergence culture: Essays on fandom and the expanding Potterverse* (pp. 55–66). McFarland.

Baker, C. (2022). Magic keys and magic mushrooms: Storylands and early dramatic play. In K. M. Jackson & M. I. West (Eds.), *Storybook worlds made real: Essays on the places inspired by children's narratives* (pp. 125–146). McFarland.

Baker, S., Istvandity, L., & Nowak, R. (2016). Curating popular music heritage: Storytelling and narrative engagement in popular music museums and exhibitions. *Museum Management and Curatorship*, *31*(4), 369–385. doi:10.1080/09647775.2016.1165141

Balbi, B., & Marasco, A. (2021). Co-designing Cultural Heritage Experiences for All with Virtual Reality: A Scenario-Based Design approach. *Umanistica Digitale*, (11), 71–98.

Ball, R. (2019, September-October). Economic development: It's about placemaking. *The European Business Review*. https://www.europeanbusinessreview.com/economic-development-its-about-placemaking/

Barrett, J., & Cocq, C. (2019). Indigenous storytelling and language learning: digital media as a vehicle for cultural transmission and language acquisition. In *Perspectives on Indigenous writing and literacies* (pp. 89–112). Brill.

Barron, P., & Leask, A. (2017). Visitor engagement at museums: Generation Y and 'Lates' events at the National Museum of Scotland. *Museum Management and Curatorship*, *32*(5), 473–490. doi:10.1080/09647775.2017.1367259

Barton, R., & Barton, G. (2014). Storytelling as an arts literacy: Use of narrative structure in Aboriginal arts practice and performance. In *Literacy in the Arts* (pp. 251–268). Springer. doi:10.1007/978-3-319-04846-8_15

Baskin, K. (2004). Complexity, stories, and knowing. *International Workshop on Complexity and Philosophy*.

Bassano, C., Barile, S., Piciocchi, P., Spohrer, J. C., Iandolo, F., & Fisk, R. (2019). Storytelling about places: Tourism marketing in the digital age. *Cities (London, England)*, *87*, 10–20. doi:10.1016/j.cities.2018.12.025

Bauman, Z. (2000). *Liquid modernity*. Polity Press.

Bec, A., Moyle, B., Schaffer, V., & Timms, K. (2021). Virtual reality and mixed reality for second chance tourism. *Tourism Management*, *83*, 104256. doi:10.1016/j.tourman.2020.104256

Bec, A., Moyle, B., Timms, K., Schaffer, V., Skavronskaya, L., & Little, C. (2019). Management of immersive heritage tourism experiencs: A conceptual model. *Tourism Management*, *72*, 117–120. doi:10.1016/j.tourman.2018.10.033

Beckman, F. (2015). Ambivalent Screens: Quentin Tarantino and the Power of Vision. *Film Philosophy*, *19*(1), 85–104. doi:10.3366/film.2015.0006

Bedford, L. (2001). Storytelling: The Real Work of Museums. *Curator (New York, N.Y.)*, *44*(1), 27–34. doi:10.1111/j.2151-6952.2001.tb00027.x

Bekele, M. K., & Champion, E. (2019). A comparison of immersive realities and interaction methods: Cultural learning in virtual heritage. *Frontiers in Robotics and AI*, *6*, 91. doi:10.3389/frobt.2019.00091 PMID:33501106

Bekele, M. K., Pierdicca, R., Frontoni, E., Malinverni, E. S., & Gain, J. (2018). A survey of augmented, virtual, and mixed reality for cultural heritage. *Journal on Computing and Cultural Heritage*, *11*(2), 1–36. doi:10.1145/3145534

Belk, R. W. (1975). Situational variables and consumer behaviour. *The Journal of Consumer Research*, *2*(3), 157–164. doi:10.1086/208627

Beltrán, M. E., Ursa, Y., de los Rios, S., Cabrera-Umpiérrez, M. F., Arredondo, M. T., Páramo, M., Prados, B., & Pérez, L. M. (2014). Engaging people with cultural heritage: Users' perspective. In C. Stephanidis & M. Antona (Eds.), *Universal Access in Human-Computer Interaction. Universal Access to Information and Knowledge* (pp. 639–649). Springer. doi:10.1007/978-3-319-07440-5_58

Ben Youssef, K., Leicht, T., & Marongiu, L. (2019). Storytelling in the context of destination marketing: An analysis of conceptualizations and impact measurement. *Journal of Strategic Marketing*, *27*(8), 696–713. doi:10.1080/0965254X.2018.1464498

Bender, S. (2019). Headset attentional synchrony: Tracking the gaze of viewers watching narrative virtual reality. *Media Practice and Education*, *20*(3), 277–296. doi:10.1080/25741136.2018.1464743

Benmayor, R. (2008). DST as a signature pedagogy for news humanities. *Arts and Humanities in Higher Education*, *7*(2), 188–204. doi:10.1177/1474022208088648

Benyon, D., Mival, O., & Ayan, S. (2013). Designing Blended Spaces. *The 26th BCS Conference on Human Computer Interaction (HCI)*, 398–403. 10.14236/ewic/HCI2012.1

Benyon, D. (2012). Presence in blended spaces. *Interacting with Computers*, *24*(4), 219–226. doi:10.1016/j.intcom.2012.04.005

Berg, C. R. (2006). A taxonomy of alternative plots in recent films: Classifying the" Tarantino effect. *Film Criticism*, *31*(1/2), 5–61.

Berrada, M. (2017). Co-Creation of the Tourist Experience via Internet: Towards Exploring a New Practice. *Journal of International Business Research and Marketing*, *2*(5), 18–23. doi:10.18775/jibrm.1849-8558.2015.25.3003

Best, G. (2017). Media memories on the move: Exploring an autoethnographic heritage of automobility and travel. *Journal of Heritage Tourism*, *12*(1), 52–66. doi:10.1080/1743873X.2016.1201088

Best, J. (2012). *Damned lies and statistics*. University of California Press. doi:10.1525/9780520953512

Bettman, J. R., & Park, C. W. (1980). Effects of prior knowledge and experience and phase of the choice process on consumer decision processes: A protocol analysis. *The Journal of Consumer Research*, *7*(3), 234–248. doi:10.1086/208812

Bhattacherjee, A., & Sanford, C. (2006). Influence processes for information technology acceptance: An elaboration likelihood model. *Management Information Systems Quarterly*, *30*(4), 805–825. doi:10.2307/25148755

Biedermann, B. (2020). The potentials of the digital cultural heritage: A best practice example of presentation and use: The digital representation of the Hans Gross Kriminalmuseum. In H. Kremers (Ed.), Digital cultural heritage (pp. 195-207). Springer.

Binkhorst, E., & Den Dekker, T. (2009). Agenda for co-creation tourism experience research. *Journal of Hospitality & Leisure Marketing*, *18*(2–3), 311–327. doi:10.1080/19368620802594193

Black, N. (2013). Patient reported outcome measures could help transform healthcare. *BMJ (Clinical Research Ed.)*, *346*(jan28 1), 346. doi:10.1136/bmj.f167 PMID:23358487

Blackwood, G. (2018). Dream factory tours: The Universal Pictures movie tour attraction in the 1960s. *Historical Journal of Film, Radio and Television*, *38*(3), 516–535. doi:10.1080/01439685.2017.1369238

Bosch, A. L. M., John, M. D. T., & Elving, W. J. L. (2005). How corporate visual identity supports reputation. *Corporate Communications*, *10*(1), 108–116. doi:10.1108/13563280510596925

Bozzelli, G., Raia, A., Ricciardi, S., De Nino, M., Barile, N., Perrella, M., & Palombini, A. (2019). An integrated VR/AR framework for user-centric interactive experience of cultural heritage: The ArkaeVision project. *Digital Applications in Archaeology and Cultural Heritage*, *15*, e00124. doi:10.1016/j.daach.2019.e00124

Bracken, C. C. (2006). Perceived source credibility of local television news: The impact of television form and presence. *Journal of Broadcasting & Electronic Media*, *50*(4), 723–741. doi:10.120715506878jobem5004_9

Bramwell, B., & Lane, B. (1999). Sustainable tourism: Contributing to the debates. *Journal of Sustainable Tourism*, *7*(1), 1–5. doi:10.1080/09669589908667323

Brechman, J. M., & Purvis, S. C. (2015). Narrative, transportation and advertising. *International Journal of Advertising*, *34*(2), 366–381. doi:10.1080/02650487.2014.994803

Brigante, R., & Elger, S. A. S. (2020). *2020 Immersive Entertainment Industry Annual Report*. HERE Institute.

BrightLocal. (2015). *Local Consumer Review Survey*. https://searchengineland.com/87-percent-customers-wont-consider-low-ratings-228607

Brodie, R. J., Ilic, A., Juric, B., & Hollebeek, L. (2013). Consumer engagement in a virtual brand community: An exploratory analysis. *Journal of Business Research*, *66*(1), 105–114. doi:10.1016/j.jbusres.2011.07.029

Brorn, M., Pope, N., & Voges, K. (2003). Buying or browsing? An exploration of shopping orientations and online PI. *European Journal of Marketing*, *37*(11/12), 1666–1684. doi:10.1108/03090560310495401

Brown, T. (2009). *Change by Design*. Harper Collins. https://www.amazon.com/Change-Design-Transforms-Organizations-Innovation/dp/0061766089

Brown, J. S., Collins, A., & Duguid, P. (2016). Situated Cognition and the Culture of Learning. *Educational Researcher*, *18*(1), 32–42. doi:10.3102/0013189X018001032

Brown, L. (2016). Treading in the footsteps of literary heroes: An autoethnography. *European Journal of Hospitality. Tourism and Recreation, 7*(2), 135–145.

Bruner, E. M. (2005). *Culture on Tour: Ethnographies of Travel*. University of Chicago.

Bryan, A. (2011). *The new digital storytelling: Creating narratives with new media*. Praeger.

Bryon, J., & Bryon, J. (2012). *Tour Guides as Storytellers – From Selling to Sharing Tour Guides as Storytellers – From Selling to Sharing*. doi:10.1080/15022250.2012.656922

Budtz, C., Fog, K., & Yakaboylu, B. (2005). *Storytelling: Branding in practice*. Springer.

Buhalis, D. (2000). Marketing the competitive destination of the future. *Tourism Management, 21*(1), 97–116. doi:10.1016/S0261-5177(99)00095-3

Buhalis, D., & Foerste, M. (2015). SoCoMo marketing for travel and tourism: Empowering co-creation of value. *Journal of Destination Marketing & Management, 4*(3), 151–161. doi:10.1016/j.jdmm.2015.04.001

Buonincontri, P., & Marasco, A. (2017). Enhancing cultural heritage experiences with smart technologies: An integrated experiential framework. *European Journal of Tourism Research, 17*, 83–101. doi:10.54055/ejtr.v17i.295

Buonincontri, P., Morvillo, A., Okumus, F., & van Niekerk, M. (2017). Managing the experience co-creation process in tourism destinations: Empirical findings from Naples. *Tourism Management, 62*, 264–277. doi:10.1016/j.tourman.2017.04.014

Burdette, M. (2016). *The Swayze effect*. https://www.oculus.com/story-studio/blog/the-swayze-effect/

Caber, M., Drori, N., Albayrak, T., & Herstein, R. (2021). *Social media usage behaviors of religious tourists: The cases of the Vatican, Mecca, and Jerusalem*. John Wiley & Sons Ltd. Doi:10.1002/jtr.2444816

Calvi, L., Hover, M., Ouwens, F., & van Waalwijk, J. (2015). Visualising Vincent's life: An engaging experience into Van Gogh's heritage. Paper presented at DeSForM 2015, Aesthetics of Interaction: Dynamic, Multisensory, Wise. 9th International Conference on Design and Semantics of Form and Movement, 14-10; 2015, Milan, Italy.

Cameron, E. (2012). New geographies of story and storytelling. *Progress in Human Geography, 36*(5), 573–592. doi:10.1177/0309132511435000

Campelo, A., Aitken, R., Thyne, M., & Gnoth, J. (2013). Sense of Place. *Journal of Travel Research, 53*(2), 154–166. doi:10.1177/0047287513496474

Campos, A. C., Mendes, J., Oom do Valle, P., & Scott, N. (2016). Co-Creation Experiences: Attention and Memorability. *Journal of Travel & Tourism Marketing*, 1–28. doi:10.1080/10548408.2015.1118424

Campos, A. C., & Almeida, S. (2022). Unique stories for unique brands: appealing to the hotel guest experience through digital storytelling. In R. Nunkoo (Ed.), *Handbook on the Tourist Experience: Design, Marketing and Management. Edward Elgar Publishing.*

Campos, A. C., Mendes, J., do Valle, P. O., & Scott, N. (2018). Co-creation of tourist experiences: A literature review. *Current Issues in Tourism, 21*(4), 369–400. doi:10.1080/13683500.2015.1081158

Camprubí, R., & Planas, C. (2020). Storytelling in tourism destination branding: The case of Girona. *Cuadernos de Turismo, 46*, 617–619. doi:10.6018/turismo.451841

Canadian Tourism Commission. (2008). CTC Debuts Edgy New Broadcast Spots, a Key Part of its 2009 Global Campaign. CTC Media Centre.

Cardoso, R. J. M. (2017). *A influência do storytelling (estrutura da narrativa) nas perceções, atitudes e comportamento dos consumidores* [Doctoral dessertation]. Instituto Politécnico de Leiria.

Carpio, R., & Birt, J. (2021). The role of the Embodiment Director in virtual reality film production. *Creative Industries Journal*, 1–10.

Carvalho, M. (2013). Construído a memória num espaço museológico: O Arquivo de Memórias da Pesca do Bacalhau do Museu Marítimo de Ílhavo. In *ARGOS – Revista do Museu Marítimo de Ílhavo* (pp. 94–101). Câmara Municipal de Ílhavo / Museu Marítimo de Ílhavo.

Casillo, M., Clarizia, F., D'Aniello, G., De Santo, M., Lombardi, M., & Santaniello, D. (2020). CHAT-Bot: A cultural heritage aware teller-bot for supporting touristic experiences. *Pattern Recognition Letters*, *131*, 234–243. doi:10.1016/j.patrec.2020.01.003

Cassel, S. H., & De Bernardi, C. (2021). Visual representations of Indigenous tourism places in social media. *Tourism, Culture & Communication*, *21*(2), 95–108. doi:10.3727/109830421X16191799471980

Cater, C., Albayrak, T., Caber, M., & Taylor, S. (2021). Flow, satisfaction and storytelling: A causal relationship? Evidence from scuba diving in Turkey. *Current Issues in Tourism*, *24*(12), 1749–1767. doi:10.1080/13683500.2020.1803221

Cavender, R., Swanson, J., & Wright, K. (2020). Transformative travel: Transformative learning through education abroad in a niche tourism destination. *Journal of Hospitality, Leisure, Sport and Tourism Education*, *27*, 100245. Advance online publication. doi:10.1016/j.jhlste.2020.100245

Central Directorate Of Rotary Capital Management. (2014). *Hierapolis (Pamukkale) Örenyeri*. Available at: http://www.dosim.gov.tr/muze/200

Chan, C. S., Peters, M., & Marafa, L. M. (2015). Public parks in city branding: Perceptions of visitors vis-à-vis residents in Hong Kong. *Urban Forestry & Urban Greening*, *14*(4), 1157–1165. doi:10.1016/j.ufug.2015.10.016

Chen, H., & Weiler, B. (2017). Heritage interpretation as a tool for sustainable management of national parks in China. *CAUTHE 2017: Time for Big ideas? Re-thinking the field for tomorrow*, 278-287.

Cheng, Q., Fang, L., & Chen, H. (2016). Visitors' brand loyalty to a historical and cultural theme park: A case study of Hangzhou Songcheng, China. *Current Issues in Tourism*, *19*(9), 861–868. doi:10.1080/13683500.2015.1006589

Chinyele, B. J., & Lwoga, N. B. (2018). Participation in decision making regarding the conservation of heritage resources and conservation attitudes in Kilwa Kisiwani, Tanzania. *Journal of Cultural Heritage Management and Sustainable Development*, *9*(5), 184–5198. doi:10.1108/JCHMSD-05-2017-0019

Chiu, H. C., Hsieh, Y. C., & Kuo, Y. C. (2012). How to Align your Brand Stories with Your Products. *Journal of Retailing*, *88*(2), 262–275. doi:10.1016/j.jretai.2012.02.001

Cho, H., Shen, L., & Wilson, K. (2012). Perceived Realism: Dimensions and Roles in Narrative Persuasion. *Communication Research*, *41*(6), 828–851. doi:10.1177/0093650212450585

Choi, H., & Murray, I. (2010). Resident attitudes toward sustainable community tourism. *Journal of Sustainable Tourism*, *18*(4), 575–594. doi:10.1080/09669580903524852

Choi, S. S. (2016). A study on effect of tourism storytelling of tourism destination brand value and tourist behavioral intentions. *Indian Journal of Science and Technology*, *9*(46), 1–6. doi:10.17485/ijst/2016/v9i46/107387

Chou, C. (2006). Research trends on Southeast Asian sea nomads. *Kyoto Review of Southeast Asia*, *7*, 1–11.

Chronis, A. (2005). Coconstructing heritage at the Gettysburg storyscape. *Annals of Tourism Research*, *32*(2), 386–406. doi:10.1016/j.annals.2004.07.009

Chronis, A. (2008). Co-constructing the narrative experience: Staging and consuming the American Civil War at Gettysburg. *Journal of Marketing Management*, *24*(1-2), 5–27. doi:10.1362/026725708X273894

Chronis, A. (2012b). Tourists as story-builders: Narrative construction at a heritage museum. *Journal of Travel & Tourism Marketing*, *29*(5), 444–459. doi:10.1080/10548408.2012.691395

Chu, J. H., & Mazalek, A. (2019). Embodied Engagement with Narrative: A Design Framework for Presenting Cultural Heritage Artifacts. *Multimodal Technologies and Interaction*, *3*(1), 1. Advance online publication. doi:10.3390/mti3010001

Chuvin, P. (2018). Introduction. In Mabeyinci Pavlos (Paulos Silentarios), Ayasofya'nın Betimi (pp. 11-26). Kırmızı Kedi Yayınevi.

Clarke, R., & Adam, A. (2010). DST in Australia: Academic perspectives and reflections. *Arts and Humanities in Higher Education*, *11*(1–2), 157–176. doi:10.1177/1474022210374223

Clark, W. C., & Dickson, N. M. (2003). Sustainability science: The emerging research program. *Proceedings of the National Academy of Sciences of the United States of America*, *100*(14), 8059–8061. doi:10.1073/pnas.1231333100 PMID:12794187

Clavé, S. A. (2007). *The global theme park industry*. CABI. doi:10.1079/9781845932084.0000

Clini, P., & Quattrini, R. (2021). Editorial: Digital cultural heritage, arts reproduction and museums systems: Languages and techniques in a Covid and post-Covid scenario for new forms of heritage against the silence of a fragile culture. *SCIRES-IT: Scientific Research and Information Technology*, *11*(1), 1–10. doi:10.2423/i22394303v11n1p1

CMI. (2008a). *10 Anos de Mudança - Município de Ílhavo 1998-2008*. Câmara Municipal de Ílhavo.

CMI. (2008b). *Boletim Informativo "Município de Ílhavo" #23 (dezembro)*. CMI.

CMI. (2012). *Boletim Informativo "Município de Ílhavo" #38 (dezembro)*. CMI.

CMI. (2014). *Boletim Informativo "Município de Ílhavo" #45 (setembro)*. CMI.

CMI. (2016). *Boletim Informativo "Município de Ílhavo" #53 (setembro)*. CMI.

CMI. (2019). *Boletim Informativo "Município de Ílhavo" #63 (dezembro)*. CMI.

CMI. (2022). *Centro de Religiosidade Marítima*. https://www.cm-ilhavo.pt/viver/areas-de-intervencao/cultura/equipamentos-culturais/museus/centro-de-religiosidade-maritima

CM-Ílhavo. (2018). *Museu Marítimo de Ílhavo ultrapassa 1 milhão de visitantes e conquista importante prémio nacional*. https://www.cm-ilhavo.pt/pages/388?news_id=3137

Cohen, E. (1988). Authenticity and Commodification in Tourism. *Annals of Tourism Research*, *15*(3), 371–386. doi:10.1016/0160-7383(88)90028-X

Cohen, J. B., Fishbein, M., & Ahtola, O. T. (1972). The nature and uses of expectancy-value models in consumer ATT research. *JMR, Journal of Marketing Research*, *9*(4), 456–460. doi:10.1177/002224377200900420

Coleman, S. (2021). Pilgrimage. In R. A. Segal & N. P. Roubekas (Eds.), *The Wiley Blackwell Companion to the Study of Religion*. John Wiley & Sons. doi:10.1002/9781119092797.ch26

Conti, E., & Cassel, S. (2020). Liminality in nature-based tourism experiences as mediated through social media. *Tourism Geographies*, *22*(2), 413–432. doi:10.1080/14616688.2019.1648544

Coral Triangle Initiative. (n.d.). *History of CTI-CFF* [Infographic]. https://coraltriangleinitiative.org/about

Cornelis, P. C. M., Strijbosch, W., & Corsius, P. (2022). Conceptualizing the dark ride experience through the dark ride cube: Evidence from the EMEA region. *Journal of Hospitality & Tourism Research (Washington, D.C.)*, 1–26. doi:10.1177/10963480221080331

Correia, R., & Brito, C. (2011). A Importância do Marketing para o Desenvolvimento Turístico: O caso de Montalegre. *Revista Turismo & Desenvolvimento*, *16*, 127–143.

Costa, N. M. (2016). A memória documental dos pescadores bacalhoeiros de Ílhavo. *Vox Musei Arte e Patrimônio*, *1*, 39–44.

Couldry, N. (2008). Mediatization or mediation? Alternative understandings of the emergent space of digital storytelling. *New Media & Society*, *10*(3), 373–391. doi:10.1177/1461444808089414

Council of Europe (2017). *Recommendation of the Committee of Ministers to member States on the European Cultural Heritage Strategy for the 21st century.* Committee of Ministers of the Council of Europe CM/Rec(2017)1.

Court, S., & Gamini, W. (2015). *People-centred approaches to the conservation of cultural heritage: Living heritage.* ICCROM. https://www.iccrom.org/sites/default/files/PCA_Annexe-2.pdf

Cross, G., & Walton, J. (2005). *The playful crowd: Pleasure places in the twentieth century.* Columbia University Press. doi:10.7312/cros12724

D'Andria, F. (2021). *Pamukkale Hierapolis.* Available at: https://denizli.ktb.gov.tr/TR-211745/pamukkale-hierapolis.html

D'Andria, F. (2010). *Hierapolis (Pamukkale).* Ege Yayınları.

Da Costa, C. (2019). *3 reasons why brand storytelling is the future of marketing.* Available at: https://www.forbes.com/sites/celinnedacosta/2019/01/31/3-reasons-why-brand-storytelling-is-the-future-of-marketing/?sh=7fd1667455ff

da Fonseca, S. (2007). *Ílhavo Ensaio Monográfico: do século X ao século XX.* Papiro Editora.

da Silva, A. J. M. (2001). Gazela Primeiro - o navio e a sua gente. In Á. Garrido (Ed.), *A Pesca do Bacalhau: História e Memória* (pp. 255–272). Editorial Notícias.

da Silva, A. J. M. (2015). The fable of the cod and the promised sea - About Portuguese traditions of Bacalhau. *Heritages and Memories from the Sea*, *48296*(January), 130–143.

Dahlstrom, M. F. (2014). Using narratives and storytelling to communicate science with nonexpert audiences. *Proceedings of the National Academy of Sciences of the United States of America*, *111*(Supplement 4), 13614–13620. doi:10.1073/pnas.1320645111 PMID:25225368

Dal Falco, F., & Vassos, S. (2017). Museum experience design: A modern storytelling methodology. *The Design Journal*, *20*(sup1), S3975–S3983. doi:10.1080/14606925.2017.1352900

Daley, B. (2021). *Storytelling with digital culture is booming - find out why.* Europeana. https://pro.europeana.eu/post/storytelling-with-digital-culture-is-booming-find-out-why

Dann, G. M. (1981). Tourism Motivations: An Appraisal. *Annals of Tourism Research*, *8*(2), 189–219. doi:10.1016/0160-7383(81)90082-7

Dantas, V. (2007). O futuro dos santuários no contexto da mobilidade humana. In Santuário de Fátima (Ed.), Atas das Jornadas Aljustrel e Valinhos – o outro pulmão de Fátima (pp. 83-100). Fátima: Santuário de Fátima.

David, M. B. (1991). Consulting and change in the storytelling organisation. *Journal of Organizational Change Management, 4*(3), 7–17. doi:10.1108/EUM0000000001193

Davidson, R., & Maitland, R. (1997). *Tourism destinations.* Hodder & Stoughton.

Davis, F. (1980). *A technology acceptance model for empirically testing new end-user information systems: Theory and results* [Doctoral dissertation]. Massachusetts Institute of Technology]

Davis, A. J., & Prentice, R. C. (1995). Conceptualising the latent visitor to heritage attractions. *Tourism Management, 16*(7), 491–500. doi:10.1016/0261-5177(95)00066-W

Davis, H., & McLeod, S. L. (2003). Why humans value sensational news: An evolutionary perspective. *Evolution and Human Behavior, 24*(3), 208–216. doi:10.1016/S1090-5138(03)00012-6

Dawson, P., & Sykes, C. (2018). Concepts of time and temporality in the storytelling and sensemaking literatures: A review and critique. *International Journal of Management Reviews, 00*, 1–18.

Dawson, P., & Sykes, C. (2018). Concepts of Time and Temporality in the Storytelling and Sensemaking Literatures: A Review and Critique. *International Journal of Management Reviews, 21*(1), 97–114. doi:10.1111/ijmr.12178

De Ascaniis, S., Mutangala, M., & Cantoni, L. (2018). ICTs in the tourism experience at religious heritage sites: A review of the literature and an investigation of pilgrims' experiences at the sanctuary of Loreto (Italy). *Church. Communication and Culture, 3*(3), 310–334. doi:10.1080/23753234.2018.1544835

Deacon, H. (2006). The 2003 Intangible Heritage Convention: keynote address. *South African Museums Association Bulletin, 32*(1), 2-9.

Debevec, K., & Iyer, E. (1988). Self-referencing as a Mediator of the Effectiveness of Sex-Role Portrayal in Advertising. *Psychology and Marketing, 5*(1), 71–85. doi:10.1002/mar.4220050106

Deffner, A., Psatha, E., Bogiantzidis, N., Mantas, N., Vlachaki, E., & Ntaflouka, P. (2015). Accessibility to Culture and Heritage: Designing for All. *Proceedings of the AESOP 2015 Conference Definite space—Fuzzy Responsibility.*

Delgadillo, Y., & Escalas, J. E. (2004). Narrative word-of-mouth communication: Exploring memory and ATT effects of consumer storytelling. *Advances in Consumer Research. Association for Consumer Research (U. S.), 31*, 186–192.

Delgado-Ballester, E., & Fernández-Sabiote, E. (2016). "Once upon a brand": Storytelling practices by Spanish brands. *Spanish Journal of Marketing, 20*(2), 115–131. doi:10.1016/j.sjme.2016.06.001

Demetrescu, E., Palombini, A., Pietroni, E., Forlani, M., & Rufa, C. (2016). Making movies: Nuove frontiere per la comunicazione audiovisiva del passato. *Archeologia e Calcolatori, 8*, 203–210.

Denizli Provincial Directorate of Culture and Tourism (n.d.a). *Tarih boyunca su ve sağlık turizmi şehri; Denizli.* Available at: https://denizli.ktb.gov.tr/TR-211794/termal-saglik-turizmi.html

Denizli Provincial Directorate of Culture and Tourism (n.d.b). *Efsaneler.* Available at: https://denizli.ktb.gov.tr/TR-211827/efsaneler.html

Denning, S. (2004). Telling tales. *Harvard Business Review, 82*(5), 122–129. PMID:15146741

Denning, S. (2005). *The leader's guide to storytelling.* Wiley.

Denning, S. (2005). *The leader's guide to storytelling: Mastering the art and discipline of business narrative.* John Wiley & Sons.

Dessart, L., & Pitardi, V. (2019). How stories generate consumer engagement: An exploratory study. *Journal of Business Research, 104*, 183–195. doi:10.1016/j.jbusres.2019.06.045

Destination Analysts. (2020, May 11). *Nature-Based Destinations and the Future of Travel. Destination Analysts.* https://www.destinationanalysts.com/blog-nature-based-destinations-and-the-future-of-travel/

Di Pietro, L., Guglielmetti Mugion, R., & Renzi, M. F. (2018). Heritage and identity: technology, values and visitor experiences. In Journal of Heritage Tourism (Vol. 13, Issue 2, pp. 97–103). Taylor and Francis Ltd. doi:10.1080/1743873X.2017.1384478

Díaz-Andreu, M. (2017). Introduction to the themed section' digital heritage and the public.'. *International Journal of Heritage Studies, 23*(5), 404–407. doi:10.1080/13527258.2017.1286780

Dionisio, M., & Nini, V. (2021). Leveraging Transmedia storytelling to engage tourists in the understanding of the destination's local heritage. *Multimedia Tools and Applications, 80*(26-27), 34813–34841. Advance online publication. doi:10.100711042-021-10949-2

Djabarouti, J. (2021). Stories of feelings and things: Intangible heritage from within the built heritage paradigm in the UK. *International Journal of Heritage Studies, 27*(4), 391–406. doi:10.1080/13527258.2020.1798271

Douglas, K. (2010). *Pictures of time beneath: science, heritage, and the uses of the deep past.* CSIRO Publishing. doi:10.1071/9780643100251

Dowling, D., Fearghail, C. O., Smolic, A., & Knorr, S. (2018). Faoladh: A case study in cinematic VR storytelling and production. In Int*ernational Conference on Interactive Digital Storytelling* (pp. 359-362). Springer. 10.1007/978-3-030-04028-4_42

Dreon, O., Kerper, R. M., & Landis, J. (2011). Digital storytelling: A tool for teaching and learning in the YouTube generation. *Middle School Journal, 42*(5), 4–10. doi:10.1080/00940771.2011.11461777

Drummond, S. (2001). Introduction to quality. In *Quality Issues in Heritage Visitor Attractions* (pp. 5–15). Butterworth-Heinemann. doi:10.1016/B978-0-7506-4675-8.50007-1

Duxbury, N., & Richards, G. (Eds.). (2019). *A research agenda for creative tourism.* Edward Elgar Publishing. doi:10.4337/9781788110723

Eagly, A. H., & Chaiken, S. (2007). The advantages of an inclusive definition of ATT. *Social Cognition, 25*(5), 582–602. doi:10.1521oco.2007.25.5.582

Eddy, R., Baker, C., Macy, R., Murray, J., & Salter, A. (2020). Hacking droids and casting spells: Locative augmented reality games and the reimagining of the theme park. *Proceedings of the 31st ACM Conference on Hypertext and Social Media*, 37-46. 10.1145/3372923.3404801

Ekinci, Y. (2003). From destination image to destination branding: An emerging area of research. *e-Review of Tourism Research (eRTR), 1*(2), 21-24.

Ekinci, Y., & Hosany, S. (2006). Destination Personality: An Application of Brand Personality to Tourism Destinations. *Journal of Travel Research, 45*(2), 127–139. doi:10.1177/0047287506291603

Elliot, S. (2009). User-Generated Videos in Tourism Destination Marketing: Using Narrative Analysis to Deconstruct Video Travel Stories. *TTRA International Conference.*

Ellis, C. (2004). *The ethnographic I: A methodological novel about autoethnography.* AltaMira Press.

Emmendoerfer, M. L., Olavo, A. V. A., Silva, A. C., Jr., & Mediotte, E. J. (2020). Innovation lab in the touristic development context: Perspectives for creative tourism. In S. M. Cabeça, A. R.Gonçalves, J. F. Marques & M. Tavares (Eds.), Creative tourism dynamics: Connecting travellers, communities, cultures and places (pp. 87-101). Grácio Editor.

Erdemoğlu, B. (2015). *Troia ve Çevresi* (E. Yazıcı, Ed.). Uranüs Fotoğraf Ajansı ve Yayıncılık.

Erdoğan, E. G. (2012). Bizans dönemi'nde Ayasofya, tarihçesi ve mimari özellikleri hakkında genel bilgiler. *İstanbul. Journal of Social Sciences*, *1*, 1–7.

Erhat, A., & Kadir, A. (2019). Introduction. In İlyada (pp. 1-84). Türkiye İş Bankası Kültür Yayınları.

Escalas, J. E. (2004). Narrative processing: Building consumer connections to brands. *Journal of Consumer Psychology*, *14*(1/2), 168–180. doi:10.120715327663jcp1401&2_19

Escalas, J. E. (2007). Imagine yourself in the product. Mental simulation, narrative transportation, and persuasion. *Journal of Advertising*, *33*(2), 37–48. doi:10.1080/00913367.2004.10639163

Escalas, J. E. (2007). Self-referencing and Persuasion: Narrative transportation Versus Analytical Elaboration. *The Journal of Consumer Research*, *33*(4), 421–429. doi:10.1086/510216

Escalfoni, R., Braganholo, V., & Borges, M. R. (2009). Applying group storytelling to capture innovation features. *Proceedings of the 13th International Conference on Computer Supported Cooperative Work in Design*, 209-214. 10.1109/CSCWD.2009.4968060

European Commission, Directorate-General for Education, Youth, Sport and Culture. (2019). *Sustainable cultural tourism*. Publications Office. https://data.europa.eu/doi/10.2766/400886

Expedia and The Center for Generational Kinetics. (2018). *Generations on the Move*. https://viwfinder.expedia.com/wp-content/uploads/2018/01/Expedia-Generations-on-the-move.pdf

Fauconnier, G., & Turner, M. (2002). *The way we think: conceptual blending and the mind's hidden complexities*. Basic Books.

Febby, A., & Hifni, A. (2018). Destination brand storytelling: Analisis naratif video the journey to a wonderful world kementerian pariwisata. *Jurnal Komunikasi Indonesia*, *7*(1), 73–86.

Feld, S., & Basso, K. H. (1996). *Senses of place*. School of American Research Press.

Ferreira, S., Alves, A., & Quico, C. (2014). The role of social media: The TravelPlot Porto case study. *Revista Turismo & Desenvolvimento*, (21/22), 151–160. doi:10.34624/rtd.v2i21/22.11407

Ferreira, T. C. (2018). Bridging planned conservation and community empowerment: Portuguese case studies. *Journal of Cultural Heritage Management and Sustainable Development*, *8*(2), 179–193. doi:10.1108/JCHMSD-05-2017-0029

Firmansyah, F., Mustofa, A., Estradivari, Damora, A., Handayani, C., Ahmadia, G., & Harris, J. (2017). *Satu dekade pengelolaan Taman Nasional Wakatobi: Keberhasilan dan tantangan konservasi laut* [A decade of Wakatobi National Park management: The achievements and challenges of sea conservation]. World Wide Fund for Nature-Indonesia.

Fischer, D., Selm, H., Sundermann, A., & Storksdieck, M. (2020). Storytelling for sustainability. *Storytelling for Sustainability in Higher Education*, (April), 38–51. doi:10.4324/9780429291111-5

Fishbein, M., & Ajzen, I. (1975). *Belief, ATT, intention, and behaviour: An introduction to theory and research*. Addison-Wesley.

Fisher, J. A., & Schoemann, S. (2019). Correction to: toward an ethics of interactive storytelling at dark tourism sites in virtual reality. In R. Rouse, H. Koenitz, & M. Haahr (Eds.), Lecture Notes in Computer Science: Vol. 11318. Interactive Storytelling. Springer. , ICIDS 2018. doi:10.1007/978-3-030-04028-4_81

Flavián, C., Ibáñez-Sánchez, S., & Orús, C. (2019). The impact of virtual, augmented and mixed reality technologies on the customer experience. *Journal of Business Research*, *100*, 547–560. doi:10.1016/j.jbusres.2018.10.050

Fonseca, F. P., & Ramos, R. A. R. (2006). O planeamento estratégico de marketing territorial em busca de potenciar o território: o caso de Almeida. *"PLURIS 2006 : Anais Do 2º Congresso Luso-Brasileiro de Planeamento Urbano Regional Integrado Sustentável"*. http://repositorium.sdum.uminho.pt/handle/1822/7224

Franco, A., Correia, R., & Meneses, R. (2019). Os 5 sentidos associados a Portugal na perspectiva dos turistas. *Revista Ibérica de Sistemas e Tecnologias de Informação*, *24*, 272–286.

Frank, A. W. (2000). The standpoint of storytellers. *Qualitative Health Research*, *10*(3), 354–365. doi:10.1177/104973200129118499 PMID:10947481

Frias, C., Silva, C., & Seabra, C. (2017). Storytelling: Motivações, autenticidade e envolvimento dos Turistas Culturais. *Revista Turismo & Desenvolvimento*, *27/28*, 1355–1366. doi:10.34624/rtd.v1i27/28.9977

Frost, W., Frost, J., Strickland, P., & Smith Maguire, J. (2020). Seeking a competitive advantage in wine tourism: Heritage and storytelling at the cellar-door. *International Journal of Hospitality Management*, *87*(102460), 1–9. doi:10.1016/j.ijhm.2020.102460

Frow, P., Nenonen, S., Payne, A., & Storbacka, K. (2015). Managing Co-creation Design: A Strategic Approach to Innovation. *British Journal of Management*, *26*(3), 463–483. doi:10.1111/1467-8551.12087

Fuller, R. B. (1975). *Synergetics: Explorations in the Geometry of Thinking*. Macmillan. https://philpapers.org/rec/FULSEI

Fu, X., Baker, C., Zhang, W., & Zhang, R. (2022). Theme park storytelling: Deconstructing immersion in Chinese theme parks. *Journal of Travel Research*. Advance online publication. doi:10.1177/00472875221098933

Gabriel, Y. (2000). *Storytelling in organizations: Facts, fictions, and fantasies*. Oxford University Press. doi:10.1093/acprof:oso/9780198290957.001.0001

Gao, J., Lin, S., & Zhang, C. (2020). Authenticity, involvement, and nostalgia: Understanding visitor satisfaction with an adaptive reuse heritage site in urban China. *Journal of Destination Marketing & Management*, *15*(100404), 1–10. doi:10.1016/j.jdmm.2019.100404

García Haro, M. Á., Martínez-Ruiz, M. P., & Martínez-Canas, R. (2014). The Effects of the Value Co-Creation Process on the Consumer and the Company. *Expert Journal of Marketing*, *2*(2), 68–81.

Garcia, J. A., Gomez, M., & Molina, A. (2012). A Destination-Branding model: An empirical analysis based on stakeholders. *Tourism Management*, *33*(3), 646–661. doi:10.1016/j.tourman.2011.07.006

Gardiner, S., & Scott, N. (2017). Experience Design. In N. Scott, M. D. Martino, & M. V. Niekerkm (Eds.), *Knowledge transfer to and within tourism* (Vol. 8, pp. 13–25). Emerald Publishing Limited. doi:10.1108/S2042-144320170000008002

Gardner, B. B., & Levy, S. J. (1955). The product and the brand. *Harvard Business Review*, *33*, 33–39.

Garrido, Á. (2016). *Portugal no Mar: Homens que foram ao bacalhau - 2.ª Edição* (CMI/MMI). Âncora Editores.

Garrido, Á. (2018). Espaço Público e memórias Inclusivas: A Experiência de Gestão do Património da Pesca do Bacalhau no Museu Marítimo de Ílhavo. In A. Polónia, C. M. Costa, & F. Mouta (Eds.), Boas prática para políticas públicas de memória, ciência e património (pp. 89–106). Media XXI – Publishing, Research and Consulting Reservados.

Garrido, Á. (2004). *O Estado Novo e a Campanha do Bacalhau*. Círculo de Leitores.

Gawronski, B., & Bodenhausen, G. V. (2006). Associative and propositional processes in evaluation: An integrative review of implicit and explicit ATT change. *Psychological Bulletin*, *132*(5), 692–731. doi:10.1037/0033-2909.132.5.692 PMID:16910748

Geia, L. K., Hayes, B., & Usher, K. (2013). Yarning/Aboriginal storytelling: Towards an understanding of an Indigenous perspective and its implications for research practice. *Contemporary Nurse*, *46*(1), 13–17. doi:10.5172/conu.2013.46.1.13 PMID:24716757

Georgescu, M., & Popescul, D. (2015). The impact of new Information and Communication Technologies on the creation and dissemination of knowledge inheritance. *Procedia: Social and Behavioral Sciences*, *188*, 122–129. doi:10.1016/j.sbspro.2015.03.347

Gertner, R. (2019). The impact of cultural appropriation on destination image, tourism, and hospitality. *Thunderbird International Business Review*, *61*(6), 873–877. doi:10.1002/tie.22068

Ghisoiu, M., Bolan, P., Gilmore, A., & Carruthers, C. (2017). "Conservation" and co-creation through film tourism at heritage sites: An initial focus on Northern Ireland. *Revista Turismo & Desenvolvimento*, *1*(27/28), 2125–2135.

Giannini, T., & Bowen, J. (2019a). Digital culture. In T. Giannini & J. Bowen (Eds.), *Museums and digital culture: New perspectives and research* (pp. 3–26). Springer. doi:10.1007/978-3-319-97457-6_1

Gibson, J. (2015). Message from Mungo. *Australian Historical Studies*, *46*(2), 308–310. doi:10.1080/1031461X.2015.1044159

Gill, P. (2015). *Toward a scientific approach to identifying and understanding indicators of radicalization and terrorist intent: Eight key problems*. Academic Press.

Gillam, S. (2017). *Spotlight VR/AR: Innovation in transformative storytelling*. MW17: MW 2017. https://mw17.mwconf.org/paper/spotlight-vrar-innovation-in-transformative-storytelling/

Gilliam, D. A., & Flaherty, K. E. (2015). Storytelling by the sales force and its effect on buyer – seller exchange. *Industrial Marketing Management*, *46*, 132–142. doi:10.1016/j.indmarman.2015.01.013

Gilliam, D. A., Flaherty, K. E., & Rayburn, S. W. (2014). The dimensions of storytelling by retail salespeople. *International Review of Retail, Distribution and Consumer Research*, *24*(2), 231–241. doi:10.1080/09593969.2013.849611

Godovykh, M., Milman, A., & Tasci, A. D. A. (2019). Theme park experience: Factors explaining amount of pleasure from a visit, time allocation for activities, perceived value, queuing quality, satisfaction, and loyalty. *Journal of Tourism and Leisure Studies (Champaign, Ill.)*, *4*(2), 1–21. doi:10.18848/2470-9336/CGP/v04i02/1-21

Godwin, V. (2017). Theme park as interface to the Wizarding (story) World of Harry Potter. *Transformative Works and Cultures, 25*.

Go, F. M., Alvarez, M. D., & Yüksel, A. (2016). Conclusions and implications for heritage tourism destinations. In M. D. Alvarez, F. M. Go, & A. Yüksel (Eds.), *Heritage tourism destinations: Preservation, communication and development* (pp. 186–193). CABI.

Goldsmith, J. (2018, July 27). *Undertold history of Native America lands on PBS with a big splash*. Current-News for people in public media. https://current.org/2018/07/undertold-history-of-native-america-lands-on-pbs-with-a-big-splash/

Gonçalves, J., Seabra, C., & Silva, C. (2018). Histórias de cultura. O poder do Storytelling em destinos de Turismo Cultural. *Caderno de Geografia*, (37), 113–120. doi:10.14195/0871-1623_37_9

Gottschall, J. (2012). *The storytelling animal: How stories make us human.* Houghton Mifflin Harcourt.

Gottschall, J. (2012). *The Storytelling Animal: How Stories Make Us Human.* Houghton Mifflin Harcourt.

Governorship of Denizli. (n.d.). *Pamukkale travertenleri.* Available at http://www.denizli.gov.tr/pamukkale

Governorship of Şanlıurfa (n.d.). *Göbeklitepe.* Available at http://www.sanliurfa.gov.tr/gobeklitepe

Gravili, S., Rosato, P., & Iazzi, A. (2017). Managing DMOs through storytelling: A model proposal for network and value co-creation in tourism. *International Business Research, 10*(7), 8. doi:10.5539/ibr.v10n7p8

Green, M. C., & Brock, T. C. (2000). The role of transportation in the persuasiveness of public narratives. *Journal of Personality and Social Psychology, 79*(5), 701–721. doi:10.1037/0022-3514.79.5.701 PMID:11079236

Green, M. C., & Brock, T. C. (2002). In the mind's eye: Transportation-imagery model of narrative persuasion. In M. C. Green, J. J. Strange, & T. C. Brock (Eds.), *Narrative impact: Social and cognitive foundations* (pp. 315–341). Lawrence Erlbaum Associates Publishers.

Green, M. C., Brock, T. C., & Kaufman, G. F. (2004). Understanding media enjoyment: The role of transportation into narrative worlds. *Communication Theory, 14*(4), 311–327. doi:10.1111/j.1468-2885.2004.tb00317.x

Green, M. C., & Dill, K. E. (2013). Engaging with Stories and Characters: Learning, Persuasion, and Transportation into Narrative Worlds. In K. Dill (Ed.), *Oxford Handbook of Media Psychology* (pp. 449–461). Oxford University Press. doi:10.1093/oxfordhb/9780195398809.013.0025

Gröppel-Wegener, A., & Kidd, J. (2019). Introduction: The seductive power of immersion. In *Critical Encounters with Immersive Storytelling.* https://www.taylorfrancis.com/books/mono/10.4324/9780429055409/critical-encounters-immersive-storytelling-alke-gröppel-wegener-jenny-kidd

Gross, M., & Zimmerman, R. (2002). *Interpretive center: The history, design, and development of nature and visitor center.* UW-SP Foundation Press, Inc.

Guazzaroni, G., & Pillai, A. S. (2020). *Virtual and augmented reality in education, art, and museums.* IGI Global. doi:10.4018/978-1-7998-1796-3

Gubrium, J. F., & Holstein, J. A. (2009). *Analyzing Narrative Reality.* Sage Publications, Inc. doi:10.4135/9781452234854

Gülyaz, M. E., & Ölmez, İ. (1998). *Kapadokya.* Dünya Kitap.

Guttentag, D. A. (2010). Virtual reality: Applications and implications for tourism. *Tourism Management, 31*(5), 637–651. doi:10.1016/j.tourman.2009.07.003

Guy, R. (2019). Will the real hillbilly please stand up? Urban Appalachian migration and culture seen through the lens of *Hillbilly Elegy.* In A. Harkins & M. McCarroll (Eds.), *Appalachian Reckoning: A Region Responds to Hillbilly Elegy* (pp. 86–104). West Virginia University Press.

Hagen, O. (2008). Driving environmental innovation with corporate storytelling: Is radical innovation possible without incoherence? *International Journal of Innovation and Sustainable Development, 3*(4), 217–233. doi:10.1504/IJISD.2008.022227

Hall, A. E., & Bracken, C. C. (2011). I really liked that movie - testing the relationship between trait empathy, transportation, perceived realism, and movie enjoyment. *Journal of Media Psychology, 23*, 90–99. doi:10.1027/1864-1105/a000036

Hall, C. M. (2000). *Tourism Planning: Policies, processes, relationships.* Prentice Hall.

Ham, S. H. (1992). *Environmental interpretation: A practical guide for people with big ideas and small budgets*. Fulcrum Publishing.

Ham, S. H. (2013). *Interpretation: Making a difference on purpose*. Fulcrum Publishing.

Han, D. I. D., Weber, J., Bastiaansen, M., Mitas, O., & Lub, X. (2019). Virtual and augmented reality technologies to enhance the visitor experience in cultural tourism. In *Augmented reality and virtual reality* (pp. 113–128). Springer, Cham. doi:10.1007/978-3-030-06246-0_9

Hanna, S. P., Potter, A. E., Modlin, E. A., Carter, P., & Butler, D. L. (2017). *Social memory and heritage tourism methodologies*. Routledge.

Hanna, S., Del Casino, V. Jr, Selden, C., & Hite, B. (2004). Representation as Work in America's Most Historic City. *Social & Cultural Geography*, *5*(3), 459–481. doi:10.1080/1464936042000252822

Hanna, S., & Rowley, J. (2013). Place brand practitioners' perspectives on the management and valuation of the brand experience. *The Town Planning Review*, *84*(4), 473–493. doi:10.3828/tpr.2013.25

Hansen, F. A., Kortbek, K. J., & GrønbÆk, K. (2012). Mobile Urban Drama: Interactive storytelling in real world environments. In *New Review of Hypermedia and Multimedia* (Vol. 18, Issues 1–2, pp. 63–89). Taylor & Francis Group. doi:10.1080/13614568.2012.617842

Harrigan, P., & Wardrip-Fruin, N. (2009). Introduction. In P. Harrigan & N. Wardrip-Fruin (Eds.), *Third person: Authoring and exploring vast narratives* (pp. 1–9). MIT Press.

Harrison, D. (2004a). Introduction: Contested narratives in the domain of World heritage. *Current Issues in Tourism*, *7*(4-5), 281–290. doi:10.1080/13683500408667985

Harrison, J. (2002). *Being a tourist – Finding meaning in pleasure travel*. UBC Press.

Hartman, S., Parra, C., & de Roo, G. (2019). Framing strategic storytelling in the context of transition management to stimulate tourism destination development. *Tourism Management, 75*, 90–98. doi:.2019.04.014 doi:10.1016/j.tourman

Hartman, S., Parra, C., & de Roo, G. (2019). Framing strategic storytelling in the context of transition management to stimulate tourism destination development. *Tourism Management*, *75*(April), 90–98. doi:10.1016/j.tourman.2019.04.014

Hassan, S. S. (2000). Determinants of market competitiveness in an environmentally sustainable tourism industry. *Journal of Travel Research*, *38*(3), 239–245.

Hatchuel, A. (2001). Towards design theory and expandable rationality: The unfinished program of Herbert Simon. *The Journal of Management and Governance*, *5*(3–4), 260–273. doi:10.1023/A:1014044305704

Hausman, A. V., & Siekpe, J. S. (2009). The effect of web interface features on consumer online PIs. *Journal of Business Research*, *62*(1), 5–13. doi:10.1016/j.jbusres.2008.01.018

Hautz, J., Fuller, J., Hutter, K., & Thurridl, C. (2014). Let users generate your video ads? The impact of video source and quality on consumers' perceptions and intended behaviors. *Journal of Interactive Marketing*, *28*(1), 1–15. doi:10.1016/j.intmar.2013.06.003

Hay, N. A., Chien, P. M., & Ruhanen, L. (2021). Tell me your story: Branding destinations through residents' (place) stories. *Journal of Vacation Marketing*, 1–16. doi:10.1177/13567667211060567

Hee, L. T. (2001). Evaluation of brand equity of tourism destinations in Korea. *Tourism Research.*, *25*(3), 171–192.

Heitor, J. (2020). *A Imagem Peregrina de Nossa Senhora de Fátima*. Cordel D'Prata.

Heitor, J. C. (2019). *A imagem peregrina de Nossa Senhora de Fátima. Uma marca na promoção do turismo religioso de Fátima?* Tese doutoramento em Turismo. Universidade de Lisboa, Instituto de Geografia e Ordenamento do Território. http://hdl.handle.net/10451/39519

Hench, J. (2009). *Designing Disney: Imagineering and the art of the show*. Disney Editions.

Henderson, K. (2006). *Dimensions of Choice: Qualitative Approaches to Parks, Recreation, Tourism, Sport, and Leisure Research* (2nd ed.). Venture Publishing, Inc.

Hensel, W. A., & Rasco, T. L. (1992). Storytelling as a method for teaching values and ATTs. *Academic Medicine*, *67*(8), 500–504. doi:10.1097/00001888-199208000-00003 PMID:1497776

Herskovitz, S., & Crystal, M. (2010). The essential brand persona: Storytelling and branding. *The Journal of Business Strategy*, *31*(3), 21–28. doi:10.1108/02756661011036673

Hertzman, E., Anderson, D., & Rowley, S. (2008). Edutainment heritage tourist attractions: A portrait of visitors' experiences at Storyeum. *Museum Management and Curatorship*, *23*(2), 155–175. doi:10.1080/09647770802012227

Heskett, J. (2005). *Design: A very short introduction*. Oxford University Press. doi:10.1093/actrade/9780192854469.001.0001

He, Y., Ma, Y. H., & Zhang, X. R. (2017). "Digital heritage": Theory and innovative practice. *The International Archives of the Photogrammetry, Remote Sensing and Spatial Information Sciences*, *42*(2), 335–342. doi:10.5194/isprs-archives-XLII-2-W5-335-2017

Hinchman, L. P., & Hinchman, S. (Eds.). (1997). *Memory, identity, community: The idea of narrative in the human sciences*. Suny Press.

Hobbs, P. (2015). *Walt's utopia: Disneyland and American mythmaking*. McFarland.

Ho, C. I., & Mi, L. E. (2008). The place marketing using storytelling: Exploratory research. *Tourism Research.*, *32*(4), 411–429.

Hollebeek, L. D., & Macky, K. (2019). Digital content marketing's role in fostering consumer engagement, trust, and value. *Journal of Interactive Marketing*, *45*, 27–41. doi:10.1016/j.intmar.2018.07.003

Holloway, S. S., van Eijnatten, F. M., Romme, A. G. L., & Demerouti, E. (2016). Developing actionable knowledge on value crafting: A design science approach. *Journal of Business Research*, *69*(5), 1639–1643. doi:10.1016/j.jbusres.2015.10.031

Hooper-Greenhill, E. (1994). *The Educational role of the museum*. Routledge.

Hover, M., & Vugts, O. (2013). Efteling and the Fairytale Tree: Storytelling between tradition and innovation. *Trendrapporttoerisme, Recreatie en Vrijetijd*, 272-275.

Howard, P. (2003). Heritage: Management, interpretation, identity. *Continuum.*

Hsiao, K. L., Lu, H. P., & Lan, W. C. (2013). The influence of the components of storytelling blogs on readers' travel intentions. *Internet Research*, *23*(2), 160–182. doi:10.1108/10662241311313303

Hsin Chang, H., & Wen Chen, S. (2008). The impact of online store environment cues on PI: Trust and perceived risk as a mediator. *Online Information Review*, *32*(6), 818–841. doi:10.1108/14684520810923953

Huertas, A. (2018). How live videos and stories in social media influence tourist opinions and behaviour. *Information Technology & Tourism*, *19*(1-4), 1–28. doi:10.100740558-018-0112-0

Hughes, G. (1995). Authenticity in Tourism. *Annals of Tourism Research*, *22*(4), 781–803. doi:10.1016/0160-7383(95)00020-X

Hung, W.-L., Lee, Y.-J., & Huang, P.-H. (2016). Creative experiences, memorability and revisit intention in creative tourism. *Current Issues in Tourism, 19*(8), 763–770. doi:10.1080/13683500.2013.877422

ICOMOS. (1994). *The Nara document on authenticity.* https://www.icomos.org/charters/nara-e.pdf

Iles, J. (2008). Encounters in the Fields: Tourism to the Battlefields of the Western Front. *Journal of Tourism and Cultural Change, 6*(2), 138–154. doi:10.1080/14766820802376299

Ingerpuu, L. (2018). Socialist architecture as today's dissonant heritage: Administrative buildings of collective farms in Estonia. *International Journal of Heritage Studies, 24*(9), 954–968. doi:10.1080/13527258.2018.1428664

Instagram. (2020). *Introducing Instagram Reels.* https://about.instagram.com/blog/announcements/introducing-instagram-reels-announcement

International Association of Amusement Parks and Attractions (IAAPA). (2018). *IAAPA history.* IAAPA.org.

ISO 9241-210: 2010 (2010). Ergonomics of human system interaction-Part 210: Human-centred design for interactive systems. International Standardization Organization (ISO).

Järv, R. (2010). Fairy tales and tourist trips. *Fabula, 51*(3–4), 281–294. doi:10.1515/FABL.2010.026

Jenkins, H. (2010). Transmedia storytelling and entertainment: An annotated syllabus. *Continuum: Journal of Media & Cultural Studies, 24*(6), 943–958. doi:10.1080/10304312.2010.510599

Jenkins, H. (2004). Game design as narrative architecture. In N. Wardrip-Fruin & P. Harrigan (Eds.), *First person: New media as story, performance, and game* (pp. 118–130). MIT Press.

Jenkins, H. (2006). *Convergence culture: Where old and new media collide.* New York University Press.

Jetter, H.-C., Geyer, F., Schwarz, T., & Reiterer, H. (2012). Blended Interaction: Toward a Framework for the Design of Interactive Spaces. *AVI 2012 : International Working Conference on Advanced Visual Interfaces.* https://kops.uni-konstanz.de/handle/123456789/43527

Jiang, S., Moyle, B., Yung, R., Tao, L., & Scott, N. (2022). Augmented reality and the enhancement of memorable tourism experiences at heritage sites. *Current Issues in Tourism*, 1–16. doi:10.1080/13683500.2022.2026303

Jones, J. (2019). Notes on a Mountain Man. In A. Harkins & M. McCarroll (Eds.), *Appalachian Reckoning: A Region Responds to Hillbilly Elegy* (pp. 270–277). West Virginia University Press.

Jones, K., & Wills, J. (2005). The invention of the park: Recreational landscapes from the Garden of Eden to Disney's Magic Kingdom. *Polity.*

Jung, T. H., & tom Dieck, M. C. (2017). Augmented reality, virtual reality and 3D printing for the co-creation of value for the visitor experience at cultural heritage places. *Journal of Place Management and Development, 10*(2), 140–151. doi:10.1108/JPMD-07-2016-0045

Jung, T., tom Dieck, M. C., Moorhouse, N., & tom Dieck, D. (2017). Tourists' experience of Virtual Reality applications. In *2017 IEEE International Conference on Consumer Electronics (ICCE)* (pp. 208-210). IEEE. 10.1109/ICCE.2017.7889287

Kang, J.-A., Hong, S., & Hubbard, G. T. (2020). The role of storytelling in advertising: Consumer emotion, narrative engagement level, and word-of-mouth intention. *Journal of Consumer Behaviour, 19*(1), 47–56. doi:10.1002/cb.1793

Kaosiri, Y. N., Fiol, L. J. C., Tena, M. A. M., Artola, R. M. R., & Garcia, J. S. (2019). User-generated content sources in social media: A new approach to explore tourist satisfaction. *Journal of Travel Research, 58*(2), 253–25. doi:10.1177/0047287517746014

Karzen, M. & Demonja, D. (2020). Importance of Storytelling: how to create more resilient cultural heritage? *Nova prisutnost, 18*(3), 653-668. doi:10.31192/np.18.3.15

Kasemsarn, K., & Nickpour, F. (2015). The challenge of digital storytelling for cultural tourism. *Journal of the Faculty of Architecture King Mongkut's Institute of Technology Ladkrabang, 20*(1), 154–164.

Katifori, A., Karvounis, M., Kourtis, V., Perry, S., Roussou, M., & Ioanidis, Y. (2018). Applying interactive storytelling in cultural heritage: opportunities, challenges and lessons learned. In R. Rouse, H. Koenitz, & M. Haahr (Eds.), *Interactive Storytelling. ICIDS 2018* (Vol. 11318, pp. 603–612). Springer. doi:10.1007/978-3-030-04028-4_70

Katifori, A., Perry, S., Vayanou, M., Antoniou, A., Ioannidis, I.-P., McKinney, S., Chrysanthi, A., & Ioannidis, Y. (2020). "Let Them Talk!": Exploring Guided Group Interaction in Digital Storytelling Experiences. *Journal on Computing and Cultural Heritage, 13*(3), 21. doi:10.1145/3382773

Katifori, A., Tsitou, F., Pichou, M., Kourtis, V., Papoulias, E., Ioannidis, Y., & Roussou, M. (2020). Exploring the Potential of Visually-Rich Animated Digital Storytelling for Cultural Heritage. In *Springer Series on Cultural Computing* (pp. 325–345). Springer., doi:10.1007/978-3-030-37191-3_17

Kavaratzis, M. (2004). From city marketing to city branding: Towards a theoretical framework for developing city brands. *Place Branding, 1*(1), 58–73. doi:10.1057/palgrave.pb.5990005

Kavaratzis, M., & Ashworth, G. J. (2006). City branding: An effective assertion of identity or a transitory marketing trick? *Place Branding, 2*(3), 183–194. doi:10.1057/palgrave.pb.5990056

Kay, I. (2013, September 17). Bi-narrative structure. *Pure Imagineering: Theme Park Narratology*. https://pureimagineering.blogspot.com/2013/09/experiential-and-presentational-stories.html

Kaya, B., & Gümüş, M. (2020). The importance of storytelling as a postmodern tool in the context of tourism. *Journal Of Applied Tourism Research, 1*(2), 115–132.

Keller, K. L. (2008). Strategic Brand Management - Building, Measuring, and Managing Brand Equity (D. Parker, Ed., 3rd ed.). Pearson Education.

Keller, K. L. (1993). Conceptualizing, Measuring, and Managing Customer-Based Brand Equity. *Journal of Marketing, 57*(1), 1–22. doi:10.1177/002224299305700101

Keller, K., & Kotler, P. (2012). *Marketing Management* (14th ed.). Pearson.

Kerr, G. (2005). From destination brand to location brand. *Journal of Brand Management, 13*(4-5), 276–283. doi:10.1057/palgrave.bm.2540271

Kerz, C. (2016). Atmosphere, immersion, and authenticity in Colonial Williamsburg. In S. Lukas (Ed.), *A reader in themed and immersive spaces* (pp. 195–205). ETC Press.

Keskin, H., Akgün, A., Zehir, C., & Ayar, H. (2016). Tales of cities: City branding through storytelling. *Journal of Global Strategic Management, 10*(1), 31–41. doi:10.20460/JGSM.20161022384

Keskin, N. (2000). *Efes*. Keskin Color.

Kidd, J. (2018). "Immersive" Heritage Encounters. *The Museum Review, 3*(1). http://orca.cf.ac.uk/id/eprint/110788

Kılıç, F., & Yolbulan Okan, E. (2020). Storytelling and narrative tools in award-winning advertisements in Turkey: An interdisciplinary approach. *Journal of Marketing Communications*, 1–16.

Kilteni, K., Groten, R., & Slater, M. (2012). The Sense of Embodiment in Virtual Reality. *Presence (Cambridge, Mass.)*, *21*(4), 373–387. doi:10.1162/PRES_a_00124

Kim, J. E., Lloyd, S., & Cervellon, M. C. (2016). Narrative-transportation storylines in luxury brand advertising: Motivating consumer engagement. *Journal of Business Research*, *69*(1), 304–313. doi:10.1016/j.jbusres.2015.08.002

Kim, J. H., & Youn, H. (2017). How to design and deliver stories about tourism destinations. *Journal of Travel Research*, *56*(6), 808–820. doi:10.1177/0047287516666720

Kim, K., Hwang, J., Zo, H., & Lee, H. (2016). Understanding user's continuance intention toward smartphone augmented reality applications. *Information Development*, *32*(2), 161–174. doi:10.1177/0266666914535119

Kim, S. H., Song, M. K., & Shim, C. (2020). Storytelling by medical tourism agents and its effect on trust and behavioural intention. *Journal of Travel & Tourism Marketing*, *37*(6), 679–694. doi:10.1080/10548408.2020.1795046

King, J. (2002). Destination marketing organizations—Connecting the experience rather than promoting the place. *Journal of Vacation Marketing*, *8*(2), 105–108. doi:10.1177/135676670200800201

Kohl, J. (2020). Four levels of interpretive development based on Tilden's principle of wholeness. *Journal of Interpretation Research*, *25*(1), 51–57. doi:10.1177/1092587220963520

Kolar, T., & Zabkar, V. (2010). A consumer-based model of authenticity: An oxymoron or the foundation of cultural heritage marketing? *Tourism Management*, *31*(5), 652–664. doi:10.1016/j.tourman.2009.07.010

Kontiza, K., Antoniou, A., Daif, A., Reboreda-Morillo, S., Bassani, M., González-Soutelo, S., Lykourentzou, I., Jones, C. E., Padfield, J., & López-Nores, M. (2020). On How Technology-Powered Storytelling Can Contribute to Cultural Heritage Sustainability across Multiple Venues—Evidence from the CrossCult H2020 Project. *Sustainability*, *12*(4), 1666. doi:10.3390u12041666

Korat, G. (2018). *Taş kapıdan taç kapıya Kapadokya*. Doğan Kitap.

Korez-Vide, R. (2017). Storytelling in sustainable tourism management: Challenges and opportunities for Slovenia. *Journal of Advanced Management Science, 5*(5).

Kotler, P., Armstrong, G., Saunders, J., & Wong, V. (1996). *Principles of marketing*. Prentice Hall Europe.

Kotler, P., Haider, D. H., & Rein, I. (1993). *Marketing Places: attracting investment, industry, ans tourism to cities, states, and nations*. Free Press.

Kotsi, F., Balakrishnan, M. S., Michael, I., & Ramsøy, T. Z. (2018). Place branding: Aligning multiple stakeholder perception of visual and auditory communication elements. *Journal of Destination Marketing & Management*, *7*, 112–130. doi:10.1016/j.jdmm.2016.08.006

Kremers, H. (2020). *Digital cultural heritage*. Springer. doi:10.1007/978-3-030-15200-0

Kreuter, M., Green, M. C., Cappella, J., Slater, M., Wise, M., Storey, D., & Woolley, S. (2007). Narrative communication in cancer prevention and control: A framework to guide research and application. *Annals of Behavioral Medicine*, *33*(3), 221–235. doi:10.1007/BF02879904 PMID:17600449

Kuflik, T., Wecker, A. J., Lanir, J., & Stock, O. (2015). An integrative framework for extending the boundaries of the museum visit experience: Linking the pre, during and post visit phases. *Information Technology & Tourism*, *15*(1), 17–47. doi:10.100740558-014-0018-4

Kusuma, P., Brucato, N., Cox, M. P., Letellier, T., Manan, A., Nuraini, C., Grangé, P., Sudoyo, H., & Ricaut, F.-X. (2017). The last sea nomads of the Indonesian archipelago: Genomic origins and dispersal. *European Journal of Human Genetics, 25*(8), 1004–1010. doi:10.1038/ejhg.2017.88 PMID:28513608

Lambert, J. (2003). *Digital storytelling cookbook and traveling companion.* Digital Diner Press.

Lambert, J. (2010). *The digital storytelling cookbook.* Digital Diner Press.

Lambert, J. (2013). *Digital storytelling: Capturing lives, creating community.* Routledge. doi:10.4324/9780203102329

Landorf, C. (2009). Managing for sustainable tourism: A review of six cultural World Heritage Sites. *Journal of Sustainable Tourism, 17*(1), 53–70. doi:10.1080/09669580802159719

Langellier, K., & Peterson, E. (2004). *Storytelling in daily life: Performing narrative.* Temple University Press.

Langhof, J. G., & Güldenberg, S. (2019). Pirates, ghosts and customer loyalty: Reviewing the dark ride experience. *Tourism Management Perspectives, 31*, 398–420. doi:10.1016/j.tmp.2019.06.004

Lapian, A. B. (2011). *Orang laut bajak laut raja laut: Sejarah kawasan laut Sulawesi abad XIX* [Sea People Sea Pirate Sea King: The history of Sulawesi sea in 19th century]. Komunitas Bambu.

Laroche, M., Kim, C., & Zhou, L. (1996). Brand familiarity and confidence as determinants of PI: An empirical test in a multiple brand context. *Journal of Business Research, 37*(2), 115–120. doi:10.1016/0148-2963(96)00056-2

Lebowitz, M. (1984). Creating characters in a story-telling universe. *Poetics, 13*(3), 171–194. doi:10.1016/0304-422X(84)90001-9

Le, D., Scott, N., & Lohmann, G. (2019). Applying experiential marketing in selling tourism dreams. *Journal of Travel & Tourism Marketing, 36*(2), 220–235. doi:10.1080/10548408.2018.1526158

Lee, S., Jeong, E., & Qu, K. (2020). Exploring theme park visitors' experience on satisfaction and revisit intention: A utilization of experience economy model. *Journal of Quality Assurance in Hospitality & Tourism, 21*(4), 474–497. doi:10.1080/1528008X.2019.1691702

Lee, T. H. (2013). Influence analysis of community resident support for sustainable tourism development. *Tourism Management, 34*, 37–46. doi:10.1016/j.tourman.2012.03.007

Leighton, D. (2007). "Step back in time and live the legend": Experiential marketing and the heritage sector. *International Journal of Nonprofit and Voluntary Sector Marketing, 12*(2), 117–125. doi:10.1002/nvsm.288

Leisen, B. (2001). Image Segmentation: The Case of a tourism destination. *Journal of Services Marketing, 15*(1), 49–66. doi:10.1108/08876040110381517

Lekić, R., Blajić, B., & Franjić, T. (2017, May). Interpretation of Mythical Landscape and Holy Geography in Creative Cultural Tourism. In *Proceedings of the 4th International Scientific Conference on ToSEE—Tourism in Southern and Eastern Europe, Opatija, Croatia* (pp. 4-6). Academic Press.

Lemelin, R. H., Whyte, K., Johansen, K., Higgins Desbiolles, F., Wilson, C., & Hemming, S. (2013). Conflicts, battlefields, indigenous peoples and tourism: Addressing dissonant heritage in warfare tourism in Australia and North America in the twenty-first century. *International Journal of Culture, Tourism and Hospitality Research, 7*(3), 257–271. doi:10.1108/IJCTHR-05-2012-0038

Lepp, A. (2007). Residents' attitudes towards tourism in Bigodi village, Uganda. *Tourism Management, 28*(3), 876–885. doi:10.1016/j.tourman.2006.03.004

Lewis, P. J. (2011). Storytelling as Research/Research as Storytelling. *Qualitative Inquiry, 17*(6), 505–510. doi:10.1177/1077800411409883

Li, C. H., & Liu, C. C. (2020). The effects of empathy and persuasion of storytelling via tourism micro-movies on travel willingness. *Asia Pacific Journal of Tourism Research, 25*(4), 382–392. doi:10.1080/10941665.2020.1712443

Lichrou, M., O'malley, L., & Patterson, M. (2010). Narratives of a tourism destination: Local particularities and their implications for place marketing and branding. *Place Branding and Public Diplomacy, 6*(2), 134–144. doi:10.1057/pb.2010.10

Liedtka, J. (2018). Why design thinking works. *Harvard Business Review, 96*(5), 72–79. https://designatdarden.org/app/uploads/2017/07/DSWP13-01rev2-1.docx

Lien, N. H., & Chen, Y. L. (2013). Narrative ads: The effect of argument strength and story format. *Journal of Business Research, 66*(4), 516–522. doi:10.1016/j.jbusres.2011.12.016

Li, J., Krishnamurthy, S., Roders, A., & Wesemael, P. (2020). Community participation in cultural heritage management: A systematic literature review comparing Chinese and international practices. *Cities (London, England), 96*, 102476. doi:10.1016/j.cities.2019.102476

Lim, Y. J., Osman, A., Salahuddin, S. N., Romle, A. R., & Abdullah, S. (2016). Factors influencing online shopping behaviour: The mediating role of PI. *Procedia Economics and Finance, 35*(1), 401–410. doi:10.1016/S2212-5671(16)00050-2

Lindstrom, M. (2013). *Brand Sense. Os Segredos Sensoriais que nos levam a comprar.* Gestãoplus Edições.

Lin, L. P. L., Huang, S. C. L., & Ho, Y. C. (2020). Could virtual reality effectively market slow travel in a heritage destination? *Tourism Management, 78*, 104027. doi:10.1016/j.tourman.2019.104027

Little, C., Patterson, D., Mitchell, L., Pappin, D., & Rosendahl, D. (2018, May). Willandra Lakes Fossil Human Trackway: Presentation and Protection. In *Proceedings of the 3r International Conference on Best Practices in World Heritage: Integral Actions Menorca, Spain* (pp. 4 - 7). Academic Press.

Little, C., Bec, A., Moyle, B. D., & Patterson, D. (2020). Innovative methods for heritage tourism experiences: Creating windows into the past. *Journal of Heritage Tourism, 15*(1), 1–13. doi:10.1080/1743873X.2018.1536709

Little, C., Patterson, D., Moyle, B., & Bec, A. (2018, January). Every footprint tells a story: 3D scanning of heritage artifacts as an interactive experience. In *Proceedings of the Australasian Computer Science Week Multiconference* (pp. 1-8). 10.1145/3167918.3167974

Liu, H., Wu, L., & Li, X. (2019). Socia media envy: How experience sharing ons ocial networking site drives millennials' aspirational tourism consumption. *Journal of Travel Research, 58*(3), 355–369. doi:10.1177/0047287518761615

Liu, L., & Minton, A. (2018). Faith-filled brands: The interplay of religious branding and brand engagement in the self-concept. *Journal of Retailing and Consumer Services, 44*, 305–314. doi:10.1016/j.jretconser.2018.07.022

Liu, Y., Dupre, K., & Jin, X. (2021). A systematic review of literature on contested heritage. *Current Issues in Tourism, 24*(4), 442–465. doi:10.1080/13683500.2020.1774516

Li, Y. M. (2014). Effects of story marketing and travel involvement on tourist behavioral intention in the tourism industry. *Sustainability, 6*(12), 9387–9397. doi:10.3390u6129387

Logan, W. S., Nic, C. M., & Kockel, U. (Eds.). (2015). *A companion to heritage studies*. Wiley-Blackwell. doi:10.1002/9781118486634

Lombardo, V., & Damiano, R. (2012). Storytelling on mobile devices for cultural heritage. *New Review of Hypermedia and Multimedia, 18*(1–2), 11–35. doi:10.1080/13614568.2012.617846

Long, G. A. (2007). *Transmedia storytelling: Business, aesthetics, and production at the Jim Henson Company* [Doctoral dissertation]. Massachusetts Institute of Technology.

Lowenthal, D. (1985). *The past is a foreign country*. Cambridge University Press.

Lowenthal, D. (2015). *The past is a foreign country: Revisited*. Cambridge University Press. doi:10.1017/CBO9781139024884

Lu, B., Fan, W., & Zhou, M. (2016). Social presence, trust, and social commerce PI: An empirical research. *Computers in Human Behavior, 56*(1), 225–237. doi:10.1016/j.chb.2015.11.057

Ludwig, T. (2021). How can interpetive experiences make people mindful? In *Proceedings of Interpret Europe Web Conference on Recreating Tourism through Heritage Interpretation* (pp. 22-44). Interpret Europe.

Ludwig, C., & Wang, Y.-W. (2020). Open-air museums and historical theme parks in the UK and China. In C. Lugwig, L. Walton, & Y.-W. Wang (Eds.), *The heritage turn in China: The reinvention, dissemination, and consumption of heritage* (pp. 131–168). Amsterdam University Press.

Lukas, S. (2012). *The immersive worlds handbook*. Focal Press. doi:10.4324/9780240820989

Lundby, K. (Ed.). (2008). *Digital storytelling, mediatized stories: Self-representations in new media*. Peter Lang.

Lund, N. F., Cohen, S. A., & Scarles, C. (2018). The power of social media storytelling in destination branding. *Journal of Destination Marketing & Management, 8*, 271–280. doi:10.1016/j.jdmm.2017.05.003

Lund, N. F., & Kimbu, A. N. (2021). Applying the Hollywood scriptwriting formula to destination branding. *Current Issues in Tourism, 24*(8), 1058–1078. doi:10.1080/13683500.2020.1739005

Lund, N., Scarles, C., & Cohen, S. (2020). The Brand Value Continuum: Countering Co-destruction of Destination Branding in Social Media through Storytelling. *Journal of Travel Research, 59*(8), 1506–1521. doi:10.1177/0047287519887234

Lundqvist, A., Liljander, V., Gummerus, J., & van Riel, A. (2012). The impact of storytelling on the consumer brand experience: The case of a firm-originated story. *Journal of Brand Management, 20*(4), 283–297. doi:10.1057/bm.2012.15

Lutz, B., & Weintke, M. (1999). Virtual Dunhuang Art Cave: A cave within a CAVE. *Computer Graphics Forum, 18*(3), 257–264. doi:10.1111/1467-8659.00346

Macgregor, C. (2022). Preserving the ancient human trackways site in the Willandra lakes world heritage area. *Studies in Conservation, 67*(sup1), 150–155. Advance online publication. doi:10.1080/00393630.2022.2050616

Madden, T. J., Ellen, P. S., & Ajzen, I. (1992). A comparison of the theory of planned behaviour and the theory of reasoned action. *Personality and Social Psychology Bulletin, 18*(1), 3–9. doi:10.1177/0146167292181001

Magnani, M., Guttorm, A., & Magnani, N. (2018). Three-dimensional, community-based heritage management of indigenous museum collections: Archaeological ethnography, revitalization and repatriation at the Sámi Museum Siida. *Journal of Cultural Heritage, 31*, 162–169. doi:10.1016/j.culher.2017.12.001

Mak, A. H. (2017). Online Destination Image: Comparing National Tourism Organisation's and Tourists' Perspectives. *Tourism Management, 60*, 280–297. doi:10.1016/j.tourman.2016.12.012

Marasco, A., & Balbi, B. (2019). Designing Accessible Experiences for Heritage Visitors through Virtual Reality. *e-Review of Tourism Research, 17*(3), 426-443.

Marasco, A. (2020). Beyond virtual cultural tourism: History-living experiences with cinematic virtual reality. *Tourism and Heritage Journal*, *2*, 1–16. doi:10.1344/THJ.2020.2.1

Marasco, A., Buonincontri, P., van Niekerk, M., Orlowski, M., & Okumus, F. (2018). Exploring the role of next-generation virtual technologies in destination marketing. *Journal of Destination Marketing & Management*, *9*, 138–148. doi:10.1016/j.jdmm.2017.12.002

Mardis, B. (2007, May 17). Looking back at the birth of Lake Cumberland. *Commonwealth Journal*. https://www.somerset-kentucky.com/news/looking-back-at-the-birth-of-lake-cumberland/article_35f3a55c-93b1-5030-8283-421648164e48.html

Marechal, G. (2010). Autoethnography. In A. Mills, G. Durepos, & E. Wiebe (Eds.), *Encyclopedia of case study research* (Vol. 2, pp. 43–45). Sage Publications.

Marques, F. (2001). O Creoula: a pesca do bacalhau no crepúsculo da nevegação à vela. In Á. Garrido (Ed.), *A Pesca do Bacalhau: História e Memória* (pp. 225–253). Editorial Notícias.

Marstine, J. (2006). Introduction. In J. Martsine (Ed.), *New Museum Theory and Practice: An Introduction* (pp. 1–36). Blackwell. doi:10.1002/9780470776230.ch

Martin, J. I. (2011). *A pesca em Portugal*. Academic Press.

Martin, J., Feldman, M. S., Hatch, M. J., & Sitkin, S. B. (1983). The uniqueness paradox in organizational stories. *Administrative Science Quarterly*, *28*(3), 438–453. doi:10.2307/2392251

Maschio, T. (2017). *Storyliving: An ethnographic study of how audiences experience VR and what that means for journalists*. Google News Lab.

Mathis, E. F., Kim, H. L., Uysal, M., Sirgy, J. M., & Prebensen, N. K. (2016). The effect of co-creation experience on outcome variable. *Annals of Tourism Research*, *57*, 62–75. doi:10.1016/j.annals.2015.11.023

Mathisen, L. I. N. E., and Chen, J. S. (2014). Storytelling in a co-creation perspective. *Creating experience value in tourism*, 157-168.

Mathisen, L. (2018). Storytelling in a co-creation perspective. In N. K. Prebensen, J. S. Chen, & M. S. Uysal (Eds.), *Creating experience value in tourism* (pp. 137–147). CABI. doi:10.1079/9781786395030.0137

Maurya, U. K., & Mishra, P. (2012). What is a brand? A perspective on brand meaning. *European Journal of Business and Management*, *4*(3), 122–133.

Maze, J. R. (1973). The concept of ATT. *Inquiry*, *16*(1-4), 168–205. doi:10.1080/00201747308601684

McAdams, D. P. (2013). The Psychological Self as Actor, Agent, and Author. *Perspectives on Psychological Science*, *8*(3), 272–295. doi:10.1177/1745691612464657 PMID:26172971

McCabe, S., & Foster, C. (2006). The role and function of narrative in tourist interaction. *Journal of Tourism and Cultural Change*, *4*(3), 194–215. doi:10.2167/jtcc071.0

Mcdonald, J. K. (2009). Imaginative instruction: What master storytellers can teach instructional designers instructional designers. *Educational Media International*, *46*(2), 111–222. doi:10.1080/09523980902933318

Mcgregor, I., & Holmes, J. G. (1999). How storytelling shapes memory and impressions of relationship events over time. *Journal of Personality and Social Psychology*, *76*(3), 403–419. doi:10.1037/0022-3514.76.3.403

McLellan, H. (2000). Experience design. *Cyberpsychology & Behavior*, *3*(1), 59–69. doi:10.1089/109493100316238

Medlik, S., & Middleton, V. T. C. (1973). The Tourist Product and its Marketing Implications. *International Tourism Quarterly, 3.*

Meehan, J. R. (1977). TALE-SPLIN: An interactive program that writes stories. *Proceedings of the Fifth International Joint Conference on Artificial Intelligence, 77*, 91-98.

Mei, X. Y., Hågensen, A. M. S., & Kristiansen, H. S. (2020). Storytelling through experiencescape: Creating unique stories and extraordinary experiences in farm tourism. *Tourism and Hospitality Research, 20*(1), 93–104. doi:10.1177/1467358418813410

Meng, B., & Cui, M. (2020). The role of co-creation experience in forming tourists' revisit intention to home-based accommodation: Extending the theory of planned behavior. *Tourism Management Perspectives, 33*(100581), 1–13. doi:10.1016/j.tmp.2019.100581

Messenger, P. M., & Smith, G. S. (Eds.). (2014). *Cultural heritage management: A global perspective*. University Press of Florida.

Michopoulou, E., & Buhalis, D. (2013). Information provision for challenging markets: The case of the accessibility requiring market in the context of tourism. *Information & Management, 50*(5), 229–239. doi:10.1016/j.im.2013.04.001

Michopoulou, E., Darcy, S., Ambrose, I., & Buhalis, D. (2015). Accessible tourism futures: The world we dream to live in and the opportunities we hope to have. *Journal of Tourism Futures, 1*(3), 179–188. doi:10.1108/JTF-08-2015-0043

Mijnheer, C. L., & Gamble, J. R. (2019). Value co-creation at heritage visitor attractions: A case study of Gladstone's Land. *Tourism Management Perspectives, 32*(100567), 1–12. doi:10.1016/j.tmp.2019.100567

Mittell, J. (2013). Forensic fandom and the drillable text. *Spreadable Media*. https://spreadablemedia.org/essays/mittell/index.html#.X0bapshKhnI

Moin, S. M. A., Hosany, S., & O'Brien, J. (2020). Storytelling in destination brands' promotional videos. *Tourism Management Perspectives, 34*, 100639. doi:10.1016/j.tmp.2020.100639

Montagud, M., Orero, P., & Matamala, A. (2020). Culture 4 all: Accessibility-enabled cultural experiences through immersive VR360 content. *Personal and Ubiquitous Computing, 24*(6), 887–905. doi:10.100700779-019-01357-3

Morgan, S. W., & Dennehy, R. F. (1995). Organizational storytelling: Telling tales in the business classroom. *Developments in Business Simulation & Experiential Exercises, 22*, 160-165.

Morgan, S. W., & Dennehy, R. F. (1997). The power of organizational storytelling: A management. *Journal of Management Development, 16*(7), 494–501. doi:10.1108/02621719710169585

Morwitz, V. G., Steckel, J. H., & Gupta, A. (2007). When do PIpredict sales? *International Journal of Forecasting, 23*(3), 347–364. doi:10.1016/j.ijforecast.2007.05.015

Moscardo, G. (2010). The shaping of tourist experience: The importance of stories and themes. *The tourism and leisure experience: Consumer and managerial perspectives, 44*, 43-58.

Moscardo, G. (2018). Tourist Experience Design: A Storytelling Framework. In Quality Services and Experiences in Hospitality and Tourism (pp. 93–107). Emerald. doi:10.1108/S2042-144320180000009007

Moscardo, G. (2021). The story turn in tourism: Forces and futures. *Journal of Tourism Futures, 7*(2), 168-173.

Moscardo, G. (2008). Understanding tourist experience through mindfulness theory. In M. Kozak & A. DeCrop (Eds.), *Handbook of Tourist Behavior* (pp. 99–115). Routledge. doi:10.4324/9780203881804.pt4

Moscardo, G. (2010). The Shaping of Tourist Experience: The Importance of Stories and Themes. In M. Morgan, P. Lugosi, & J. Ritchie (Eds.), *The Tourism and Leisure Experience* (pp. 43–58). Channel View. doi:10.21832/9781845411503-006

Moscardo, G. (2017a). Exploring mindfulness and stories in tourist experiences. *International Journal of Culture, Tourism and Hospitality Research, 11*(2), 111–124. doi:10.1108/IJCTHR-11-2016-0108

Moscardo, G. (2017b). Stories as a Tourist Experience Design Tool. In D. R. Fesenmaier & Z. Xiang (Eds.), *Design Science in Tourism: Foundations of Destination Management* (pp. 97–124). Springer International Publishing. doi:10.1007/978-3-319-42773-7_7

Moscardo, G. (2020a). The story turn in tourism: Forces and futures. *Journal of Tourism Futures, 7*(2), 168–173. doi:10.1108/JTF-11-2019-0131

Moscardo, G. (2020b). Stories and design in tourism. *Annals of Tourism Research, 83*, 102950. doi:10.1016/j.annals.2020.102950

Mossberg, L., Therkelsen, A., Huijbens, E. H., Björk, P., & Olsson, A. K. (2010). *Storytelling and Destination Development.* Nordic Council of Ministers. www.handels.gu.se

Mossberg, L., Therkelsen, A., Huijbens, E., Björk, A-K., & Olsson, P. (2010). *Storytelling and Destination Development.* Nordic Innovation Centre (NICe).

Mossberg, L., Therkelsen, A., Huijbens, E., Björk, P., & Olsson, A. (2010). *Storytelling and destination development.* Nordic Innovation Centre (NICe) project number: 08041. Available at: https://www.nordicinnovation.org/2010/storytelling-and-destination-development

Mossberg, L., Therkelsen, A., Huijbens, H. E., Björk, A-K., & Olsson, P. (2010). Storytelling and Destination Development. Nordic Innovation Centre (NICe) Project Number: 08041.

Mossberg, L. (2008). Extraordinary experiences through storytelling. *Scandinavian Journal of Hospitality and Tourism, 8*(3), 195–210. doi:10.1080/15022250802532443

Mossberg, L., & Eide, D. (2017). Storytelling and meal experience concepts. *European Planning Studies, 25*(7), 1184–1199. doi:10.1080/09654313.2016.1276156

Muenster, S., Apollonio, F. I., Bell, P., Kuroczynski, P., Di Lenardo, I., Rinaudo, F., & Tamborrino, R. (2019). Digital cultural heritage meets digital humanities. In D. Gonzalez-Aguilera, F. Remondino, I. Toschi & ICOMOS/ISPRS Committee for Documentation of Cultural Heritage (Eds.), *27th CIPA International Symposium "Documenting the past for a better future" 1–5 September 2019, Ávila, Spain* (pp. 812-820). Copernicus Publications.

Muljono-Larue, F., & Pranoto, I. B. (2008). *Orang Bajo Suku Pengembara Laut: Pengalaman Seorang Antropolog* [Nomadic people of the sea: The Badjos of Indonesia] (F. Zacot, Trans.). Kepustakaan Populer Gramedia. (Original work published 2002)

Munar, A. M. (2016). The house of tourism studies and the systemic paradigm. In *Tourism Social Science Series* (Vol. 22, pp. 131–153). Emerald Group Publishing Ltd. doi:10.1108/S1571-504320150000022014

Muñiz, A. M. Jr, & Schau, H. J. (2007). Vigilante marketing and consumer-created communications. *Journal of Advertising, 36*(3), 35–50. doi:10.2753/JOA0091-3367360303

Mura, P., & Sharif, S. P. (2017). Narrative analysis in tourism: A critical review. *Scandinavian Journal of Hospitality and Tourism, 17*(2), 194–207. doi:10.1080/15022250.2016.1227276

Mura, P., Tavakoli, R., & Sharif, S. P. (2017). 'Authentic but not too much': Exploring perceptions of authenticity of virtual tourism. *Information Technology & Tourism, 17*(2), 145–159. doi:10.100740558-016-0059-y

Murray, J. H. (2017). *Hamlet on the Holodeck: The Future of Narrative in Cyberspacr*. MIT Press. https://mitpress.mit.edu/books/hamlet-holodeck-updated-edition

Murzyn-Kupisz, M. (2012). Cultural, economic and social sustainability of heritage tourism: Issues and challenges. *Economic and Environmental Studies, 12*(2), 113–133.

Nabi, R. L., & Green, M. C. (2015). The Role of a Narrative's Emotional Flow in Promoting Persuasive Outcomes. *Media Psychology, 18*(2), 137–162. doi:10.1080/15213269.2014.912585

Naef, P., & Ploner, J. (2016). Tourism, conflict and contested heritage in former Yugoslavia. *Journal of Tourism and Cultural Change, 14*(3), 181–188. doi:10.1080/14766825.2016.1180802

Naniopoulos, A., & Tsalis, P. (2015). A methodology for facing the accessibility of monuments developed and realised in Thessaloniki, Greece. *Journal of Tourism Futures, 1*(3), 240–253. doi:10.1108/JTF-03-2015-0007

Nash, K. (2011). Documentary-for-the-Other: Relationships, Ethics and (Observational) Documentary. *Journal of Mass Media Ethics, 26*(3), 224–239. doi:10.1080/08900523.2011.581971

Nasser, N. (2003). Planning for urban heritage places: Reconciling conservation, tourism, and sustainable development. *Journal of Planning Literature, 17*(4), 467–479. doi:10.1177/0885412203017004001

National Association for Interpretation. (2021). *What is interpretation?* https://www.interpnet.com/ nai/interp/About/What_is_Interpretation_/nai/_About/what_is_interp.aspx?hkey=b5ddeff3-03a8-4000bf7 3-433c37c8a7af

Naversen, N. (2017). *The sliced onion technique: How to theme a space*. Themed Attraction.

Nezakati, H., Amidi, A., Jusoh, Y. Y., Moghadas, S., Aziz, Y. A., & Sohrabinezhadtalemi, R. (2015). Review of social media potential on knowledge sharing and collaboration in tourism industry. *Procedia: Social and Behavioral Sciences, 172*, 120–125. doi:10.1016/j.sbspro.2015.01.344

Nielsen, J. K. (2017). Museum communication and storytelling: Articulating understandings within the museum structure. *Museum Management and Curatorship, 32*(5), 440–455. doi:10.1080/09647775.2017.1284019

Noy, C. (2007). The Poetics of Tourist Experience: An Autoethnography of a Family Trip to Eilat. *Journal of Tourism and Cultural Change, 5*(3), 141–157. doi:10.2167/jtcc085.0

Nozdreva, R., & Churakova, A. (2021). Digital Tools in Empirical Marketing Strategies of Global Hotel Companies. In N. Konina (Ed.), *Digital Strategies in a Global Market: Navigating the Fourth Industrial Revolution* (pp. 213–227). Palgrave Macmillan. doi:10.1007/978-3-030-58267-8_16

O'Keefe, B., Benyon, D., Chandwani, G., Menon, M., & Duke, I. I. R. (2014). A Blended Space for Heritage Storytelling. *Proceedings of the 28th International BCS Human Computer Interaction Conference: Sand, Sea and Sky - Holiday HCI, HCI 2014*, 120–129. 10.14236/ewic/HCI2014.13

O'Leary, J. T., & Fesenmaier, D. (2017). Concluding Remarks: Tourism Design and the Future of Tourism. In *Design Science in Tourism: Foundations of Destination Management* (pp. 265–272). Springer. doi:10.1007/978-3-319-42773-7_16

OECD. (2014). *Tourism and creative economy*. OECD. doi:10.1787/9789264207875-

OECD. (2014). *Tourism and the creative economy, OECD Studies on Tourism*. OECD Publishing.

Ohler, J. (2006). The world of digital storytelling. *Educational Leadership, 63*(4), 44–47.

Okech, R. N. (2010). Socio-cultural impacts of tourism on World Heritage sites: Communities' perspective of Lamu (Kenya) and Zanzibar Islands. *Asia Pacific Journal of Tourism Research, 15*(3), 339–351. doi:10.1080/10941665.2010.503624

Oliveira, T., Araujo, B., & Tam, C. (2020). Why do people share their travel experiences on social media? *Tourism Management, 78*, 1–14. doi:10.1016/j.tourman.2019.104041 PMID:32322615

Olsen, D. H., & Timothy, D. J. (2002). Contested religious heritage: Differing views of Mormon heritage. *Tourism Recreation Research, 27*(2), 7–15. doi:10.1080/02508281.2002.11081215

Olson, J. M., & Zanna, M. (1993). ATTs and ATT change. *Annual Review of Psychology, 44*(1), 117–154. doi:10.1146/annurev.ps.44.020193.001001

Olsson, A. K., Therkelsen, A., & Mossberg, L. (2016). Making an effort for free – volunteers' roles in destination-based storytelling. *Current Issues in Tourism, 19*(7), 659–679. doi:10.1080/13683500.2013.784242

Ooi, C. S. (2002). *Cultural tourism and tourism cultures: The business of mediating experiences in Copenhagen and Singapore*. Copenhagen Business School Press DK.

Overend, D. (2012). Performing sites: Illusion and authenticity in the spatial stories of the guided tour. *Scandinavian Journal of Hospitality and Tourism, 12*(1), 44–54. doi:10.1080/15022250.2012.678070

Oye, N. D. (2013). Reducing corruption in African developing countries: The relevance of E-Governance. *Greener Journal of Social Sciences, 3*(1), 6–13. doi:10.15580/GJSS.2013.1.103112183

Ozdemir, M. A. (2021). Virtual Reality (VR) and Augmented Reality (AR) technologies for accessibility and marketing in the tourism industry. In ICT Tools and Applications for Accessible Tourism (pp. 277-301). IGI Global.

Pace, S. (2008). YouTube: An Opportunity for Consumer Narrative Analysis? *Qualitative Market Research, 11*(2), 213–226. doi:10.1108/13522750810864459

Pachucki, C., Grohs, R., & Scholl-Grissemann, U. (2021). No story without a storyteller: The impact of the storyteller as a narrative element in online destination marketing. *Journal of Travel Research*. doi:10.1177/00472875211046052

Pagano, A., Palombini, A., Bozzelli, G., De Nino, M., Cerato, I., & Ricciardi, S. (2020). Arkaevision VR game: User experience research between real and virtual Paestum. *Applied Sciences (Basel, Switzerland), 10*(9), 3182. doi:10.3390/app10093182

Page, R. E. (2012). Stories and social media: Identities and interaction. In *Stories and Social Media: Identities and Interaction*. Taylor and Francis. doi:10.4324/9780203148617

Palombini, A. (2017). Storytelling and telling history. Towards a grammar of narratives for cultural heritage dissemination in the digital era. *Journal of Cultural Heritage, 24*, 134–139. doi:10.1016/j.culher.2016.10.017

Pan, B., MacLaurin, T., & Crotts, J. C. (2007). Travel Blogs and the Implications for Destination Marketing. *Journal of Travel Research, 46*(August), 35–45. doi:10.1177/0047287507302378

Park, H., Kim, J., Bang, S., & Woo, W. (2018). The effect of applying film-induced tourism to virtual reality tours of cultural heritage sites. In *3rd Digital Heritage International Congress (DigitalHERITAGE)* (pp. 1-4). IEEE. 10.1109/DigitalHeritage.2018.8810089

Pasquinelli, C. (2013). Competition, cooperation and co-opetition: Unfolding the process of inter-territorial branding. *Urban Research & Practice, 6*(1), 1–18. doi:10.1080/17535069.2012.727579

Pawaskar, P., & Goel, M. (2014). A Conceptual Model: Multisensory Marketing and Destination Branding. *Procedia Economics and Finance, 11*, 255–267. doi:10.1016/S2212-5671(14)00194-4

Pearce, J., & Moscardo, G. (2019). The role of stories in travel posts to social media. In M. Mansson, A. Buchmann, C. Cassinger, & L. Eskilsson (Eds.), *Routledge Companion to Media and Tourism*. Routledge.

Pekin, F. (2018). *Kapadokya: kayalardaki şiirsellik*. İletişim Yayınları.

Pellowski, A. (1977). *The World of Storytelling*. R. R. Bowker Company.

Peralta, E. (2003). O mar por tradição: O património e a construção das imagens do turismo. *Horizontes Antropológicos*, *9*(20), 83–96. doi:10.1590/S0104-71832003000200005

Peralta, E. (2010). "We're all seafarers": An ethnography of the (in)visibilities of power in the social representation of the local past in Ílhavo. *Etnográfica (Lisboa)*, *14*(3), 443–464. doi:10.4000/etnografica.249

Pera, R. (2017). Empowering the new traveller: Storytelling as a co-creative behavior. *Current Issues in Tourism*, *20*(4), 331–338. doi:10.1080/13683500.2014.982520

Pera, R., & Viglia, G. (2016). Exploring how video DST builds relationship. *Psychology and Marketing*, *33*(12), 1142–1150. doi:10.1002/mar.20951

Pérez, E., Merchán, P., Merchán, M. J., & Salamanca, S. (2020). Virtual reality to foster social integration by allowing wheelchair users to tour complex archaeological sites realistically. *Remote Sensing*, *12*(3), 419. doi:10.3390/rs12030419

Perovic, Đ., Moric, I., Pekovic, S., Stanovcic, T., Roblek, V., & Pejic Bach, M. (2018). The antecedents of tourist repeat visit intention: Systemic approach. *Kybernetes*, *47*(9), 1857–1871. doi:10.1108/K-12-2017-0480

Perry, S., Roussou, M., Economou, M., Young, H., & Pujol, L. (2017). Moving beyond the virtual museum: Engaging visitors emotionally. In *2017 23rd International Conference on Virtual System & Multimedia (VSMM)* (pp. 1-8). IEEE.

Peterson, R. A. (1994). Meta-analysis of Alpha Cronbach's Coefficient. *The Journal of Consumer Research*, *21*(2), 381–391. doi:10.1086/209405

Petrelli, D., Dulake, N., Marshall, M., Kockelkorn, H., & Pisetti, A. (2016, April 7). Do it together: the effect of curators, designers, and technologists sharing the making of new interactive visitors' experiences. *Museums and the Web 2016*. https://mw2016.museumsandtheweb.com/paper/do-it-together-the-effect-of-curators-designers-and-technologists-sharing-the-making-of-new-interactive-visitors-experiences/

Petty, R. E., & Wegener, D. T. (1996). ATT change: Multiple roles for persuasion variables. In D. Gilbert, S. Fiske, & G. Lindzey (Eds.), *The handbook of social psychology* (Vol. 1, pp. 323–390). McGraw-Hill.

Petty, R., Cacioppo, J., & Schumann, D. (1983). Central and Peripheral Routes to Advertising Effectiveness: The Moderating Role of Involvement. *The Journal of Consumer Research*, *10*(Sept), 135–146. doi:10.1086/208954

Philips, D. (1999). Narrativised spaces: The functions of stories in theme parks. In D. Crouch (Ed.), *Leisure/Tourism geographies: Practices and geographical knowledge* (pp. 92–108). Routledge.

Pieraccini, M., Guidi, G., & Atzeni, C. (2001). 3D digitizing of cultural heritage. *Journal of Cultural Heritage*, *2*(1), 63–70. doi:10.1016/S1296-2074(01)01108-6

Pietroni, E., Forlani, M., & Rufa, C. (2015). Livia's Villa Reloaded: An example of re-use and update of a pre-existing Virtual Museum, following a novel approach in storytelling inside virtual reality environments. 2015 Digital Heritage, 511-518.

Pietroni, E., & Adami, A. (2014). Interacting with virtual reconstructions in museums: The Etruscanning Project. *Journal on Computing and Cultural Heritage*, *7*(2), 1–29. doi:10.1145/2611375

Pike, S. (2007). Consumer-based brand equity for destinations. *Journal of Travel & Tourism Marketing*, *22*(1), 51–61. doi:10.1300/J073v22n01_04

Pikkemaat, B., & Schukert, M. (2007). Success factors of theme parks: An exploratory study. *Tourism (Zagreb)*, *55*(2), 197–208.

Pine, B. J., II, & Gilmore, J. H. (1998, July-August). Welcome to the experience economy. *Harvard Business Review*. https://hbr.org/1998/07/welcome-to-the-experience-economy

Pine, B. J., & Gilmore, J. H. (2011). *The experience economy* (2nd ed.). Harvard Business School Publishing.

Pine, J., & Gilmore, J. H. (1998). Welcome to the experience economy. *Harvard Business Review*, *76*(4), 97–105. PMID:10181589

Pine, J., & Gilmore, J. H. (1998). Welcome to the Experience Economy. *Harvard Business Review*, *76*(4), 97–105. PMID:10181589

Pinto, R., & Neto, D. C. (2017). The Brazilian demand for dried and salted cod - from 1989 to 2014. *Smart and Inclusive Development in Rural Areas - Iberian Conference on Rural Studies.*

Pinzaru, F. (2012). From conjunctural urban storytelling to city branding: An empirical model for bucharest. *Management Research And Practice*, *4*(2), 33–44.

Pires, A. da R., Castro, E. A. de, Esteves, C., & Santos, C. (2000). *Ilhavo 2020: construir uma Visão Partilhada das Opções e dos Percursos de Desenvolvimento*. Academic Press.

Piscitelli, M. (2019). Promotion of heritage tourism through branding and graphic communication. In S. Srivastava (Ed.), *Conservation and promotion of heritage tourism* (pp. 87–106). IGI Global. doi:10.4018/978-1-5225-6283-2.ch004

Polletta, F., Chen, P., Gardner, B., & Motes, A. (2011). The sociology of storytelling. *Annual Review of Sociology*, *37*(1), 109–130. doi:10.1146/annurev-soc-081309-150106

Pollock, S., Quesenberry, K. A., & Coolsen, M. K. (2019). Drama goes viral: Effects of story development on shares and views of online advertising videos. *Journal of Interactive Marketing*, *48*, 1–16. doi:10.1016/j.intmar.2019.05.001

Pompílio, M. (2019). *O storytelling como recurso na promoção do turismo em Fortaleza*. Master Dissertation, Universidade Estadual do Ceará.

Popova, Y. B. (2015). *Stories, meaning, and experience: Narrativity and enaction* (1st ed.). Routledge., doi:10.4324/9781315880488

Poria, Y., Butler, R., & Airey, D. (2003). The core of heritage tourism. *Annals of Tourism Research*, *30*(1), 238–254. doi:10.1016/S0160-7383(02)00064-6

Poulios, I. (2014). *Past in the present: A living heritage approach*. Ubiquity Press.

Poux, F., Valembois, Q., Mattes, C., Kobbelt, L., & Billen, R. (2020). Initial user-centered design of a virtual reality heritage system: Applications for digital tourism. *Remote Sensing, 12*(16), 1-32.

Prahalad, C. K., & Ramaswamy, V. (2004). Co-creating unique value with customers. *Strategy and Leadership*, *32*(3), 4–9. doi:10.1108/10878570410699249

Prahalad, C., & Ramaswamy, V. (2004). Co-creation experiences: The next practice in value creation. *Journal of Interactive Marketing*, *18*(3), 5–14. doi:10.1002/dir.20015

Prebensen, N. K., Chen, J. S., & Uysal, M. S. (2017). Tourist experience creation An overview. *Co-Creation in Tourist Experiences,* 1–9.

Prebensen, N. K., Chen, J. S., & Uysal, M. S. (2018). Co-creation of tourist experience: Scope, definition and structure. In N. K. Prebensen, J. S. Chen, & M. S. Uysal (Eds.), *Creating experience value in tourism* (pp. 1–10). CABI. doi:10.1079/9781786395030.0001

Prebensen, N. K., & Foss, L. (2011). Coping and co-creating in tourist experiences. *International Journal of Tourism Research, 13*(1), 54–67. doi:10.1002/jtr.799

Prebensen, N. K., Vittersø, J., & Dahl, T. I. (2013). Value Co-creation significance of tourist resources. *Annals of Tourism Research, 42,* 240–261. doi:10.1016/j.annals.2013.01.012

Prebensen, N. K., Woo, E., & Uysal, M. S. (2014). Experience value: Antecedents and consequences. *Current Issues in Tourism, 17*(10), 910–928. doi:10.1080/13683500.2013.770451

Prebensen, N. K., & Xie, J. (2017). Efficacy of co-creation and mastering on perceived value and satisfaction in tourists' consumption. *Tourism Management, 60,* 166–176. doi:10.1016/j.tourman.2016.12.001

Prentice, R. (1993). *Tourism and heritage attractions.* Routledge.

Prentice, R. (2005). Heritage: The key sector in the 'new' tourism. In G. Corsane (Ed.), *Heritage, museums and galleries: An introductory reader* (pp. 267–278). Routledge. doi:10.4324/9780203326350_chapter_19

Price, H. (2004). *Walt's revolution! By the numbers.* Ripley Entertainment.

Providência, F. (2004). *Manual de Identidade Gráfica - Câmara Municipal de Ílhavo.* Academic Press.

Puczko, L., & Ratz, T. (2000). Tourist and resident perceptions of the physical impacts of tourism at Lake Balaton, Hungary: issues for sustainable tourism management. *Journal of Sustainable Tourism, 8*(6), 458-478.

Puhakka, R. (2011). *Matkailukysynnän trendit vuoteen 2030 mennessä.* http://www.kulmat.fi/images/tiedostot/Artikkelit/tuleva_trendit_2030.pdf

Rabinovitz, L. (2006). From Hale's Tours to Star Tours: Virtual voyage, travel ride films, and the delirium of the hyper-real. In J. Ruoff (Ed.), *Virtual voyages: Cinema and travel* (pp. 42–60). Duke University Press. doi:10.2307/j.ctv1131b73.6

Rach, M., & Scott, R. (2020). The Use of Virtual Reality in Marketing: Exploring the Need for Technology and Language Adaptation to Create High Quality Immersive Experiences. In *Marketing and Smart Technologies* (pp. 67–77). Springer. doi:10.1007/978-981-15-1564-4_8

Radić, M., Ravasi, D., & Munir, K. (2021). Privatization: Implications of a shift from state to private ownership. *Journal of Management.*

Rahn, S. (2010). The dark ride of Snow White: Narrative strategies at Disneyland. In K. M. Jackson & M. I. West (Eds.), *Disneyland and culture: Essays on the parks and their influence* (pp. 87–100). McFarland.

Ramos, C. & Rodrigues, J. (2017). Turismo e Tecnologia: Qualificando a Experiência do Turista Religioso. *Revista Rosa dos Ventos – Turismo e Hospitalidade, 9*(1), 32-48. doi:10.18226/21789061.v9i1p032

Rasoolimanesh, S. M., Seyfi, S., Hall, C. M., & Hatamifar, P. (2021). Understanding memorable tourism experiences and behavioural intentions of heritage tourists. *Journal of Destination Marketing & Management, 21,* 100621. doi:10.1016/j.jdmm.2021.100621

Rasoolimanesh, S. M., Seyfi, S., Rastegar, R., & Hall, C. M. (2021). Destination image during the COVID-19 pandemic and future travel behaviour: The moderating role of past experience. *Journal of Destination Marketing & Management*, *21*, 100620. doi:10.1016/j.jdmm.2021.100620

Rauhala, M & Vikström, T. (2014). *Storytelling työkaluna*. Talentum Media.

Reamy, T. (2002). *Imparting knowledge through storytelling*. Available at: https://www.kmworld.com/Articles/Editorial/Features/Imparting-knowledge-throughstorytelling-Part-1-of-a-two-part-article-9358.aspx

Reed, A. (2002). City of Details: Interpreting the Personality of London. *Journal of the Royal Anthropological Institute*, *8*(1), 127–141. doi:10.1111/1467-9655.00102

Reinermann, J. L., Lubjuhn, S., Bouman, M., & Singhal, A. (2014). Entertainment-education: Storytelling for the greater, greener good. *International Journal of Sustainable Development*, *17*(2), 176–191. doi:10.1504/IJSD.2014.061781

Republic of Türkiye Ministry of Culture and Tourism. (n.d.). *Hierapolis archaeological site*. Available at: https://muze.gov.tr/muze-detay?SectionId=PMK01&DistId=PMK

Reyes, M. C. (2018). Measuring user experience on interactive fiction in cinematic virtual reality. In *International Conference on Interactive Digital Storytelling* (pp. 295-307). Springer. 10.1007/978-3-030-04028-4_33

Rhodes, N., Toole, J., & Arpan, L. M. (2016). Persuasion as reinforcement: Strengthening the pro-environmental attitude-behavior relationship through ecotainment programming. *Media Psychology*, *19*(3), 455–478. doi:10.1080/15213269.2015.1106322

Richards, G. (1996). *Cultural tourism in Europe*. Cab International.

Richards, G. (2020). Designing creative places: The role of creative tourism. *Annals of Tourism Research*, *85*, 102922. Advance online publication. doi:10.1016/j.annals.2020.102922

Richards, G. (2021). Making places through creative tourism? In N. Duxbury (Ed.), *Cultural sustainability, tourism and development: (Re)articulations in tourism contexts* (pp. 36–48). Routledge. doi:10.4324/9780367201777-4

Richards, G., & Cooper, C. (2018). The creative economy, entertainment and performance. In C. Cooper, S. Volo, W. C. Gartner, & N. Scott (Eds.), *SAGE handbook of tourism management: Applications of theories and concepts to tourism* (Vol. 2, pp. 315–327). SAGE Reference. doi:10.4135/9781526461490.n22

Riessman, C. (1993). *Narrative Analysis*. Sage Publications, Inc.

Ritchie, J. R., & Ritchie, J. B. (1998). The branding of tourism destinations. *Annual Congress of the International Association of Scientific Experts in Tourism.*

Rizvic, S., Boskovic, D., Okanovic, V., Ivkovic-Kihic, I., & Sljivo, S. (2019). Virtual Reality Experience of Sarajevo War Heritage. In *EUROGRAPHICS Workshop on Graphics and Cultural Heritage* (pp. 1-6). Academic Press.

Rizvic, S., Djapo, N., Alispahic, F., Hadzihalilovic, B., Cengic, F. F., Imamovic, A., Okanovic, V., & Boskovic, D. (2017). Guidelines for interactive digital storytelling presentations of cultural heritage. In *9th International Conference on Virtual Worlds and Games for Serious Applications (VS-Games)* (pp. 253-259). IEEE. 10.1109/VS-GAMES.2017.8056610

Rizvic, S., Okanovic, V., & Boskovic, D. (2020). Digital storytelling. In *Visual computing for cultural heritage* (pp. 347–367). Springer. doi:10.1007/978-3-030-37191-3_18

Robiady, N. D., Windasari, N. A., & Nita, A. (2020). Customer engagement in online social crowdfunding: The influence of storytelling technique on donation performance. *International Journal of Research in Marketing*. Advance online publication. doi:10.1016/j.ijresmar.2020.03.001

Robin, B. R., & McNeil, S. G. (2019). Digital storytelling. In *The international encyclopedia of media literacy,* 1-8. doi:10.1002/9781118978238.ieml0056

Robin, B. R. (2008). Digital storytelling: A powerful technology tool for the 21st century classroom. *Theory into Practice, 47*(3), 220–228. doi:10.1080/00405840802153916

Robinson, M., & Picard, D. (2006). *Tourism, culture and sustainable development.* UNESCO.

Robinson, P. B., Stimpson, D. V., Huefner, J. C., & Hunt, H. K. (1991). An ATT approach to the prediction of entrepreneurship. *Entrepreneurship Theory and Practice, 15*(4), 13–32. doi:10.1177/104225879101500405

Rogers, H. (2019, December 17). *Rogers, McConnell Secure Kentucky Wildlands National Heritage Area Study Act in Final Federal Funding Package for 2020* [Press Release]. United States House of Representatives, Office of Congressman Hal Rogers. https://halrogers.house.gov/2019/12/rogers-mcconnell-secure-kentucky-wildlands-national-heritage-area-study-act-in-final-federal-funding-package-for-2020

Rohde, J. (2017, May 5). Details, the inner berm. *Instagram.* Available at: https://www.instagram.com/p/BTugAK_AoZN/

Rohde, J. (2007). From myth to mountain: Insights into virtual placemaking. *Computer Graphics, 41*(3), 1. doi:10.1145/1281324.1281325

Rojek, C., & Urry, J. (1997). *Touring cultures: Transformations of travel and theory.* Routledge.

Romagosa, F. (2020). The COVID-19 crisis: Opportunities for sustainable and proximity tourism. *Tourism Geographies: An International Journal of Tourism Space, Place, and Environment, 22*(3), 690–694. doi:10.1080/14616688.2020.1763447

Roque, M. I., & Forte, M. J. (2017b). The festivities of the Light of Carnide: A persistent authenticity. In S. Lira, R. Amoêda & C. Pinheiro (Eds.), *Sharing cultures 2017: Proceedings of the 5th international conference on intangible heritage* (pp. 375-383). Green Lines Institute for Sustainable Development.

Roque, M. I., & Forte, M. J. (2017a). Digital strategies to a local cultural tourism development: Project e-Carnide. In V. Katsoni, A. Upadhya, & A. Stratigea (Eds.), *Tourism, culture and heritage in a smart economy* (pp. 365–383). Springer International Publishing. doi:10.1007/978-3-319-47732-9_24

Ross, D. (2020). Towards meaningful co-creation: A study of creative heritage tourism in Alentejo, Portugal. *Current Issues in Tourism, 23*(22), 2811–2824. doi:10.1080/13683500.2020.1782355

Ross, D., & Saxena, G. (2019). Participative co-creation of archaeological heritage: Case insights on creative tourism in Alentejo, Portugal. *Annals of Tourism Research, 79*(102790), 1–14. doi:10.1016/j.annals.2019.102790

Rossiter, M., & Garcia, P. A. (2010). Digital storytelling: A new player on the narrative field. *New Directions for Adult and Continuing Education, 126*(126), 37–48. doi:10.1002/ace.370

Roussos, M., & Bizri, H. (1998). Mitologies: Medieval labyrinth narratives in virtual reality. In *International Conference on Virtual Worlds* (pp. 373-383). Springer. 10.1007/3-540-68686-X_36

Roussou, M. (2001). The interplay between form, story, and history: The use of narrative in cultural and educational virtual reality. In *International Conference on Virtual Storytelling* (pp. 181-190). Springer. 10.1007/3-540-45420-9_20

Roussou, M. (2008). The components of engagement in virtual heritage environments. In *New Heritage* (pp. 241–257). Routledge.

Rubin, J. (Ed.). (2020). *Global attractions attendance report 2019.* TEA/AECOM.

Ruszev, S. (2021). Tracing Embodied Narrative in VR Experiences. *Cinergie – Il Cinema E Le Altre Arti, 10*(19), 9–20.

Rutledge, P. (2019). Transmedia Psychology: Creating Compelling and Immersive Experiences. In M. Freeman & R. R. Gambarato (Eds.), *Routledge Companion to Transmedia Studies* (pp. 350–363). Toutledge.

Ryan, M.-L. (2015). *Narrative as Virtual Reality II: Revisiting Immersion and Interactivity*. JHU Press. https://www.academia.edu/16494139/Narrative_as_Virtual_Reality_II_Revisiting_Immersion_and_Interactivity

Ryan, M. L. (2016). Transmedia narratology and transmedia storytelling. *Artnodes*, *2016*(18), 1–10. doi:10.7238/a.v0i18.3049

Ryu, K., Lehto, X. Y., Gordon, S. E., & Fu, X. (2018). Compelling brand storytelling for luxury hotels. *International Journal of Hospitality Management*, *74*, 22–29. doi:10.1016/j.ijhm.2018.02.002

Sabate, F., Berbegal-Mirabent, J., Cañabate, A., & Lebherz, P. (2014). Factors influencing popularity of branded content in Facebook fan pages. *European Management Journal*, *32*(6), 1001–1011. doi:10.1016/j.emj.2014.05.001

Salazar, N. (2005). Tourism and Glocalization: "Local" Tour Guiding. *Annals of Tourism Research*, *32*(3), 628–646. doi:10.1016/j.annals.2004.10.012 PMID:32572282

Salazar, N. B. (2012). Tourism imaginaries: A conceptual approach. *Annals of Tourism Research*, *39*(2), 863–882. doi:10.1016/j.annals.2011.10.004

Salvado Alves, A. J. F. (2008). A competitividade dos territórios num contexto de globalização : Uma utopia ou uma realidade? *VI Congresso Português de Sociologia - Mundos Sociais: Saberes e Práticas.*

Salvado, J., & Joukes, V. N. P. M. (2019). Bacalhau, um marcador simbólico da identidade alimentar portuguesa: Um olhar sobre o território de Trás-os- Montes e Alto Douro. *Revista Portuguesa de Estudos Regionais*, *53*, 55–68.

Sanctuary, F. (2022). *Identidade e Missão do Santuário de Fátima*. https://www.fatima.pt/pt/pages/identidade-e-missao

Sandvik, K. (2010). *Crime scenes as augmented reality: Models for enhancing places emotionally by means of narratives, fictions and virtual reality*. Channel View Publications.

Sarica, H. Ç., & Usluel, Y. K. (2016). The effect of DST on visual memory and writing skills. *Computers & Education*, *94*, 298–309. doi:10.1016/j.compedu.2015.11.016

Schiavottiello, N., Carpetudo, C., Correia, M., Viscomi, P., Cabral, R., Alarcón, S.P. (2016). CGI Storytelling in Archaeological and Cultural Heritage Public Interpretation: Scientific Facts or Hollywood Movies? *Entre Ciência E Cultura. Da Interdisciplinaridade a Transversalidade da Arqueologia*, 569-592.

Schmidt, K. (2010). *Göbekli Tepe Kazısı 2010 Yılı Raporu*. Directorate General of Cultural Assets and Museums, Publishing Number: 155-3. Available at: http://www.kulturvarliklari.gov.tr/Eklenti/4653,33kazi3.pdf?0

Schmidt, K. (2018). *Göbekli Tepe: en eski tapınağı yapanlar* (R. Arslan, Trans.). Arkeoloji ve Sanat Yayınları. (Original book published 2006)

Schmitt, B. (1999). *Experiential marketing: How to get customers to sense, feel, think, act, and relate to your company and brands*. Free Press.

Scolari, C. A. (2009). Transmedia storytelling: Implicit consumers, narrative worlds, and branding in contemporary media production. *International Journal of Communication*, *3*, 586–606.

Scott, N., Gao, J., & Ma, J. (2017). Introduction: creating memorable experiences. In N. N. Scott, J. Gao, & J. Ma (Eds.), Visitor Experience Design (pp. 3–12). CABI Publishing.

Seide, B., & Slater, B. (2020). Virtual Cinematic Heritage for the Lost Singaporean Film Pontianak (1957). In *International Conference on Human-Computer Interaction* (pp. 396-414). Springer. 10.1007/978-3-030-50267-6_30

Selmanović, E., Rizvic, S., Harvey, C., Boskovic, D., Hulusic, V., Chahin, M., & Sljivo, S. (2020). Improving accessibility to intangible cultural heritage preservation using virtual reality. *Journal on Computing and Cultural Heritage, 13*(2), 1–19. doi:10.1145/3377143

Sestir, M., & Green, M. C. (2010). You are who you watch: Identification and transportation effects on temporary self-concept. *Social Influence, 5*(4), 272–288. doi:10.1080/15534510.2010.490672

Shankar, A., & Goulding, C. (2001). Interpretive Consumer Research: Two More Contributions to Theory and Ppractice. *Qualitative Market Research, 4*(1), 7–16. doi:10.1108/13522750110364523

Shankar, A., & Goulding, C. (2001). Interpretive Consumer Research: Two More Contributions to Theory and Practice. *Qualitative Market Research, 4*(1), 7–16.

Sheppard, B. H., Hartwick, J., & Warshaw, P. R. (1988). The theory of reasoned action: A meta-analysis of past research with recommendations for modifications and future research. *The Journal of Consumer Research, 15*(3), 325–343. doi:10.1086/209170

Shinde, K. (2021). Sacred Sites, Rituals, and Performances in the Ecosystem of Religious Tourism. *Religions, 12*(7), 523. doi:10.3390/rel12070523

Shishido, H., Ito, Y., Kawamura, Y., Matsui, T., Morishima, A., & Kitahara, I. (2017, December). Proactive preservation of world heritage by crowdsourcing and 3D reconstruction technology. In *2017 IEEE International Conference on Big Data (Big Data)* (pp. 4426-4428). IEEE.

Shrestha, R. K., Stein, T. V., & Clark, J. (2007). Valuing nature-based recreation in public natural areas of the Apalachicola River region, Florida. *Journal of Environmental Management, 85*(4), 977–985. doi:10.1016/j.jenvman.2006.11.014 PMID:17275161

Sil, C. H. (2006). *Creating a culture of collaboration: The International Association of Facilitators handbook* (Vol. 4). John Wiley & Sons.

Simmons, A. (2006). *The Story Factor: Secrets of Influence through the Art of Storytelling*. Basic Books.

Simon, H. A. (1969). *The sciences of the artificial*. MIT Press.

Skift. (2020). Available at: https://skift.com/destination2020/

Skift. (2020). *Destination storytelling 2020: A guide to telling great stories*. Available at: https://skift.com/destination2020/

Sklar, M. (2015). *One little spark!: Mickey's ten commandments and the road to Imagineering*. Disney Editions.

Škola, F., Rizvić, S., Cozza, M., Barbieri, L., Bruno, F., Skarlatos, D., & Liarokapis, F. (2020). Virtual Reality with 360-Video Storytelling in Cultural Heritage: Study of Presence, Engagement, and Immersion. *Sensors (Basel), 20*(20), 5851. doi:10.339020205851 PMID:33081154

Slater, M. (2009). Place illusion and plausibility can lead to realistic behaviour in immersive virtual environments. *Philosophical Transactions of the Royal Society of London. Series B, Biological Sciences, 364*(1535), 3549–3557. doi:10.1098/rstb.2009.0138 PMID:19884149

Smeda, N., Dakich, E., & Sharda, N. (2014). The effectiveness of DST in the classrooms: A comprehensive study. *Smart Leaning Environments, 1*(6), 1–21. doi:10.1109/ICALT.2013.157

Smith, K., & Wintrob, M. (2013). *Brand storytelling: a framework for activation.* Available at: https://onlinelibrary.wiley.com/doi/full/10.1111/drev.10227

Smith, L. (2014). Visitor Emotion, Affect and Registers of Engagement at Museums and Heritage Sites. *Conservation Science in Cultural Heritage, 14*(2), 125–131. doi:10.6092/issn.1973-9494/5447

Smith, S. (2015). A sense of place: Place, culture and tourism. *Tourism Recreation Research, 40*(2), 220–233. doi:10.1080/02508281.2015.1049814

Snyder, H. (2019). Literature review as a research methodology: An overview and guidelines. *Journal of Business Research, 104*, 333–339. doi:10.1016/j.jbusres.2019.07.039

Sobral, J. M. (2016). O Bacalhau: de Alimento de Penitência a lcone Português. *Argos - Revista Do Museu Marítimo de Ílhavo, 4*, 27–33.

Sobral, J. M., & Rodrigues, P. (2013). O "fiel amigo": O bacalhau e a identidade portuguesa. *Etnográfica (Lisboa), 17*(3), 619–649. doi:10.4000/etnografica.3252

Sole, D., & Wilson, D. (2002). *Storytelling in Organizations.* LILA: Harvard, Graduate School of Education.

Sole, D., & Wilson, D. G. (2002). Storytelling in organizations: The power and traps of using stories to share knowledge in organizations. LILA, Harvard, Graduate School of Education, 1-12.

Sole, D., & Wilson, D. G. (2002). Storytelling in organizations: The power and traps of using stories to share knowledge in organizations. *Training & Development, 53*, 44–52.

Solima. (2012). *The listening museum. New communications strategies for public museums.* Rubbettino.

Soria, A., & Molendowska-Ruiz, J. (2017). Creative Storytelling for European Pilgrimage route's audience development. In Book of Abstracts, Cultural Management and Governance for European Pilgrimage routes and religious tourism. Academic Press.

Souiden, N., Ladhari, R., & Chiadmi, N. E. (2017). Destination personality and destination image. *Journal of Hospitality and Tourism Management, 32*, 54–70. doi:10.1016/j.jhtm.2017.04.003

Sousa, C. S. M. (2019). *A influência do DST no comportamento do consumidor: O caso da Lemon Jelly* [Doctoral dissertation]. Instituto Politécnico De Leiria.

Southall, H., Marmion, M., & Davies, A. (2019). Adapting Jake Knapp's design sprint approach for AR/VR applications in digital heritage. In *Augmented reality and virtual reality. The Power of AR and VR for Business, edited by M. Claudia tom Dieck and Timothy Jung* (pp. 59–70). Springer. doi:10.1007/978-3-030-06246-0_5

Spears, N., & Singh, S. (2004). Measuring ATT toward the brand and PIs. *Journal of Current Issues and Research in Advertising, 26*(2), 53–66. doi:10.1080/10641734.2004.10505164

Spinks, R. (2020, May 21). *What is the future of vacations?* Skift. https://skift.com/2020/05/21/what-is-the-future-of-vacations/

Spreen, M. (1992). Rare populations, hidden populations, and link-tracing designs: What and why? Bulletin of Sociological Methodology. *Bulletin de Methodologie Sociologique, 36*(1), 34–58. doi:10.1177/075910639203600103

Stacey, N. (2007). *Boats to Burn: Bajo fishing activity in the Australian fishing zone.* ANU E Press.

Stanley, N. (2018). Digital storytelling. *The TESOL encyclopedia of English language teaching*, 1–7. doi:10.1002/9781118784235.eelt040

Statista. (2021). *Number of Monthly Active Facebook Users Worldwide as of 2nd Quarter 2021(in Millions)*. Available at: https://www. statista.com/statistics/264810/number-of-monthly-active-facebook-users-worldwide/

Steinemann, S., Iten, G., Opwis, K., Forder, S., Frasseck, L., & Mekler, E. (2017). Interactive narratives affecting social change. *Journal of Media Psychology*, *29*(1), 54–66. doi:10.1027/1864-1105/a000211

Stephens, J., & Tiwari, R. (2015). Symbolic estates: Community identity and empowerment through heritage. *International Journal of Heritage Studies*, *21*(1), 99–114. doi:10.1080/13527258.2014.914964

Stern, B. B. (2006). What does brand mean? Historical-analysis method and construct definition. *Journal of the Academy of Marketing Science*, *34*(2), 216–223. doi:10.1177/0092070305284991

Stevenson, N. (2019). Developing cultural understanding through story-telling. *Journal of Teaching in Travel & Tourism*, *19*(1), 8–21. doi:10.1080/15313220.2018.1560528

Streimikiene, D., Svagzdiene, B., Jasinskas, E., & Simanavicius, A. (2021). Sustainable tourism development and competitiveness: The systematic literature review. *Sustainable Development*, *29*(1), 259–271. doi:10.1002d.2133

Strielkowski, W. (2017). Promoting tourism destination through film-induced tourism: the case of Japan. *Market-Tržište*, *29*(2), 193-203. doi:10.22598/mt/2017.29.2.193

Sugathan, P., & Ranjan, K. R. (2019). Co-creating the tourism experience. *Journal of Business Research*, *100*, 207–217. doi:10.1016/j.jbusres.2019.03.032

Su, L., Cheng, J., & Swanson, S. (2020). The impact of tourism activity type on emotion and storytelling: The moderating roles of travel companion presence and relative ability. *Tourism Management*, *81*, 104138. doi:10.1016/j.tourman.2020.104138

Sundin, A., Andersson, K., & Watt, R. (2018). Rethinking communication: Integrating storytelling for increased stakeholder engagement in environmental evidence synthesis. In Environmental Evidence (Vol. 7, Issue 6, pp. 1–6). BioMed Central Ltd. doi:10.118613750-018-0116-4

Suvantola, J. (2018). *Tourist's experience of place*. Routledge. doi:10.4324/9781315184821

Swords, J., Nally, C., Rogage, K., Watson, R., Charlton, J., & Kirk, D. (2021). Colliding epistemologies, productive tensions and usable pasts in the generation of heritage-led immersive experiences. *International Journal of Heritage Studies*, *27*(2), 186–199. doi:10.1080/13527258.2020.1780462

Sylaiou, S., & Dafiotis, P. (2020). Storytelling in virtual museums: Engaging a multitude of voices. In F. Liarokapis, A. Voulodimos, N. Doulamis, & A. Doulamis (Eds.), *Visual computing for cultural heritage* (pp. 369–388). Springer. doi:10.1007/978-3-030-37191-3_19

Taçon, P. S., & Baker, S. (2019). New and emerging challenges to heritage and well-being: A critical review. *Heritage*, *2*(2), 1300–1315. doi:10.3390/heritage2020084

Tang, S. (2016). Digital storytelling approach in a multimedia feature writing course. *Journal of Language Teaching and Research*, *7*(3), 572–578. doi:10.17507/jltr.0703.19

Tasci, A., & Milman, A. (2019). Exploring experiential consumption dimensions in the theme park context. *Current Issues in Tourism*, *22*(7), 853–876. doi:10.1080/13683500.2017.1321623

Teixeira Costa, P., & Veríssimo, M. (2020). The Sea by Tradition: From the Slogan to the Brand – Exploring a Successful Case of Territorial Strategy Anchored in the Sea and Tourism The Sea by Tradition. In M. Morais de Brito, A. Dias, & M. Patuleia (Eds.), *Managing, Marketing, and Maintaining Maritime and Coastal Tourism* (pp. 264–283). IGI Global. doi:10.4018/978-1-7998-1522-8.ch016

Teles, F., & Anselmo de Castro, E. (2015). *Plano Estratégico do Município de Ílhavo.* Academic Press.

The Economic Times. (2017). *Do people buy goods and services, or stories and magic?* Available at: https://economictimes.indiatimes.com/do-people-buy-goods-and-services-or-stories-and-magic/articleshow/57016683.cms

Therkelsen, A., Halkier, H., & Jensen, O. B. (2010). Branding Aalborg: Building community or selling place? In *Towards effective place brand management.* Edward Elgar Publishing. doi:10.4337/9781849806398.00014

Throsby, D. (2009). Tourism, heritage and cultural sustainability: Three 'golden rules. In L. F. Girard & P. Nijkamp (Eds.), *Cultural tourism and sustainable local development* (pp. 13–29). Routledge.

Tiago, F., Couto, J., Faria, S., & Borges-Tiago, S. (2018). Cruise tourism: Social media content and network structures. *Tourism Review, 73*(4), 433–447. doi:10.1108/TR-10-2017-0155

Tilden, F. (1977). *Interpreting our heritage.* University of North Carolina Press.

Timothy, D. J. (1994). Environmental impacts of heritage tourism: Physical and sociocultural perspectives. *Manusia dan Lingkungan, 11*(4), 37-49.

Timothy, D. J., & Boyd, S. W. (2003). *Heritage tourism.* Prentice Hall.

Tivers, J., & Rakić, T. (Eds.). (2012). *Narratives of travel and tourism.* Ashgate.

Tobin, P. (2006). Teaching IT through storytelling. *Proceedings of The Conference on Information Technology in Tertiary Education.*

tom Dieck, M. C., Jung, T., & Michopoulou, E. (2019). Experiencing Virtual Reality in Heritage Attractions: Perceptions of Elderly Users. In Augmented Reality and Virtual Reality. The power of AR and VR for Business (pp. 89-98), Springer.

Torraco, R. J. (2005). Writing integrative literature reviews: Guidelines and examples. *Human Resource Development Review, 4*(3), 356–367. doi:10.1177/1534484305278283

Tribe, J., & Liburd, J. (2016). The tourism knowledge system. *Annals of Tourism Research, 57*, 44–61. doi:10.1016/j.annals.2015.11.011

Trunfio, M., & Della Lucia, M. (2018). Engaging destination stakeholders in the digital era: The best practice Of Italian regional DMOs. *Journal of Hospitality & Tourism Research (Washington, D.C.), 43*(3), 349–373. doi:10.1177/1096348018807293

Tsai, H. (2021). Exploring the motivation-based typology of religious tourists: A study of Welcome Royal Lord Festival. *Journal of Destination Marketing & Management, 21*, 100623. doi:10.1016/j.jdmm.2021.100623

Tscheu, F., & Buhalis, D. (2016). Augmented reality at cultural heritage sites. In *Information and Communication Technologies in Tourism* (pp. 607–619). Springer.

Tuerlan, T., Li, S., & Scott, N. (2021). Customer emotion research in hospitality and tourism: Conceptualization, measurements, antecedents and consequences. *International Journal of Contemporary Hospitality Management, 33*(8), 2741–2772. doi:10.1108/IJCHM-11-2020-1257

Turkey Culture Portal. (2018). *Troya antik kenti.* Available at: https://www.kulturportali.gov.tr/portal/troyaantikkenti

Turkey Culture Portal. (n.d.). *Göbeklitepe.* Available at: https://www.kulturportali.gov.tr/portal/gobeklitepe

Turkey Tourkeys. (2020). About Cappadocia. Available at: https://www.turkeytourkeys.com/about-cappadocia/

Türkoğlu, S. (2015). *Efes'in öyküsü.* Arkeoloji ve Sanat Yayınları.

Tussyadiah, I. P. (2014). Toward a Theoretical Foundation for Experience Design in Tourism. *Foundations of Tourism Research: A Special Series, 53*(5), 543–564. doi:10.1177/0047287513513172

Tussyadiah, I. P., & Fesenmaier, D. R. (2008). Marketing places through first-person stories—An analysis of Pennsylvania roadtripper blog. *Journal of Travel & Tourism Marketing, 25*(3-4), 299–311. doi:10.1080/10548400802508358

Tussyadiah, I. P., & Fesenmaier, D. R. (2009). Mediating Tourist Experiences. Access to Places via Shared Videos. *Annals of Tourism Research, 36*(1), 24–40. doi:10.1016/j.annals.2008.10.001

Udeaja, C., Mansuri, L. E., Makore, B. C. N., Awuah, K. G. B., Patel, D. A., Trillo, C., & Jha, K. N. (2021). Digital storytelling: The integration of intangible and tangible heritage in the city of Surat, India. In M. Rauterberg (Ed.), *Culture and computing; Interactive cultural heritage and arts* (pp. 152–168). Springer. doi:10.1007/978-3-030-77411-0_11

Ülkü, A. (2021). *Dijital dönemde hikaye/hikaye anlatıcılığı ve destinasyonlar* [Conference Presentation]. 2nd International Congress of New Generations and New Trends in Tourism, Sakarya, Türkiye.

UNEP & WTO. (2005). *Making Tourism More Sustainable. A Guide for Policy Makers.* Available at: http://www.unep.fr/shared/publications/pdf/DTIx0592xPATourismPolicyEN.pdf

UNESCO Man & the Biosphere Programme. (2012, September 30). *Wakatobi Biosphere Reserve, Indonesia.* https://en.unesco.org/biosphere/aspac/wakatobi

UNESCO World Heritage Centre. (n.d.). *Willandra Lakes Region.* Retrieved September 8, 2021, from https://whc.unesco.org/en/list/167/video

UNESCO. (1976). *Charter of cultural tourism.* http://www.lacult.unesco.org/lacult_en/docc/ChartCulturTour1976.doc

United Nation Education, Scientific and Cultural Organization. (2006). *Wakatobi National Park.* https://whc.unesco.org/en/tentativelists/2006/

United Nations World Tourism Organization (UNWTO). (2019, December 10). *Product Development.* https://www.unwto.org/tourism-development-products

UNWTO. (1995). *Global code of ethics for tourism.* https://cf.cdn.unwto.org/sites/all/files/docpdf/gcetbrochureglobalcodeen.pdf

Uysun, M. (2011). *Kapadokya kaya kiliseleri ikonografisi.* Doğu Ofset.

Vallerand, R. J., Deshaies, P., Cuerrier, J.-P., Pelletier, L. G., & Mongeau, C. (1992). Ajzen and Fishbein's theory of reasoned action as applied to moral behaviour: A confirmatory analysis. *Journal of Personality and Social Psychology, 62*(1), 98–109. doi:10.1037/0022-3514.62.1.98

Van Der Heijden, H., Verhagen, T., & Creemers, M. (2003). Understanding online PIs: Contributions from technology and trust perspectives. *European Journal of Information Systems, 12*(1), 41–48. doi:10.1057/palgrave.ejis.3000445

Van Der Linde, S., & Mans, J. (2015). Visualising values in the caribbean: A creative approach to value assessment. *Conservation and Management of Archaeological Sites, 17*(3), 257–269. doi:10.1080/13505033.2015.1129800

Van Laer, T., De Ruyter, K., Visconti, L. M., & Wetzels, M. (2014). The extended transportation-imagery model: A meta-analysis of the antecedents and consequences of consumers' narrative transportation. *The Journal of Consumer Research, 40*(5), 797–817. doi:10.1086/673383

Van Limburg, B. (2009). The anatomy of a digital storytelling system: The architecture. *Information Technology & Tourism, 11*(1), 31–42. doi:10.3727/109830509788714604

Vargo, S. L., & Lusch, R. F. (2004). Evolving to a New Dominant Logic for Marketing. *Journal of Marketing, 68*(1), 1–17. doi:10.1509/jmkg.68.1.1.24036

Vermeulen, M. (2002). The Netherlands, Holiday Country. In V. Patteeuw (Ed.), *City Branding: Image Building and Building Images*. NAI Uitgevers.

Victoria University of Wellington. (n.d.). *Study sites* [Infographic]. https://www.wgtn.ac.nz/sbs/research/ marine-biology-research/sponge-marine-ecology/research/study-sites

Vrettakis, E., Kourtis, V., Katifori, A., Karvounis, M., Lougiakis, C., & Ioannidis, Y. (2019). Narralive: Creating and experiencing mobile digital storytelling in cultural heritage. *Digital Applications in Archaeology and Cultural Heritage, 15*, e0120513. Advance online publication. doi:10.1016/j.daach.2019.e00114

Wall, G. (1989). An international perspective on historic sites, recreation, and tourism. *Recreation Research Review, 14*(4), 10–14.

Wang, N. (1999). Rethinking authenticity in tourism experience. *Annals of Tourism Research, 26*(2), 349–370. doi:10.1016/ S0160-7383(98)00103-0

Wansink, B., Kent, R. J., & Hoch, S. J. (1998). An anchoring and adjustment model of purchase quantity decisions. *JMR, Journal of Marketing Research, 35*(1), 71–81. doi:10.1177/002224379803500108

Waterton, E., & Smith, S. (2010). The recognition and misrecognition of community heritage. *International Journal of Heritage Studies, 16*(1-2), 4–15. doi:10.1080/13527250903441671

Waysdorf, A., & Reijnders, S. (2018). Immersion, authenticity and the theme park as social space: Experiencing the Wizarding World of Harry Potter. *International Journal of Cultural Studies, 21*(2), 173–188. doi:10.1177/1367877916674751

WECD. (1987). *Our Common Future: The World Commission on Environment and Development*. Oxford University Press.

Weiler, B., & Black, R. (2015). The changing face of the tour guide: One-way communicator to choreographer to co-creator of the tourist experience. *Tourism Recreation Research, 40*(3), 364–378. doi:10.1080/02508281.2015.1083742

Wells, J. C., & Stiefel, B. L. (Eds.). (2019). *Human-centered built environment heritage preservation: Theory and evidence-based practice*. Routledge.

Wiederhold, B. K. (2018). The next level of virtual reality isn't technology—It's storytelling. *Cyberpsychology, Behavior, and Social Networking, 21*(11), 671–672. doi:10.1089/cyber.2018.29129.bkw PMID:30421994

Wilkins, A. L. (1984). The creation of company cultures: The role of stories and human resource systems. *Human Resource Management, 23*(1), 41–60. doi:10.1002/hrm.3930230105

Williams, R. (2020). *Theme park fandom: Spatial transmedia, materiality, and participatory cultures*. Amsterdam University Press.

Wilson, K., & Desha, C. (2016). Engaging in design activism and communicating cultural significance through contemporary heritage storytelling. *Journal of Cultural Heritage Management and Sustainable Development, 6*(3), 271–286. doi:10.1108/JCHMSD-10-2015-0039

Wilson, M., & Zipes, J. (2005). *Storytelling and theatre: Contemporary professional storytellers and their art*. Palgrave Macmillan.

Wisniewsk, L. (2018). Religious tourism in Christian sanctuaries - the implications of mixed interests for the communication of the faith. *Church, Communication and Culture, 3*(3), 199–220. doi:10.1080/23753234.2018.1537674

Wiyonoputri, W. T., Anindita, D., & Masita, A. (2021). *Pedoman Interpretasi Daya Tarik Wisata* [Guideline for Developing Interpretation in Tourist Attractions]. Kementerian Pariwisata dan Ekonomi Kreatif / Badan Pariwisata dan Ekonomi Kreatif Republik Indonesia.

Wolf, M. J. P. (2012). *Building imaginary worlds: The theory and history of subcreation*. Routledge.

Wong, J. Y., Lee, S. J., & Lee, W. H. (2015). 'Does it really affect me?' Tourism destination narratives, destination image, and the intention to visit: Examining the moderating effect of narrative transportation. *International Journal of Tourism Research, 18*(4), 458–468.

Wood, N. T., & Muñoz, C. L. (2007). 'No rules, just right' or is it? The role of themed restaurants as cultural ambassadors. *Tourism and Hospitality Research, 7*(3–4), 242–255. doi:10.1057/palgrave.thr.6050047

Woodside, A. G. (2010). Brand-consumer storytelling theory and research: Introduction to a psychology & marketing special issue. *Psychology and Marketing, 27*(6), 531–540. doi:10.1002/mar.20342

Woodside, A. G., & Megehee, C. M. (2009). Travel storytelling theory and practice. *Anatolia, 20*(1), 86–99.

Woodside, A., & Martin, D. (2015). Introduction: The tourist gaze 4.0: Uncovering non-conscious meanings and motivations in the stories tourists tell of trip and destination experiences. *International Journal of Tourism Anthropology, 4*(1), 1–12.

Wyman, B., Smith, S., Meyers, D., & Godfrey, M. (2011). Digital Storytelling in Museums: Observations and Best Practices. *Curator (New York, N.Y.), 54*(4), 461–468. doi:10.1111/j.2151-6952.2011.00110.x

Xiang, Z., & Gretzel, U. (2010). Role of social media in online information search. *Tourism Management, 31*(2), 179–188. doi:10.1016/j.tourman.2009.02.016

Yavuz, M., Sumbul, M., Ergec, N., Derdiyok, C. (2016). Storytelling in Destination Brand Communication: A Qualitative Analysis. *International Interdisciplinary Business-Economics Advancement Journal, 1*, 63–72.

Yavuz, M. C., Sümbül, M., Ergeç, N. E., & Derdiyok, Ç. (2016). Storytelling in destination brand communication: A qualitative analysis. *International Interdisciplinary Business-Economics Advancement Journal, 1*(2), 63–72. doi:10.5038/2640-6489.1.2.1008

Yerasimos S. (1993). *Türk metinlerinde Konstantiniye ve Ayasofya Efsaneleri* (Ş. Tekeli, Trans). İletişim Yayınları. (Original book published 1990)

Yılmaz, R., & Ciğerci, F. M. (2019). A brief history of storytelling: From primitive dance to digital narration. In *Handbook of research on transmedia storytelling and narrative strategies* (pp. 1–14). IGI Global. doi:10.4018/978-1-5225-5357-1.ch001

Yin, R. (2003). *Case Study Research: Design and Methods* (3rd ed.). Sage Publications Ltd.

Yoder-Wise, P. S., & Kowalski, K. (2003). The power of storytelling. *Nursing Outlook, 51*(1), 37–42. doi:10.1067/mno.2003.2 PMID:12595827

Yoon, S. J. (2002). The antecedents and consequences of trust in online-purchase decisions. *Journal of Interactive Marketing, 16*(2), 47–63. doi:10.1002/dir.10008

Younger, D. (2016). *Theme park design and the art of themed entertainment*. Inklingwood Press.

Yu, H., & Chang, Y. (2013). How to influence the brand attitude of the audience by micro-films. *Journal of Promotion Management, 19*(5), 674–686.

Yuxiang, Z., & Jingwen, L. (2021). Review of cultural heritage crowdsourcing in the domain of digital humanities. *Data Analysis and Knowledge Discovery, 5*(1), 36–55. doi:10.11925/infotech.2096-3467.2020.0906

Zainal, Z. (2007). The Case Study as a Research Method. *Jurnal Kemanusiaan, 9*, 1–6.

Zarezadeh, Z. Z., Rastegar, H. R., & Gretzel, U. (2018). Reviewing the Past to Inform the Future: A Literature Review of Social Media in Tourism. *Czech Journal of Tourism, 7*(2), 115–131. doi:10.1515/cjot-2018-0006

Zavattaro, S. M., Daspit, J. J., & Adams, F. G. (2015). Assessing managerial methods for evaluating place brand equity: A qualitative investigation. *Tourism Management, 47*, 11–21.

Zebregs, S., Putte, B., Neijen, P., & Graaf, A. (2014). The differential impact of statistical and narrative evidence on beliefs, ATT, and intention: A meta-analysis. *Health Communication, 30*(3), 282–289. doi:10.1080/10410236.2013.842528 PMID:24836931

Zeng, B., & Gerritsen, R. (2014). What do we know about social media in tourism? A review. *Tourism Management Perspectives, 10*, 27–36. doi:10.1016/j.tmp.2014.01.001

Zevi, F., Demma, F., Nuzzo, E., Rescigno, C., & Valeri, C. (2008). *Museo Archeologico dei Campi Flegrei. Catalogo generale. Cuma*. Electa Napoli.

Zhang, K. Z. K., Zhao, S. J., Cheung, C. M. K., & Lee, M. K. O. (2014). Examining the influence of online reviews on consumers' decision-making: A heuristic-systematic model. *Decision Support Systems, 67*, 78–89. doi:10.1016/j.dss.2014.08.005

Zhang, Y., Gao, J., Cole, S., & Ricci, P. (2021). How the spread of user-generated contents (UGS) shapes international tourism distribution: Using agent-based modeling to inform strategic UGC marketing. *Journal of Travel Research, 60*(7), 469–491. doi:10.1177/0047287520951639

Zika, J. (2018). Dark rides and the evolution of immersive media. *Journal of Themed Experience and Attractions Studies, 1*, 54–60.

Zlatanov, S., & Popesku, J. (2021). Promotion of Religious Tourism Sites on Facebook by NTOs. *International Journal of Religious Tourism and Pilgrimage, 9*(3), 70–81. doi:10.21427/5gct-vb53

About the Contributors

Ana Cláudia Campos holds a PhD in Tourism, a MA in Tourism Management, and a BA in Philosophy. Currently she is Assistant Researcher at CinTurs, the Research centre for Tourism, Sustainability and Well-being, Universidade do Algarve, Portugal. Her research interests are in tourism marketing, tourism experience, tourist psychology and co-creation.

Sofia Almeida holds a PhD in Tourism and a degree in Business Communication. She is Assistant Professor at the Faculty of Tourism and Hospitality (FTH), Universidade Europeia. Her research focuses on the areas of hotel management, networks, coopetition, hotel marketing consortia, hotel branding and digital marketing. Sofia has been publishing in hospitality studies with several of her colleagues, as in Cooperation in the Hospitality of the Center: Myth or Reality? with C. Costa and J. M. Simões (Edições Actual, Almedina, 2018) and the network cooperation as a lever factor in Hospitality (Journal of Tourism and Development, 2015). She is a research member of the Centre for Geographical Studies, IGOT, Universidade Lisboa.

* * *

Guido Acampa attended the Artistic High School and the Academy of Fine Arts. A painter for passion, he exhibits from a very young age in some contemporary art galleries. He soon began working in the world of advertising and communication having founded his own company. He approached cinema as a self-taught artist, making videos for publishing and the art circuit and experimenting with the cinematographic language applied to new media for the fruition of cultural heritage.

Lorna Kankor Adu-Debrah holds an MBA in Finance and is currently a second year PhD student at the University of Ghana. She also runs and teaches at a high school in Accra, the capital of Ghana. Her research interests are in co-creation, domestic tourism and ecotourism.

Luís José Andrade holds a PhD in Labor Sciences, a MA in Social and Organizational Psychology and a BA in Psychology. He is Professor, Manager, Coacher and Senior Consultant in Management areas. Currently, he is Managing Director in the Customer Service area in a leading Business Group in Portugal. His career has been consolidated by holding different positions in business organisations, including Human Resources and Organizational Consulting. He has lectured and coordinated curricular units in Human Resources, Organizational Behaviour, and Psycho-sociology, Marketing, Market Research, and Research Methods. He regularly publishes scientific and opinion articles and is a co-author of the Practical Handbook on Human Resource Management by Verlag Dashofer publishers.

Carissa Baker is an Assistant Professor at the University of Central Florida's Rosen College of Hospitality Management. She holds a Ph.D. in Texts and Technology as well as an MA and BA in Literature. Her interdisciplinary scholarship focuses on aspects of the themed entertainment industry, especially narrative and culture.

Barbara Balbi, restorer at the Superintendency of the city of Naples of Italian Ministry of Culture, specialized in art history. Barbara holds a PhD in Industrial Design (digital humanities) and a Master Degree in New Technologies for Social and Human Sciences. She has participated in teaching activities of the University Suor Orsola Benincasa of Naples and has had a research grant on Design of New Technologies for the Fruition of Cultural Heritage. She has lectured in national and international conferences and she is the author of articles on the topics of UX of interaction in digital fruition of cultural heritage and restoration techniques. Since 2019, on behalf of the Superintendency, she is in charge of the conservation of the fine arts heritage of some district of Naples, supervising and directing many restoration works.

Alexandra Bec, PhD, is currently a Senior Economic Research and Data Analyst with City of Gold Coast. She has experience working in academic, industry and local government on strategic research projects. Her research interests include socio-economic development, resilience, investment evaluation, and tourism.

Meghan Beevor holds a MSc in Management with a specialization in Tourism Management from the University of Algarve and a BSocSc in Anthropology from the University of Ottawa. After her years of working as an archaeologist in Canada, Meghan has become passionate about the management of cultural heritage and experiences. Having specialized in the implementation and impact of storytelling within the context of archaeological museums and heritage attractions, she hopes to use a combination of her academic knowledge and management skills to assist institutions in creating engaging, interactive, and memorable cultural experiences. Her research interests include cultural heritage management, new museology, storytelling, experience design, visitor experience, and co-creation in heritage tourism.

Stephen Mahama Braimah is a senior lecturer at the University of Ghana Business School in the Department of Marketing and Entrepreneurship. In 1992, Dr Braimah worked as an Assistant Administrative Officer with the Soil Research Institute of the Council for Scientific and Industrial Research (CSIR). Later in 1993, he joined the private sector, where he worked in various capacities for different companies in different industries. He has held various senior management positions, including marketing, commercial, and general manager. His experiences include project management, procurement, international business, marketing and sales, customer service, and management and administration. Stephen has worked in the pharmaceutical industry, the petroleum sector, the automobile industry, the shipping sector, and the ICT sector, among others. His experiences cover project management, international business, marketing, retailing, sales and sales management, customer care, and administration. Dr Braimah has executed projects worth several million dollars. Stephen is also an experienced speaker and human capital development consultant. In collaboration with various organisations, he has designed and implemented different management capacity building and executive development programmes for public and private sector organisations, covering organisations in and outside Ghana. The organisations covered in these training programmes range from SMEs, private firms, banks, petroleum companies, government ministries, departments, and agencies. Dr Stephen Mahama Braimah's research interest is in marketing and service marketing, strategy, branding, tourism, and hospitality.

João Caldeira Heitor holds a PhD in Tourism from the University of Lisbon. He is Professor and Secretary General at the Instituto Superior de Gestão, Lisbon, Portugal. He is the Scientific Director of the Degree in Tourism Management at the Instituto Superior de Gestão. Integrated Researcher at CiTUR - Leira, Portugal, develops research, publishes and participates in several international congresses and conferences. Additionally, effect consultancy in religious, cultural and destination management tourism.

Statia Elliot is Director of the School of Hospitality, Food and Tourism Management at University of Guelph, Canada. Statia teaches Contemporary Issues in Tourism at the graduate level, and researches Tourism Destination Image and Place Branding. She is Chair of the Ontario Tourism Education Corporation Board, member of the Travel and Tourism Research Association international board, and a Fellow of the Ontario Hostelry Institute.

Buntu Marannu Eppang holds a PhD in management, a MA in management, a BA in English Literature and a Diploma in Tourism. He is now an Assistant Professor at Politeknik Pariwisata Makassar (Makassar Tourism Polytechnic) and has worked for this institution for about 20 years teaching marketing, tour planning, tour guiding, MICE and destination management. He is also a researcher and the head of center of excellent for marine tourism development at this institution. His research interests are in tourism marketing, tourism experience and destination management especially for marine tourism destination.

Günay Erol is working as Associate Professor in Tourism Faculty at Nevşehir Hacı Bektaş Veli University in Turkey. He has experience in different tourism businesses and holds various certificates about Ticketing and Reservation and Online Reservation. His research areas are Tourism and Hotel Management, Tourist Behaviour and Mobility, Organizational Behaviour in Tourism, Social Media and Tourism, Tourism and Gastronomy. He has many publications in international journals and also book chapters and projects in tourism field.

Manuela Guerreiro is an Assistant Professor in the Faculty of Economics, University of Algarve (Portugal). She is Co-coordinator and integrated researcher at the Research Centre for Tourism, Sustainability and Well-Being (CinTurs). She is the Programme Leader of the Master of Marketing Management at this institution. Her current research interests include Marketing and Design, Consumer Behaviour, Consumer Experiences, Brand Management, Social Media Marketing, Arts and Events Marketing, Destination Marketing, Branding, and Image.

Taylor Johannigman is currently an undergraduate student at the University of Kentucky, pursuing dual degrees in Marketing and Hospitality Management and Tourism. Her scholarly interests include tourism marketing, social media content creation, and tourism storytelling.

Michael W. Lever is an Assistant Professor in the International School of Hospitality, Sports, and Tourism Management at Fairleigh Dickinson University in Vancouver, Canada. His research incorporates his background in the areas of consumer behaviour, management, and marketing and applies it to concepts in tourism, particularly those of destination image and place branding. In so doing, he hopes to understand how these related areas may influence both destination marketing practices and tourism and place theory. He currently serves as a reviewer for the Journal of Travel Research, and as a Director on the Travel and Tourism Research Association (Canada Chapter) Board.

Chris Little is a Technical Team Leader at the Department of Defence and a PhD student at Griffith University. He has extensive experience in 3D scanning and modelling across industry, academia, and government. Mr Little's research interests include using 3D technology for heritage preservation, as well as using the 3D data to bring heritage sites, monuments and artefacts back to life through digital technologies.

Alessandra Marasco is senior researcher at the National Research Council of Italy - Institute for Research on Innovation and Services for Development (CNR-IRISS). Her research focuses on the management of innovation in services, particularly in creative services, cultural heritage and tourism. Specific interests include collaborative innovation, service experience design and digital innovation, virtual reality in visitor experiences. She served as lead researcher in several funded projects on innovation in cultural heritage tourism and as co-editor of the section on Destination Competitiveness in the annual Report on Tourism in Italy. Her research has been published in leading journals and conference proceedings.

Mariana Inácio Marques holds a PhD in Communication Sciences, a Master degree in tourism and a degree in Organizational Communication – Marketing, Public Relations and Publicity. She is currently an Auxiliar Professor and researcher at Instituto Superior de Gestão, ISG Business & Economics School. She is an enthusiast about tourism and marketing. These are her main research areas. She is also a researcher from GOVCOPP, Research Unit on Governance, Competitiveness and Public Policies, University of Aveiro, Portugal. Currently she is a doctoral student in Tourism at the University of Aveiro.

Brent Moyle is an Associate Professor in the Department of Tourism, Sport and Hotel Management, Griffith University. He takes an interdisciplinary approach to tourism research, partnering with researchers and communities to maximise outcomes. Brent takes pride in conducting research at the interface between theory and practice, engaging extensively with industry partners to complete research with impact. His passion for sustainable regional development laid the foundations for a number of long-term collaborations with local government and parks agencies.

Francesca Nicolais holds a PhD in Communication Sciences, and is adjunct professor at the University Suor Orsola Benincasa of Naples, where she develops research projects on design, gamification, packaging and communication and on the powerful relationship between Humanities and Technologies. She is coordinator of cultural innovation projects for Pej's kitchen, a multimedia production company, and co-founder of Riot studio (www.riotstudio.it), a space dedicated to creativity, cultural research and innovation. She loves to be co-organizer of TEDxNapoli since 2012, content curator of the format

Who is the designer, co-inventor of a European patent on the preparation of fluorescent nanoparticles for corporate security. "Cross fertilization and knowledge integration are the keywords that guide my research projects".

Dale Patterson, PhD, is currently an Associate Professor (Digital Design, QCA). Dale teaches into a number of programs (including the Visualization & Immersive design major of the Bachelor of Design) and has a strong interest in innovative teaching approaches. Dale leads a number of externally funded projects including the development of innovative interfaces/systems, using technologies such as wearable interfaces (Watch), AR (Augmented Reality) and VR (Virtuial Reality) to bring heritage & past events/cultures to life, as well as visualisation and interaction techniques for play based interactions.

Maria Isabel Roque, PhD in History with specialisation in Museology, is Assistant Professor of Museum Studies at Universidade Católica Portuguesa, Lisbon. She is integrated researcher at CIDEHUS-UÉ at Universidade de Évora. She integrated several religious art exhibitions curatorial team. Her research focuses on heritage cultural mediation linked to theoretical research on art history, museum studies, cultural tourism, and digital humanities, which has been published in books, journals, and international conferences proceedings.

Noel Scott, PhD, is Adjunct Professor in the School of Business and Law at Edith Cowan University. He provides training and consulting in a number of countries in Asia and the Pacific. Clients include the United Nations World Tourism Organization (UNWTO), Organization for Economic Co-operation and Development, Association of Southeast Asian Nations (ASEAN), and Government of Fiji. Professor Scott has led eight Australia Awards Sustainable Tourism Short Courses in Indonesia and Sri Lanka for the Australian Department of Foreign Affairs and Trade. He has held positions as Research Professor at Griffith University, Associate Professor at The University of Queensland, and prior to his academic career has held management research positions in several leading companies including Queensland Tourist and Travel Corporation. He holds a PhD and Bachelor of Science with Honours from The University of Queensland, and Masters degrees from Warwick University and Queensland University of Technology. His research interests include the study of tourism experiences, and destination management and marketing. He is a frequent speaker at academic and industry conferences. He has over 300 academic publications including 17 books. He has supervised 30 doctoral students to successful completion of their theses. He is on the Editorial Board of 10 journals, a member of the International Association of China Tourism Scholars (IACST) and a Fellow of the Council for Australasian Tourism and Hospitality Education (CAUTHE). He holds a number of other Adjunct and Visiting Professor positions at University of the Sunshine Coast, Mataram University, Indonesia and Taylor's University, Malaysia.

Liubov Skavronskaya, PhD, is a Lecturer at Griffith College, Brisbane, Australia. Liubov studied tourism and hospitality management in China at the South China University of Technology and Chinese Studies at the Sun Yat-sen University. Liubov specialising in pathway education and currently teach tourism, hospitality, events management, and academic skills. Her research interests include emotions, memorable tourism experiences cognitive psychology, and China tourism.

Emmanuel Nii Ayi Solomon holds an HND in Tourism, Bsc. in Business Administration (Marketing option), MPhil in Marketing and is currently a PhD candidate at the University of Ghana Business School. He is also an Assistant lecturer at Accra Technical University where he lectures at the Marketing department. His recent focus is on how to use netnography to examine experiences told through blogs, photography and social media portals by tourists by combining principles of storytelling. His area of research includes destination marketing and competitiveness, destination branding, and issues bordering on destination sustainability. He is an award-winning Ghanaian Tour Guide, Tour Planner, and Tourism Consultant. As part of marketing Ghana's tourism products, he has attended several international trade exhibitions and conferences to promote Ghana's travel products. He has over a decade of experience in the tourism industry and has guided tourists from all walks.

Jason Swanson is currently an Associate Professor in the Retailing and Tourism Management Department at the University of Kentucky. He has a Ph.D. in Parks, Recreation and Tourism Management and a Master of Management in Hospitality. His research and creative interests include community-based tourism storytelling, travel media production, the transformative nature of travel, and tourism-related public policy.

Paulo Teixeira Costa is a coordinator professor in polytechnic higher education, researcher, trainer, and consultant in the fields of Tourism, Territorial Marketing, Planning and Management of Major Events. He is also a doctoral student in the Tourism PhD Program at the University of Aveiro, Portugal. He is member of the UNIDCOM/IADE – Design and Communication Research Unit, and member (as PhD student) of GOVCOPP, Research Unit in Governance, Competitiveness and Public Policy, University of Aveiro. His topics of interest include tourism experience, territorial marketing, events management, and co-creation.

Ana Sofia Torres Pereira completed her PhD on "Women Screenwriters: A dynamic definition of language in the feminine in Portuguese cinema" at Faculdade de Ciências Sociais e Humanas, Universidade Nova de Lisboa. After completing her degree in Sound and Image at Universidade Católica Portuguesa in 2006, she founded Cimbalino Filmes, a production company. Around the same time, she also started her teaching career at her Alma Mater, focusing on topics such as screenwriting, copywriting, storytelling, media and communication. Currently, besides her activity as a lecturer at Instituto Superior de Ciências da Informação e da Administração (Aveiro) and researcher at CITAR, Centro de Investigação em Ciência e Tecnologia das Artes (Universidade Católica Portuguesa), she works with numerous national and international production companies as a copywriter, screenwriter, and script-doctor.

Abdullah Ülkü is an Assistant Professor in the Tourist Guide department of Harran University, Tourism Faculty in Türkiye. He is also a tourist guide. He has experience in different tourism businesses. His research areas are destination marketing, digital transformation in tourism and tourist guide. He has publications in international journals and also book chapters.

Wiwien T. Wiyonoputri has been working in the tourism field since 1992 when she joined the Centre for Tourism Research – Institut Teknologi Bandung (ITB). Her field of interest includes community-based tourism and interpretation development for the last fifteen years. She holds a bachelor degree in Regional and City Planning, a master degree in Development Study focusing on Tourism from ITB. She is a PATA Duncan Sandys program fellow, Leadership for Environment and Development (LEAD) Fellow Cohort 8, an international member of the National Associations for Interpretation of the United States and European Association for Heritage Interpretation. In 2018, she initiated the Indonesian Society for Interpretation, a forum for those who are concerned and would like to learn, share, and discuss about interpretation.

Index

T

U

V

www.ingramcontent.com/pod-product-compliance
Lightning Source LLC
LaVergne TN
LVHW062143300926
849161LV00010B/2607